MATERIAL PROPERTIES OF STEEL IN FIRE CONDITIONS

MATERIAL PROPERTIES OF STEEL IN FIRE CONDITIONS

WEIYONG WANG
Chongqing University, P. R. China

VENKATESH KODUR
Michigan State University, United States

ELSEVIER

ACADEMIC PRESS
An imprint of Elsevier

Academic Press is an imprint of Elsevier
125 London Wall, London EC2Y 5AS, United Kingdom
525 B Street, Suite 1650, San Diego, CA 92101, United States
50 Hampshire Street, 5th Floor, Cambridge, MA 02139, United States
The Boulevard, Langford Lane, Kidlington, Oxford OX5 1GB, United Kingdom

Library of Congress Cataloging-in-Publication Data
A catalog record for this book is available from the Library of Congress

British Library Cataloguing-in-Publication Data
A catalogue record for this book is available from the British Library

ISBN: 978-0-12-813302-6

For information on all Academic Press publications visit our website at https://www.elsevier.com/books-and-journals

Publisher: Matthew Deans
Acquisition Editor: Glyn Jones
Editorial Project Manager: Peter Adamson
Production Project Manager: Sojan P. Pazhayattil
Cover Designer: Matthew Limbert

Typeset by TNQ Technologies

Contents

Author profiles

Weiyong Wang

Dr. Weiyong Wang is a professor and director in the Institute of Safety and Disaster Prevention Engineering of College of Civil Engineering at the Chongqing University (CQU). He received his PhD from Tongji University, China, in June 2008. From March to July in 2010, he worked as a visiting professor sponsored by Japan Society for the Promotion of Science (JSPS) in Tokyo University of Science (TUS), Japan. From January to November in 2011, he worked for Professor Kodur Venkatesh as a research associate in Michigan State University (MSU), USA. From May 2014 to September in 2015, he worked as a visiting scholar sponsored by China Scholarship Council (CSC) in University of Texas at Austin (UT Austin), USA.

Dr. Wang has supervised 30 PhD and MS students over the last ten years. Dr. Wang's research has focused on the experimental behavior and analytical modeling of steel structures and composite structures under fire conditions; constitutive modeling of material properties at elevated temperatures, developing guidelines for fire resistance design of steel and composite structures; and evaluating fire performance of high strength steel materials and performance-based fire safety design. He has secured more than 20 competitive research grants from diverse funding agencies and published 1 book and over 100 peer-reviewed papers in international and Chinese journals and conferences in structural, material, and fire resistance areas and has given 30 oral presentations in conferences.

He is a professional structural engineer, member of Council of Stability and Fatigue Branch of China Steel Structure Association, member of Housing Construction Steel Structure Branch of China Steel Structure Association, committee member of Professional Committee of Building Fire Prevention of China Fire Fighting Association, committee member of Professional Committee of Structures in Fire of China Architecture Society of China, committee member of Engineering Fire Prevention Technology Branch of Chinese Society of Civil Engineering, standing committee of Academic Committee on Human Settlement Environment of High-Rise Buildings of China and Architecture Society of China. He serves as associate editor of Journal of Progress in Steel Building Structures, Member of Editorial Board of Journal of Civil and Environmental Engineering, as well as reviewers for 30 international and Chinese Journals. He has won many awards including Excellent Young Teachers in Chongqing University (2012), Huang's Youth Innovation Award (2014), Advanced Faculty of Chongqing University (2018), and Tang's Excellent Research Teacher Award (2018). In 2018, as a major group member, he won Science Research Famous Achievement Award in Higher Institution (the first prize).

Venkatesh Kodur

Dr. Venkatesh Kodur is a University Distinguished Professor in the Department of Civil and Environmental Engineering at Michigan State University. He also serves as Director of the Centre on Structural Fire Engineering and Diagnostics at Michigan State University.

Dr. Kodur is an internationally recognized scholar for his contributions in structural fire engineering field. His research has focused on the experimental behavior and analytical modeling of structural systems under extreme fire conditions, constitutive modelling of material properties at elevated temperatures, developing guidelines for fire resistance design of structural systems, and building collapse investigations. He conducts unique research in fire safety field and has developed fundamental understanding on the behavior of materials and structural systems under extreme fire conditions. His research contributions have led to the development of fundamental understanding on the fire behavior of material and structural systems and also resulted in numerous design approaches and innovative and cost-effective solutions for enhancing fire resistance of structural systems.

Dr. Kodur has advised more than 25 postdoctoral researchers, 25 PhD students, 26 MS students, and number of undergraduate students over the last 25 years. Many of his (former) PhD and postdoctoral students are currently faculty members in reputed universities throughout the world. Dr. Kodur, together with his students and collaborators, has published results from his research in 425+ peer-reviewed papers in journals and conferences and has given numerous keynote presentations in major international conferences. He is one of the highly cited authors in Civil Engineering and Fire Protection Engineering disciplines, and as per Google Scholar, he has more than 10100 citations with an "h" index of 55. The methodologies, techniques, and design guidelines, resulting from his research, have been incorporated in to various codes and standards, as well as practical applications, in the United States and around the world, and are instrumental in minimizing the destructive impact of fire in the built infrastructure.

Prof. Kodur is a Fellow of the Canadian Academy of Engineering and a Foreign Fellow of Indian National Academy of Engineering. He is a professional engineer, Fellow of American Society of Civil Engineers, Fellow of Structural Engineering Institute, Society of Civil Engineers, Fellow of American Concrete Institute, Associate Editor of Journal of Structural Engineering, Past Chairman of ACI Fire Protection Committee, Chairman of ASCE-29 (Fire) Standards Committee and a member of UK-EPSRC College of Reviewers. He has won many awards including MSU University Distinguished Professor, AISC Faculty Fellowship Award, MSU Distinguished Faculty Award, NRCC (Government of Canada) Outstanding Achievement Award and NATO Award for collaborative research. Dr. Kodur was part of the FEMA/ASCE Building Performance Assessment Team that studied the collapse of WTC buildings as a result of September 11 incidents.

Preface

About the book

This book is a new contribution aimed at providing a comprehensive understanding of properties of steel under elevated temperature exposure, as encountered in fire situations. The application of new types of steel in constructions has increased dramatically in recent years, accompanied by the development of new provisions in codes of practice, such as specification for design of high strength steel structures, for evaluating fire resistance. For applying these provisions and to calculate fire resistance of steel structures, the basic material properties of steel are needed. There is dearth of information on property data for high strength steels under elevated temperature exposure. Most fire design codes, including Eurocode 3 and Eurocode 4, GB51249-2017, provide only properties of mild steel. Recent research clearly shows that variation of properties of new types of steel is quite different, especially at elevated temperatures. In recent years, the authors of this book, as well as few other researchers around the world, have carried out number of material-level tests to evaluate high temperature properties of several types of high strength steels, such as Q460 steel, S460 steel, Q690 steel, S690 steel, and A572 steels, etc. However, there is no compilation of these data in a single source. This book is aimed at providing high temperature properties of high strength steels, together with relations for various properties.

Contents and coverage

The contents of this book mainly deal with the high temperature mechanical properties of high strength steels, including strength, elastic modulus, and creep behavior, as well as their application in fire resistance design of steel structures. The book is organized into six chapters to facilitate smooth flow for readers with different levels of background in structural engineering, material science/engineering, and fire protection engineering. Each chapter starts with introductory concepts to provide a fundamental understanding and then presents test methods and property data, together with relations for these properties as a function of temperature.

Chapter 1 provides a brief introduction to high strength steels and their application in building structures and background to fire hazard together with its adverse effect on performance of steel structures. In addition, Chapter 1 also introduces test methods on thermal and mechanical properties of steel, including creep strain, at elevated temperatures. Chapter 2 provides information relating to thermal properties of steel at elevated temperature, namely thermal expansion, specific heat, and thermal conductivity. Chapters 3 to 4 present tension test results on mechanical property of high strength steels at elevated temperature and after fire exposure (residual properties). Furthermore, the property relations as a function of temperature to determine yield strength, tensile strength, and elastic modulus of several types of high strength steels, including Q420 steel, Q460 steel, Q690 steel, S460 steel, S690 steel, S960 steel, BISPLATE 80 steel, and ASTM A572 steel, etc., are given in these chapters. Chapter 5 deals with the evolution of creep strain of high strength steels at elevated temperatures. Creep test data, creep model to predict creep strain, and creep buckling of high strength steel columns are covered in this chapter. The application of proposed property relations of high strength steels in evaluating fire resistance is presented in Chapter 6, as well as lays out the knowledge gap and needed research for advancing the knowledge relating to high temperature properties of high strength steels. In each chapter, the subject matter is explained with the aid of numerous illustrations, figures, and tables, and step-by-step numerical examples are presented to help readers grasp the subject matter with ease.

The content presented in the book are basic properties needed to undertake fire resistance analysis of high strength steel structures. We believe the information presented in this book is useful to design professionals and also to researches or students to undertake fire resistance analysis and design on high strength steel structures.

Target audience

The target audience for this book is engineering professionals, students, and teachers in civil, mechanical and architectural engineering disciplines, and building officials and regulators. A good knowledge on mechanics of materials and structures is essential when reading this book, while general background on the fire safety design of steel structures is an advantage. It is hoped that the book will enable researchers, practitioners and students to develop greater insight of structural fire engineering, so that safer steel structures could be designed through enhancing fire resistance.

Although care has been taken to present the research results, some errors might have crept in inadvertently. We would high appreciate if these errors and suggestions for improvement are brought to my notice. If needed, an errata list will be placed on www.wangfire.net.

Acknowledgments

In the last 10 years, the authors have been devoted to the study of structural behaviors and analysis of high strength steel structures in a fire. It is unimaginable that the work could have been carried out, year-by-year, without financial support. Therefore, we gratefully acknowledge the Chinese National Program on Key Research and Development Project (Grant No. 2016YFC0701200), the National Natural Science Foundation of China (Grant Nos. 51678090, 51878096), and the Fundamental Research Funds for the Central Universities of China (Grant Nos. cqu2018CDHB1B02, 2019CDQYTM027) for the financial support. We are also grateful to the authors of the references cited by this book, especially Ac. Prof. Xuhong Qiang at Tongji University and Prof. Hongbo Liu at Tianjin University, for their assistance in writing this book. Any opinions, findings and conclusions or recommendations expressed in this paper are those of the authors and do not necessarily reflect the views of the sponsors.

Weiyong Wang
Chongqing University, People's Republic of China
Venkatesh Kodur
Michigan State University, United States

List of symbols

Δl	thermal expansion due to temperature elevation
l	length of steel member
T	temperature of steel
α_s	thermal expansion coefficient
C_s	specific heat of steel
λ_s	thermal conductivity of steel
m	mass of the steel member
d	diameter
f_1	fundamental frequency of the steel specimen
T_1	modification factor of the fundamental frequency
μ	Poisson's ratio of steel
f_{yT}	yield strength of steel at elevated temperature
f_y	yield strength of steel at room temperature
f_{uT}	tensile strength of steel at elevated temperature
f_u	tension strength of steel at room temperature
E_T	elastic modulus of steel at elevated temperature
E	elastic modulus of steel at room temperature
f'_{yT}	yield strength of steel after fire exposure
f'_{uT}	tensile strength of steel after fire exposure
E'_T	elastic modulus of steel after fire exposure
$f_{0.2}$	0.2% nominal yield strength of steel at elevated temperature
$f_{1.0}$,	yield strengths of steel at the strain levels of 1.0%
$f_{2.0}$,	yield strengths of steel at the strain levels of 2.0%
ε	tension or compressive strain of steel
ε_{cr}	creep strain of steel
ε_p	plastic strain of steel
σ	normal stress of steel
$\sigma_{0.2}$	static 0.2% tensile proof stress
n	strain-hardening exponent
n_0	original strain-hardening exponent
K	modular coefficient in Ramberg–Osgood expression
$\sigma_{0.01}$	static 0.01% tensile proof stress
m	exponent for the modified Ramberg–Osgood model
ε_{pu}	plastic strain at ultimate strength
t	time
a, b and c	temperature-dependent material properties
$c_1 \sim c_7$	coefficients that are dependent on type of steel
λ	slenderness ratio of the column
e_0	ratio of equivalent initial eccentricity
φ	stable factor of steel column at room temperature
f	design strength of steel at room temperature
γ_R	resistance factor of steel

η	ratio of maximum residual stress value at ambient condition to nominal yield strength of steel
α	residual stress reduction coefficient
H	the height of cross section
B	width of cross section
t_f	thickness of flange
t_w	thickness of web
A	area of cross section
I	moment of inertia of an area
i	radius of gyration
L	length of steel column
P_{cr}	load bearing capacity of columns
φ'_{T}	stability factor of steel columns after fire exposure
φ_{T}	stability factor of steel column at elevated temperature
L_R	load level
δ	deflection of steel member
N	axial force
d_i	thickness of fire protection
λ_i	thermal conductivity of fire protection
W_p	plastic section modulus
β_r	rotational restraint stiffness ratio
$\dot{\varepsilon}_{cr}$	creep strain rate

CHAPTER 1

Introduction

Contents

1.1 Initial consideration

Fire represents one of the most severe conditions encountered during the lifetime of a steel structure, and therefore, the provision of appropriate fire safety measures for steel structural members is a major safety requirement in building design. The basis for this requirement can be attributed to the fact that when other measures for containing the fire fail, structural integrity and stability is the last line of defense. To provide rational fire designs of steel structures, in the last three decades, there have been important research endeavors devoted to developing better understanding of steel structural behavior under fire conditions and to develop rational design approaches for evaluating fire resistance of steel structures. However, all the design

Material Properties of Steel in Fire Conditions
ISBN 978-0-12-813302-6
https://doi.org/10.1016/B978-0-12-813302-6.00001-1

approaches are based on the high-temperature material properties of mild steels.

In recent years, high strength steels (HSSs) have been paid more attention in construction applications. There is very limited knowledge in fire resistance design of HSS structures due to lack of material properties of these steels under fire conditions. Before establishing design methods for HSS structures, material properties of HSSs under fire conditions are necessaries. In the last 10 years, the authors carried out a great deal of tests on mechanical properties of HSSs in fire conditions and after fire exposure, as well as fire response of HSS structures in fire and after fire exposure. The main objective of this book is to present the test results of the material properties of HSSs under fire conditions by the authors as well as other researchers. In addition, the application of mechanical properties of HSSs in structural fire engineering, namely, fire resistance investigations on HSS structures, is also introduced in this book.

1.2 High strength steel

As ideal building materials, steels are widely used in constructions and infrastructures for many years. They can often be classified into many types, namely mild steels, cold-formed thin-walled section steels, and alloy steels. Stainless steels, fire resistance steels, and HSSs are all alloy steels and often called as special steels. HSS generally exhibits improved strength, good ductility, and better corrosion resistance. When formed into structural elements, significant benefits arise in terms of stiffness, strength, and potential energy absorption. In recent years, HSS has found wide applications in modern construction because of the many advantages it offers over conventional mild steel, such as high strength and strength to weight ratio (Wang et al., 2010).

The definition for classifying structural steel based on strength varies in different design standards. American Society for Testing and Materials (ASTM) specifies a nominal yield stress of 690 N/mm^2 for quenched and tempered steel, which is the equivalent to HSS used in plate form (Bjorhovde, 2004). The Hong Kong standard defines HSS as having a nominal yield stress between 460 and 690 N/mm^2 (Hong Kong Buildings Department, 2005). The European standard Eurocode 3 (European Committee for Standardization, 2005a) has an implied delineation point of 460 N/mm^2 to distinguish between mild and HSS. In Australia, high strength structural steel is defined as steel that has a maximum yield stress exceeding 450 N/mm^2 (Standards Australia, 1998). In China, steels with a yield strength higher than 420 N/mm^2 and ultimate strength greater than 590 N/mm^2 are classified as HSS (Li et al., 2006).

The chemical compositions have great effect on the material properties of steels, not only at room temperature but also at elevated temperatures. The primary chemical compositions in conventional carbon (mild) steel are iron, carbon, manganese, sulfur, and phosphorous. The HSSs are made by adding some alloys, such as nickel, titanium, boron, and chromium, and these alloys influence the mechanical properties of the HSSs. For example, molybdenum, chromium, and niobium can increase the fire resistance property of steel, whereas chrome and nickel can enhance the corrosion resistance of steel (Li et al., 2006). The chemical composition of HSS is different from that of mild steel, and therefore, the high-temperature behavior of HSS is envisioned to be different. Current design rules on fire resistance of steel structures (European Committee for Standardization, 2005b; British Standards Institution, 1990; GB51249, 2017) are primarily based on the experimental results from mild steel and do not account for specific property variations in HSSs.

As stated above, steel is often characterized based on yield strength and elastic modulus property. Cottrell atmosphere is easily formed in low carbon (mild) steel, so the low carbon steel at room temperature has a clear and well-defined yield plateau in tension. However, alloy steels, such as HSSs, do not have a clearly demarked yield plateau at room or high temperatures, as shown in Fig. 1.1. Generally, the yield strength of mild steel and HSS at elevated temperature is evaluated at strain levels of 0.2% offset or 0.5, 1.5, or 2.0%. The 0.2% offset yield strength ($f_{0.2,\ T}$) is the intersection point of the stress–strain curve and the proportional line offset by 0.2% strain. The yield strengths of $f_{0.5,\ T}$, $f_{1.5,\ T}$, and $f_{2.0,\ T}$ are those corresponding to the strain levels of 0.5%, 1.5%, and 2.0%, respectively. Because there is no clearly demarked yield point, the definition for determining the yield strength of steel at elevated temperatures varies in different design codes of practice. The European Convention for Constructional Steelwork (2001) specifies that the yield strength at room temperature corresponds to 0.2% offset strain level, and for temperatures exceeding 400°C, yield strength is assumed at a nominal strain level of 0.5%. When temperature is in the range of 25–400°C, yield strength is assumed to correspond to the interpolation between the strain level of 0.2% offset and 0.5% nominal strain. BS5950 (BSI, 1990) specifies that the yield strength at elevated temperature can be taken at the strain level of nominal strain 0.5, 1.5, or 2% for different types

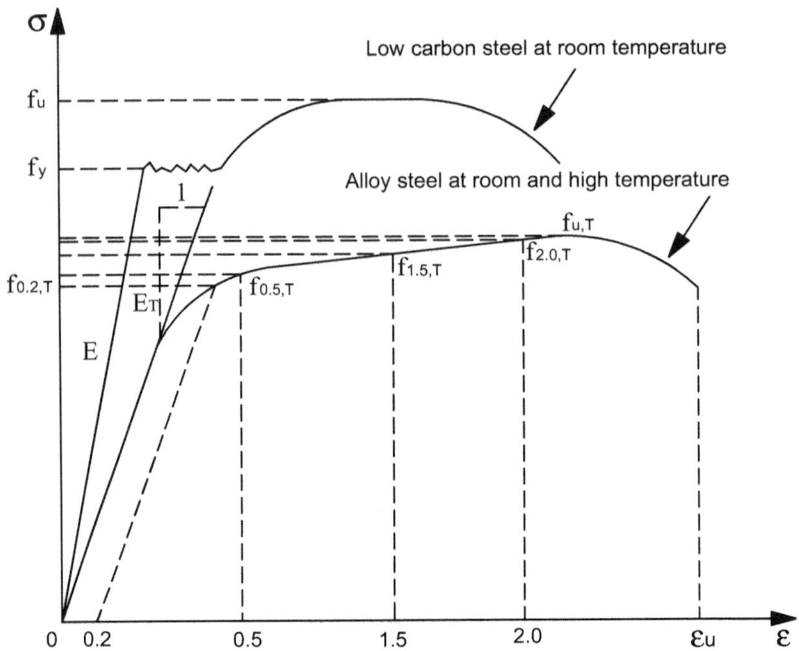

Figure 1.1 Typical representation of stress–strain curves of steel at elevated temperature.

of structural members. The yield strength at elevated temperature in Eurocode 3 (European Committee for Standardization, 2005b) and Eurocode 4 (European Committee for Standardization, 2006) is based on the strain level of 0.2% offset. Kirby and Preston (1988) recommended the use of 1.0% nominal strain for evaluating the yield strength at high temperature. AISC (2005) recommends that the properties for steel at elevated temperatures are adopted from the European Convention for Constructional Steelworks (2001). In current studies on strength properties of HSSs, the yield strengths of $f_{0.2,\,T}$, $f_{0.5,\,T}$, $f_{1.0,\,T}$, $f_{1.5,\,T}$, and $f_{2.0,\,T}$ are often obtained and presented.

The elastic modulus is the ratio of stress to strain of material in the elastic stage. It is often measured as the tangent modulus to the initial linear elastic portion of the stress–strain curve. Sometimes, the strain at the elastic stage is not accurately recorded due to the limitation of displacement extensometer used in the test. The dynamic method is also utilized, which can compute the elastic modulus of materials at elevated temperatures. Steel specimen possess specific mechanical resonant frequencies that are determined by the elastic modulus, mass, and geometric property of test specimen. Therefore, the dynamic elastic properties of steel can be evaluated if the geometry, mass, and mechanical resonant frequencies of a steel specimen are measured. Dynamic elastic modulus is determined using the resonant frequency in either the flexural or the longitudinal mode of vibration. The detailed information related to test methods on material properties of steels at elevated temperature will be presented and discussed in Section 1.6.

1.3 Applications of high strength steel in building structures

As discussed above, in different countries, different definitions on HSS are selected; therefore, various types of HSS are used around the world. The most common types of HSS in the China are Q460 steel, Q550 steel, Q690 steel, Q890 steel, and Q960 steel, which have yield strength of 460, 550, 690, 890, and 960 N/mm^2, respectively. One of the primary applications for high strength Q460 steel is in hollow structural sections that are widely used in the structural members of high-rise and large-span buildings. Examples of well-known buildings utilizing Q460 steel include Beijing National Stadium (also known as the Bird's Nest) (shown in Fig. 1.2) and the China Central Television building (shown in Fig. 1.3). Another type of HSS in China is Q690 steels, which has a yield strength of 690 N/mm^2. As to now, Q690 steel has been widely used in transmission tower. In building structures, Q690 has not widely applied because of lack of design code. Recently, a lot of researches have been addressed on design of Q690 steel structures. In the near future, Q690 will be used commonly in constructions in China.

Figure 1.2 China national stadium.

Figure 1.3 China Central Television (CCTV) tower.

In Europe, most common types of HSSs are S460, S690, and S960. To reduce the sectional dimension of steel members, S460 steel and S690 steel have been used in Sony Center located at Berlin, Germany (shown in Fig. 1.4). Brittle fracture behavior of these steels has been tested at low temperature to assure the structural safety of this building in cold winter. In Australia, BISPLATE 80 steel is a typical HSS with nominal yield strength of 690 MPa. HSSs with yield strength of 650 and 690 MPa (BISPLATE 80) are used in Star City of Australia.

In Japan, HSSs with nominal yield stress of 500–600 MPa were firstly applied to bridge structures in 1950s. In 1960s, HSSs with nominal yield stress of 800 MPa were begun to use in bridge. For example, Minato Bridge (Fig. 1.5) constructed in 1974 consumed 1073t 700 MPa steels and 4195t

Figure 1.4 Sony Tower in Berlin.

Figure 1.5 Minato Bridge in Osaka.

800 MPa steel. In addition, HSSs are also widely used in high-rise buildings. For example, to reduce the structural mass, BISPLATE 80 steel has been applied in Latitude Mansion located in Japan, and good economic benefits have been obtained. Landmark Tower located in Yokohama (shown in Fig. 1.6) is the first building in Japan using HSS of 600 MPa in I-shaped steel columns. Japan Railways East Japan and Nippon Television Tower in Tokyo also used HSSs in key steel members.

In the United States, HSSs were also firstly applied to bridge structures, such as New River Gorge Bridge (shown in Fig. 1.7), build during 1974–78. In building structures, a large amount of HSSs is consumed every year.

Figure 1.6 Sony Tower in Berlin.

Figure 1.7 New River Gorge Bridge in Virginia.

1.4 Fire hazard and effect of fire on steel structures

As we all know, steel is a noncombustion material, but its yield strength and Young's modulus degrade quickly at high temperature, which makes the steel structure have a low fire resistance. At the temperature of 600°C, the normal steel will lose most of its strength and stiffness. The fire in the building may easily cause global collapse to the steel structure or severe damage to structural components.

Steel structures exhibit relatively lower fire resistance because of faster loss of strength and stiffness properties of steel with temperature and also because of rapid rise in temperatures resulting from the high thermal conductivity and the low specific heat of steel. The fire resistance of a steel structural member depends mainly on the thermal and mechanical properties of steel. Thermal properties that influence fire response of steel structures include thermal conductivity, specific heat, and thermal expansion. Mechanical properties, including strength, elastic modulus, and creep, have significant influence on the fire response of steel structures. With the degrading of strength and stiffness of steel at elevated temperatures, structural components may lose their load-bearing capacities, which leads to global collapse of the building.

In 1990, the Broadgate Street building had a fire during the construction phase, and steel columns and steel beams were seriously damaged. Fig. 1.8 illustrates the buckled steel column caused by the fire. Redistribution of the load to less severely heated components of the building leads to improved fire resistance performance of the structure.

Figure 1.8 Buckling of steel column in fire.

In 2001, a severe fire happened at the Taipei Oriental Science District Building. Although the overall structure of the building did not collapse, a large number of steel structural components were severely damaged. The damage to the steel structure included (a) fracture of the beam-to-column connection, (b) local buckling of the steel beam, and (c) large deflection in the steel beam and the floor slab. Fig. 1.9 illustrates a steel portal frame industrial building collapsed in a fire. And the collapse of the World Trade Center is another disaster caused by fire (shown in Figs. 1.10 and 1.11).

1.5 State of the art

Because of great influence of fire on HSS structures, it is very necessary to investigate the structural behavior of HSS structures under fire conditions or at elevated temperatures. As a basic knowledge, material properties of HSSs are key factors to understand the fire resistance of HSS structures. As to now, there are some research focuses on material properties of HSSs, including of mechanical properties of HSS at elevated temperatures, mechanical properties of HSS after fire exposure, and creep behavior of HSS at elevated temperatures. The following section gives a detail review of these investigations.

1.5.1 Mechanical properties of high strength steel at elevated temperatures

A review of literature indicates that many experiments have been conducted to evaluate the effect of temperature on strength properties of

Figure 1.9 Collapse of a steel portal frame in fire.

Figure 1.10 Fire in World Trade Center caused by aircraft attack.

Figure 1.11 Collapse of World Trade Center tower.

conventional mild steel (Kirby and Preston, 1988; Twilt, 1988; Chen and Young, 2004; Kodur and Harmathy, 2008). Most of these studies were conducted to assess the variation of strength and stiffness properties of structural steel with yield strength in the range of 250–350 N/mm^2. Very few studies are reported in the literature on the temperature effects on strength and stiffness properties of high strength structural steel.

High strength bolts are often used in connections, and they are made of HSSs. Li et al. (2003) conducted experiments to evaluate the properties of high strength bolt steel, made of 20MnTiB (SAC, 1988). The test data include yield strength, ultimate strength, modulus of elasticity, and elongation of 20MnTiB steel at elevated temperature. Based on the data from the tests, the mathematical models for predicting the mechanical properties of 20MnTiB steel have been established. The study found that the chemical composition of steel has a great effect on high-temperature properties.

In China, high strength structural steels are fabricated according to GB/T 1591–2008 (2008) and marked as Q, such as Q460 steel and Q690 steel. Wang et al. (2013, 2018b) carried out comprehensive studies on material properties of Q460 and Q690 steels. Strength and stiffness properties of these steels were measured at various temperatures in the range of 20–800°C. A relative comparison of measured data indicates that HSS experiences a slower loss of strength and stiffness with temperature than that of conventional steel. Data generated from the experiments, namely, stress–strain relationships, were utilized to develop relations for yield strength, tensile strength, and elastic modulus of Q460 and Q690 steels as a function of temperature.

BISPLATE 80 steel is a type of HSS made in Australia and has yield strength of 690 MPa. Chen et al. (2006, 2008) studied the mechanical properties of BISPLATE 80 steel at elevated temperature and proposed relations for yield and ultimate strength and elastic modulus as a function of temperature. The elastic modulus and yield strengths were obtained at different strain levels, and the ultimate strength and thermal elongation were evaluated at different temperatures. It is shown that the reduction factors of yield strength and elastic modulus of high strength BISPLATE 80 steel and mild steel are quite similar for the temperature ranging from 22 to 540°C. The test results were compared with the predictions obtained from the American, Australian, British, and European standards.

In Europe, HSS was fabricated according to EN10025-3 and marked as S, such as S460, S690, and S960. Qiang et al. (2012a,b,c) conducted tensile tests to evaluate high-temperature mechanical properties of S460, S690, and

S690 steel. It is found that the deterioration of mechanical properties of steels under steady-state fire condition is dependent on steel grades. The comparison of test data with design standards reveals the necessity of including specified recommendations on the deterioration of mechanical properties in fire for these high strength structural steels in current leading design standards. Further, some unique predictive equations are proposed for the elevated-temperature mechanical properties of these steels.

1.5.2 Mechanical properties of high strength steel after fire exposure

Wang et al. (2015) and Zhou et al. (2019) carried out tensile tests on postfire mechanical properties of Q460 steels and Q690 steels. Tensile coupons taken from high strength Q460 and Q690 steel sheets were exposed to different temperatures up to 900°C and were then allowed to cool down to ambient temperature before they were tested to failure. Two cooling methods were considered, namely, natural air cooling and cooling by water. Tensile coupon tests were conducted to obtain their postfire stress—strain curves and associated mechanical properties (yield strength, elastic modulus, ultimate strength, and elongation). It was found that the postfire mechanical properties of high strength Q460 and Q690 steel are significantly reduced below the original ambient temperature mechanical properties if they had been exposed to temperatures exceeding 600°C. Hence, a new set of equations is proposed to predict the postfire mechanical properties of high strength Q460 and Q690 steel.

Qiang et al. (2012b, 2013) carried out experiments on postfire mechanical properties of high strength S460, S690, and S960 steels and proposed predictive equations. The study revealed that all the steel grades (S460, S690, and S960) almost fully regained their elastic modulus when experienced temperatures less than 600°C in steel. However, a significant degradation of elastic modulus was observed for specimen experiencing temperatures exceeding 600°C. Furthermore, the higher strength steels (S690 and S960) experienced greater losses than the lower strength S460 specimen. It was also found that all steel grades regained more than 75% of its elastic modulus when experiencing temperatures up to 800°C and more than 60% when experiencing temperatures up to 1000°C.

Chiew et al. (2014) presented an investigate to the mechanical properties of reheated, quenched, and tempered (RQT) HSS plate (grade S690) at elevated temperatures as well as its residual strength after heating and cooling down. It is found that while RQT-S690 HSS showed good resistance to temperature (below 400°C) heating, serious deterioration in strength would occur at higher temperatures, and its postfire mechanical properties are different from those of traditional hot-rolled normal strength steels.

1.5.3 Creep strain of high strength steel at elevated temperatures

When exposed to elevated temperatures, steel structures undergo increasing permanent deformations with time even when the applied stress level is below that of yield stress, and this time-dependent deformation is referred to as creep. Thus, creep can be defined as an increase in strain in a solid material under constant stress over a period of time. Many studies (Kodur and Dwaikat, 2010; Wang et al., 2018b) have shown that creep strain in steel has great influence on fire response of steel structures.

A comprehensive study on the high-temperature creep in high strength Q460 and Q690 steel was undertaken by Wang et al. (2017, 2018a). A set of creep tests were carried out on Q460 and Q690 steel coupons at various stress levels in 300–900°C range. Using these high-temperature creep data, two creep models based on the creep power law and unified theory on plasticity and creep strains are proposed for Q460 and Q690 steel.

1.5.4 Fire resistance investigation on high strength steel structures

To develop a methodology for evaluating fire resistance of high strength Q460 steel columns, the load-bearing capacity of high strength Q460 steel columns was investigated by Wang and Li (2014). The current approach of evaluating load-bearing capacity of mild steel columns at room temperature is extended to high strength Q460 steel columns with due consideration to high-temperature properties of high strength Q460 steel. The critical temperature of high strength Q460 steel column was presented and compared with mild steel columns. The proposed approach was validated by comparing the predicted load capacity with that evaluated through finite element analysis and test results. In addition, parametric studies were carried out by employing the proposed approach to study the effect of residual stress and geometric imperfections. Results from parametric studies show that, only for a long column (slenderness higher than 75), the magnitude and distribution mode of residual stress have little influence on ultimate load-bearing capacity of high strength Q460 steel columns, but the geometric imperfections have significant influence on any columns. At a certain slenderness ratio, the stability factor first decreases and then increases with temperature rise.

An investigation on the behavior of restrained HSS columns at elevated temperature obtained from full-scale fire tests and finite element analyses was presented by Wang et al. (2018c). In the fire tests, applied load and restraint stiffness were two key factors to be examined. Column responses such as the axial displacement, deflection at column middle height, and axial force induced by thermal expansion associated with temperature evolution were reported. Column buckling and failure temperatures were determined based on the criteria of the axial displacement and lateral deflection of the specimen at elevated temperatures. The test results showed that both the applied load and restraint stiffness have considerable influences on fire resistances of HSS columns. It was observed that the columns with only axial restraints failed by flexure buckling about the weak axis, whereas the columns with both axial and rotational restraints and subjected to large magnitude of the applied load failed by flexural torsional buckling. Finite element analyses were conducted to simulate the fire responses of the test specimen, and the obtained numerical results are found to be reasonably agree with the test data. Parametric studies via finite element analysis were carried out to quantitatively determine the effect of applied load, restraint stiffness, and slenderness ratio on fire resistance of HSS columns.

An analytical investigation on fire resistance of high strength Q460 steel beam subjected to nonuniform temperature distribution was presented by Wang et al. (2018d). Based on the critical bending moment associated with overall flexural stability and results obtained from the previous experimental investigation on the mechanical properties of Q460 steel at elevated temperature, an equivalent stiffness method (ESM) was established to evaluate the fire resistance of the beam with the consideration of the influence of temperature gradient across the section of the beam. Lateral torsional buckling resistance, critical temperature, and overall stability coefficient were obtained for flexure of high strength Q460 steel beams at elevated temperature. A three-dimensional (3D) nonlinear finite element model (FEM), which was capable of accounting for temperature gradient and predicting critical bending moment of Q460 steel beam at elevated temperature, was developed. Results from the finite element simulations were compared with the results determined by the proposed ESM, and there was a good agreement between the results of the two methods with the maximum difference of 6%. Using the ESM, parametric studies were carried out to investigate effects of steel grade and temperature distribution pattern on fire resistance of high strength Q460 steel beams. Accordingly, a simplified design approach was proposed to predict the critical temperature and overall stability coefficient of Q460 steel beams subjected to nonuniform temperature distribution. The simplified approach was applicable to high strength Q460 steel beams with cross-sectional dimensions ranging from 175 to 350 mm and 250 to 500 mm for flange width and section height, respectively.

Wang et al. (2018b) presents a FEM for accessing the fire resistance of restrained high strength Q460 steel beams by taking high-temperature Norton creep model of steel into consideration. The validation of the established model was verified by comparing the axial force and deflection of restrained beams obtained by finite element analysis with test results. To explore the creep effect on fire response of restrained Q460 steel beams, the thermal axial force and deflection of the beams were also analyzed excluding creep effect. Results from comparison inferred that creep plays a crucial role in fire response of restrained steel beam and neglecting the effect of creep may lead to unsafe design. A set of parametric studies were accomplished by using the calibrated FEM to evaluate the governed factors influencing fire response of restrained Q460 steel beams. The parametric studies indicated that load level, rotational restraint stiffness, span-depth ratio, heating rate, and temperature distribution pattern are key factors in determining fire resistance of restrained Q460 steel beam. A simplified design approach to determine the moment capacity of restrained Q460 steel beams was proposed based on the parametric studies by considering creep effect.

Experimental study and finite element modeling on the structural behavior of high strength Q460 steel columns after being exposed to ISO 834 standard fire was presented by Wang and Liu (2016). Two section shapes were considered, namely welded H-shaped section and welded box section. The experiment composes of two stages: heating and cooling stage of the specimen and compressive test on the specimen after fire exposure. The temperature experienced in the specimen was recorded during both the heating and cooling phases, as well as the load-deflection curves, load-axial displacement curves, load-rotation curves, load-strain curves, and failure modes in the steel columns were measured during the compressive tests. 3D FEM s were also established by employing the program ANSYS to study the postfire behavior of high strength Q460 steel columns. The FEMs were used to investigate the effects of key parameters on the residual load-bearing capacity of high strength Q460 steel columns, following exposure to the

ISO 834 standard fire. The investigated parameters included heating time, cooling methods, and slenderness ratio. Finally, a simplified design approach was proposed for predicting the residual load-bearing capacity of high strength Q460 steel columns after being exposed to ISO 834 standard fire.

A study on creep buckling of Q460 steel columns subjected to elevated temperatures via experimental and numerical investigations was presented by Wang et al. (2019). Six high strength Q460 steel columns with welded H shape were tested for the determination of creep buckling strength at elevated temperatures. The effects of the slenderness ratio, load ratio, and elevated temperature on failure time corresponding to creep buckling (also called as creep buckling time) were taken into consideration in the investigations. The evolutions of furnace and column temperatures and column lateral and axial displacements associated with fire duration were measured. The creep buckling time of the columns was obtained based on the magnitudes of axial and lateral displacement of the columns subjected to constant elevated temperature and compressive load. The test results showed that the creep deformation of high strength Q460 steel columns became significant at a temperature of 800°C even when the applied load level was as low as 10%. The creep buckling time of Q460 steel columns are sensitive to load ratio, slenderness, and temperature. Finite element analysis program ANSYS was employed to establish a structural model of Q460 steel column subjected to the constant elevated temperature and compressive load, with mechanical properties at elevated temperature, creep strains, and residual stresses induced by welding being considered. The FEM was validated by the test results. Parametric studies were carried out to quantitatively determine the effects of the slenderness ratio, load ratio, temperature, and geometric imperfections on the creep buckling time of Q460 steel columns.

1.6 Test methods

For structural fire engineering, two sets of data are very important, which are the material thermal properties for thermal analysis of structural components, including the thermal conductivity, specific heat, and density, and the material mechanical properties for structural analysis, including the yield strength, elastic modulus, stress–strain relationship, and the expansion coefficient. This section describes the test methods recommended in standards on thermal and mechanical properties of structural steel, including HSS at high temperature.

1.6.1 Thermal properties tests

1.6.1.1 Thermal conductivity

According to the second law of thermodynamics, heat will flow from the hot environment to the cold one to equalize the temperature difference. This is quantified in terms of a heat flux q, which gives the rate, per unit area, at which heat flows in a given direction. Thermal conductivity is the ability of a material to conduct heat. In the International System of Units, thermal conductivity is measured in watts per meter-kelvin (W/(m•K)). Some papers report in watts per centimeter-kelvin (W/(cm•K)). Currently, there is not an accepted standard for testing thermal conductivity of material, and the test results varied significantly by using different test methods. In the world, there are some standards widely used and the commonly employed methods include guarded hot plate method, hot wire method, heat flow meter method, and transient line-source method. More details about the test method and requirements can refer to the following standards.

(1) ISO 8302: Thermal insulation—Determination of steady-state thermal resistance and related properties—Guarded hot plate apparatus.

(2) ASTM C1113: Standard Test Method for Thermal Conductivity of Refractories by Hot Wire.

(3) ASTM C518: Standard Test Method for Steady-State Thermal Transmission Properties by Means of the Heat Flow Meter Apparatus.

(4) ASTM D5930: Standard Test Method for Thermal Conductivity of Plastics by Means of a Transient Line-Source Technique.

(5) GB/T 10295: Thermal Insulation—Determination Of Steady-State Thermal Resistance And Related Properties—Heat Flow Meter Apparatus.

(6) GB/T 10297 Test Method For Thermal Conductivity Of Nonmetal Solid Materials By Hot-Wire Method.

According to the methods specified in standards, some test setup for measuring thermal conductivity of material are designed, as shown in Fig. 1.12.

1.6.1.2 Specific heat

Specific heat is a particular type of heat capacity. Specific heat is the thermodynamic property, which states the amount of heat required for a single unit of mass of a substance to be raised by 1° of temperature. Energy in the form of heat is expressed in joules (J) or kilojoules (kJ), which are the most common units associated with energy. One unit of mass is measured

Figure 1.12 Thermal conductivity apparatus.

in grams or kilograms with regard to specific heat. Per 1 gram is the standard form used in tables of specific heat values, but references using 1 kilogram are sometimes seen. One degree of temperature is measured on either the Celsius or Kelvin scale, but usually Celsius. The most commonly associated units then for specific heat are J/(g•°C).

The methods to measure the specific heat of solid material include differential scanning calorimetry and mixing method. More details about the testing procedures and specimen preparation are available in the following standards:

(1) ISO 11357-4 that specifies methods for determining the specific heat capacity of plastics by differential scanning calorimetry.

(2) ASTM E1269: Standard Test Method for Determining Specific Heat Capacity by Differential Scanning Calorimetry.

1.6.2 Tensile strength test at elevated temperature

1.6.2.1 Brief introduction on tensile testing

Tensile testing, also known as tension testing, is a fundamental materials science and engineering test in which a sample is subjected to a controlled tension until failure. Properties that are directly measured via a tensile test are ultimate tensile strength, breaking strength, maximum elongation, and reduction in area. From these measurements the following properties can also be determined: Young's modulus, Poisson's ratio, yield strength, and strain-hardening characteristics. Uniaxial tensile testing is the most commonly used for obtaining the mechanical characteristics of isotropic materials. Some materials use biaxial tensile testing.

The preparation of test specimen depends on the purposes of testing and on the governing test method or specification. A tensile specimen is usually a standardized sample cross section. It has two shoulders and a gage (section) in between. The shoulders are large so they can be readily gripped, whereas the gage section has a smaller cross section so that the deformation and failure can occur in this area. The shoulders of the test specimen can be manufactured in various ways to mate to various grips in the testing machine. Each system has advantages and disadvantages; for example, shoulders designed for serrated grips are easy and cheap to manufacture, but the alignment of the specimen is dependent on the skill of the technician. On the other hand, a pinned grip assures good alignment. Threaded shoulders and grips also assure good alignment, but the technician must know to thread each shoulder into the grip at least one diameter's length, otherwise the threads can strip before the specimen fractures.

Tensile testing is most often carried out at a material testing laboratory. The ASTM D638 is among the most common tensile testing protocols. The ASTM D638 measures plastics tensile properties including ultimate tensile strength, yield strength, elongation, and Poisson's ratio.

The most common testing machine used in tensile testing is the universal testing machine (Fig. 1.13). This type of machine has two crossheads; one is adjusted for the length of the specimen and the other is driven to apply tension to the test specimen. There are two types: hydraulic powered and electromagnetically powered machines.

The machine must have the proper capabilities for the test specimen being tested. There are four main parameters: force capacity, speed, precision, and accuracy. Force capacity refers to the fact that the machine must be able to generate enough force to fracture the specimen. The machine must be able to apply the force quickly or slowly enough to properly mimic the actual application. Finally, the machine must be able to accurately and precisely measure the gage length and forces applied; for instance, a large machine that is designed to measure long elongations may not work with a brittle material that experiences short elongations before fracturing.

Alignment of the test specimen in the testing machine is critical because if the specimen is misaligned, either at an angle or offset to one side, the machine will exert a bending force on the specimen. This is especially bad

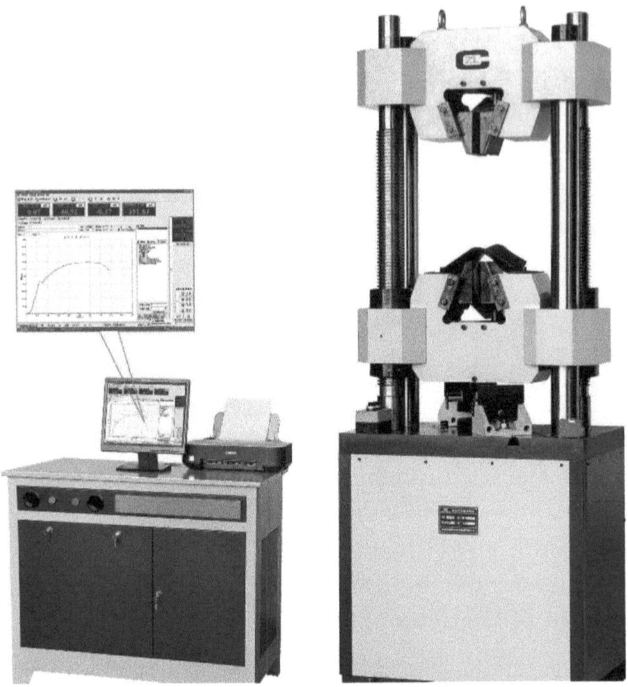

Figure 1.13 Tension test setup.

for brittle materials because it will dramatically skew the results. This situation can be minimized by using spherical seats or U joints between the grips and the test machine. If the initial portion of the stress—strain curve is curved and not linear, it indicates the specimen is misaligned in the testing machine.

The strain measurements are most commonly measured with an extensometer, but strain gages are also frequently used on small test specimen or when Poisson's ratio is being measured. Newer test machines have digital time, force, and elongation measurement systems consisting of electronic sensors connected to a data collection device (often a computer) and software to manipulate and output the data. However, analog machines continue to meet and exceed ASTM, NIST, and ASM metal tensile testing accuracy requirements, continuing to be used today.

The test process involves placing the test specimen in the testing machine and slowly extending it until it fractures. During this process, the elongation of the gage section is recorded against the applied force. The data are manipulated so that it is not specific to the geometry of the test sample. The elongation measurement is used to calculate the engineering strain, ε. The force measurement is used to calculate the engineering stress, σ. The machine does these calculations as the force increases, so that the data points can be graphed into a stress–strain curve.

1.6.2.2 Tensile test at elevated temperature

Tensile test at elevated temperature is a method to evaluate the mechanical properties of metallic materials at elevated temperature. Compared with tensile test at room temperature, the temperature control and measurement system have been added to the tensile test at elevated temperature (Fig. 1.14), and the influence factors of the test results are also more complex. More details about the specimen preparation, test setup, test procures, tensile strain rate requirements, and test results processing method are available in the following standards:

(1) ISO 6892-2: Metallic materials—Tensile testing—Part 2: Method of test at elevated temperature.

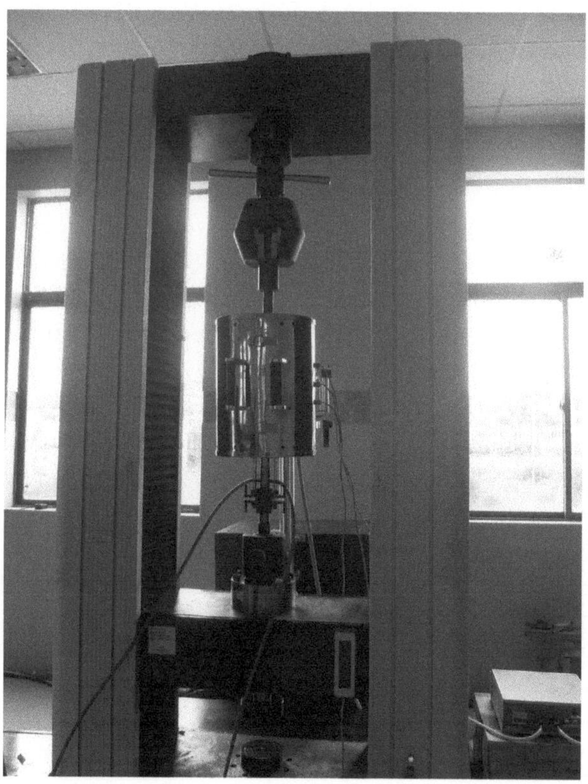

Figure 1.14 High-temperature tension test setup.

(2) GB/T 228.2: Metallic materials—Tensile testing—Part 2: Method of test at elevated temperature.
(3) ASTM E21-5 Standard Test Methods for Elevated Temperature Tension Tests of Metallic Materials.

1.6.3 Creep test at elevated temperature

1.6.3.1 Creep phenomenon

In materials science, creep (sometimes called cold flow) is the tendency of a solid material to move slowly or deform permanently under the influence of mechanical stresses. It can occur as a result of long-term exposure to high levels of stress that are still below the yield strength of the material. Creep is more severe in materials that are subjected to heat for long periods and generally increases as they are near their melting point.

The rate of deformation is a function of the material's properties, exposure time, exposure temperature, and the applied structural load. Depending on the magnitude of the applied stress and its duration, the deformation may become so large that a component can no longer perform its function. For example, creep of a turbine blade will cause the blade to contact the casing, resulting in the failure of the blade. Creep is usually of concern to engineers and metallurgists when evaluating components that operate under high stresses or high temperatures. Creep is a deformation mechanism that may or may not constitute a failure mode. For example, moderate creep in concrete is sometimes welcomed because it relieves tensile stresses that might otherwise lead to cracking. Unlike brittle fracture, creep deformation does not occur suddenly on the application of stress. Instead, strain accumulates as a result of long-term stress. Therefore, creep is a "time-dependent" deformation.

Generally, the creep−time curve includes three stages, as shown in Fig. 1.15. In the initial stage, or primary creep, or transient creep, the strain rate is relatively high but decreases with increasing time and strain due to a process analogous to work hardening at lower temperatures. For instance, the dislocation density increases and, in many materials, a dislocation subgrain structure is formed and the cell size decreases with strain. The strain rate diminishes to a minimum and becomes near constant as the secondary stage begins. This is due to the balance between work hardening and annealing (thermal softening). The secondary stage referred to as "steady-state creep," is the most understood. The microstructure is invariant during this stage, which means that recovery effects are concurrent with deformation. No material strength is lost during these first two stages of creep.

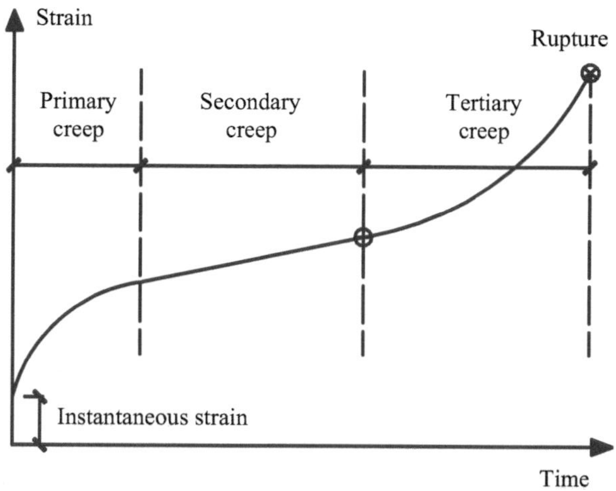

Figure 1.15 Strain as a function of time due to constant stress over an extended period.

The characterized "creep strain rate" typically refers to the constant rate in this second stage. Stress dependence of this rate depends on the creep mechanism. In tertiary creep, the strain rate exponentially increases with stress because of necking phenomena, or internal cracks or voids decrease the effective area of the specimen. Strength is quickly lost in this stage while the material's shape is permanently changed. The acceleration of creep deformation in the tertiary stage eventually leads to material fracture.

1.6.3.2 Creep test at elevated temperature

Creep strain of steel at room temperature is too small and almost has no influence on the behavior of steel structures. At high temperature and high stress level, the creep in steel is far more significant. The measurement of creep strain in steel at elevated temperature is often carried out by a universal creep test setup, as shown in Fig. 1.16.

More details about the creep testing procedures and specimen preparation are available in the following standards:

(1) ISO 204: Metallic materials—Uniaxial creep testing method in tension—Method of test.

(2) GB/T 2039: Metallic materials—Uniaxial creep testing method in tension.

(3) ASTM E139: Test Methods for Conducting Creep, Creep-Rupture, and Stress-Rupture Tests of Metallic Materials.

Figure 1.16 High-temperature creep test setup.

1.7 Book scope

This chapter provides a brief introduction on the HSSs and their application in building structures and also fire hazard and its influence on response of steel structures. State of the art is also reviewed in this chapter. This chapter also introduces test methods on mechanical properties of steel at room and elevated temperature, as well as creep strain measurement. Chapters 2–4 present thermal properties of steel, test results on tensile tests at elevated temperatures, and after fire exposure, respectively. With the information of test provided, it is possible to propose design equations as function of temperature to predict or determine the yield strength, tensile strength, and elastic modulus of several types of HSSs, such as Q460 steel, Q690 steel, S460 steel, S690 steel, S960 steel, and BISPLATE 80 steel, at elevated temperature or after fire exposure. The comparisons and discussions on mechanical properties between various HSSs are presented to understand the difference and relations of different HSSs. Chapter 5 deals with the creep strain of HSSs at elevated temperatures. Creep test data, creep model to predict creep strain, and creep buckling of HSS columns are covered in this chapter. The application of material properties of these steels is presented in Chapter 6, including fire test, temperature distribution, and fire resistance evaluation of structures made in these steels. The last chapter also prospects the knowledge gap and research needs.

References

AISC, 2005. Specification for structural steel buildings. ANSI/AISC2005:360-05, Chicago.

ASTM C1113: Standard Test Method for Thermal Conductivity of Refractories by Hot Wire.

ASTM C518: Standard Test Method for Steady-State Thermal Transmission Properties by Means of the Heat Flow Meter Apparatus.

ASTM D5930: Standard Test Method for Thermal Conductivity of Plastics by Means of a Transient Line-Source Technique.

ASTM E1269: Standard Test Method for Determining Specific Heat Capacity by Differential Scanning Calorimetry.

ASTM E139: Test Methods for Conducting Creep, Creep-Rupture, and Stress-Rupture Tests of Metallic Materials.

Bjorhovde, R., 2004. Development and use of high-performance steel. J. Constr. Steel Res. 60 (3–5), 393–400.

British Standards Institution, 1990. BS5950: The Structural Use of Steelwork in Buildings, Part 8: Code of Practice for Fire Resistant Design, London.

Chen, J., Young, B., 2004. Mechanical properties of cold-formed steel at elevated temperatures. In: Proc., 17th Int. Special Conf. on Cold-Formed Steel Structures. Univ. of Missouri-Rolla, Rolla, MO, pp. 437–466.

Chen, J., Young, B., Uy, B., 2006. Behavior of high strength structural steel at elevated temperatures. J. Struct. Eng. 132 (12), 1948–1954.

Chen, J., Young, B., 2008. Design of high strength steel columns at elevated temperatures. J. Constr. Steel Res. 64 (6), 689–703.

Chiew, S.P., Zhao, M.S., Lee, C.K., 2014. Mechanical properties of heat-treated high strength steel under fire/post-fire conditions. J. Constr. Steel Res. 98, 12–19.

European Committee for Standardization, 2005a. Eurocode 3: Design of Steel Structures—Part 1.1: General Rules and Rules for Buildings. BS EN1993-1-1, Brussels, Belgium.

European Committee for Standardization, 2005b. Eurocode 3: Design of Steel Structures—Part 1.2: General Rules-Structural fire design. BS EN1993-1-2, Brussels, Belgium.

European Committee for Standardization, 2006. Eurocode 4: Design of Composite Steel and Concrete Structures: Part 1.1: General Rules and Rules for Buildings. BS EN1994-1-1, Brussels, Belgium.

European Convention for Constructional Steelworks, 2001. European Recommendations for the Fire Safety of Steel Structures, Section III.2. ECCS General Secretariat, Brussels, Belgium.

GB/T 10295: Thermal Insulation — Determination of Steady-State Thermal Resistance and Related Properties — Heat Flow Meter Apparatus.

GB/T 10297 Test Method for Thermal Conductivity of Nonmetal Solid Materials-Hot Wire Method.

GB/T 2039: Metallic Materials- Uniaxial Creep Testing Method in Tension.

GB/T 228.2: Metallic Materials — Tensile Testing — Part 2: Method of Test at Elevated Temperature.
Hong Kong Buildings Dept, 2005. Code of Practice for the Structural Use of Steel. Government of the Hong Kong Special Administrative Region, Kowloon, Hong Kong.
pecifies Methods for Determining the Specific Heat Capacity of Plastics by Differential Scanning Calorimetry.
ISO 204: Metallic materials—Uniaxial creep testing method in tension—Method of test.
ISO 6892-2: Metallic Materials — Tensile Testing — Part 2: Method of Test at Elevated Temperature.
ISO8302: Thermal Insulation — Determination of Steady-State Thermal Resistance and Related Properties — Guarded Hot Plate Apparatus.
Kirby, B.R., Preston, P.R., 1988. High temperature properties of hot-rolled structural steels for use in fire engineering design studies. Fire Saf. J. 13 (1), 27—37.
Kodur, V.K.R., Harmathy, T.Z., 2008. In: DiNenno, P.J. (Ed.), Properties of Building Materials. SFPE Handbook of Fire Protection Engineering, fourth ed. National Fire Protection Association, Quincy, MA. 1-167—1-195.
Kodur, V.K.R., Dwaikat, M., 2010. Effect of high temperature creep on the fire response of restrained steel beams. Mater. Struct. 43 (10), 1327—1341.
Li, G.Q., Han, L.H., Lou, G.B., Jiang, S.C., 2006. Fire Resistance Design of Steel Structure and Composite Steel and Concrete Structure. Chinese Architecture and Building Press, Beijing.
Li, G.Q., Jiang, S.C., Yin, Y.Z., 2003. Experimental studies on the properties of constructional steel at elevated temperatures. J. Struct. Eng. 129 (12), 1717—1721.
Qiang, X.H., Bijlaard, F.S.K., Kolstein, H., 2012a. Deterioration of mechanical properties of high strength structural steel S460N under steady state fire condition. Mater. Des. 36, 438—442.
Qiang, X.H., Bijlaard, F.S.K., Kolstein, H., 2012b. Post-fire mechanical properties of high strength structural steels S460 and S690. Eng. Struct. 35, 1—10.
Qiang, X.H., Bijlaard, F.S.K., Kolstein, H., 2012c. Dependence of mechanical properties of high strength steel S690 on elevated temperatures. Constr. Build. Mater. 30, 73—79.
Qiang, X., Bijlaard, F.S.K., Kolstein, H., 2013. Post-fire performance of very high strength steel S960. J. Constr. Steel Res. 80, 235—242.
Standardization Administration of the People's Republic of China (SAC), 2008. GB/T 1591-2008: High Strength Low Alloy Structural Steel. China Standard Press, Beijing.
Standards Australia, 1998. AS 4100-Steel Structures. Sydney, Australia.
State Standard of the P.R. China, 2017. Code for Fire Safety of Steel Structures in Buildings. GB51249-2017. China Planning Press, Beijing, China.
Twilt, L., 1988. Strength and deformation properties of steel at elevated temperatures. Fire Saf. J. 13 (1), 9—15.
Wang, W.Y., Li, G.Q., Dai, G.X., 2010. Fire-resistance performance of high-strength-steel H shaped columns under the axial compression. J. Chongqing Univ. 33 (10), 76—82 (in Chinese).
Wang, W.Y., Liu, B., Kodur, V.K.R., 2013. Effect of temperature on strength and elastic modulus of high strength steel. J. Mater. Civ. Eng. 25 (2), 174—182.
Wang, W.Y., Liu, T.Z., 2016. Experimental and numerical study on post-fire behavior of high-strength Q460 steel columns. Adv. Struct. Eng. 19 (12), 1—16.
Wang, W.Y., Wang, K., Kodur, V.K.R., Wang, B., 2018a. Mechanical properties of high strength Q690 steel at elevated temperature. J. Mater. Civ. Eng. 30 (5), 04018062.
Wang, W.Y., Yan, S.H., Liu, J.P., 2017. Test on temperature induced creep in high strength Q460 steel. Mater. Struct. 50, 68.

Wang, W.Y., Zhang, L.B., Ge, Y., Xu, L., 2018c. Behaviour of restrained high strength steel columns at elevated temperature. J. Constr. Steel Res. 148, 251–264.

Wang, W.Y., Zhang, L.B., He, P.Z., 2018b. A numerical investigation on restrained high strength Q460 steel beams including creep effect. Int. J. Steel Struct. 18 (5), 1497–1507.

Wang, W.Y., Zhou, H.Y., Xu, L., 2019. Creep buckling of high strength Q460 steel columns at elevated temperatures. J. Constr. Steel Res. 157, 414–425.

Wang, W.Y., Zhou, H.Y., Zhou, Y.C., Xu, L., 2018d. An approach for predicting fire resistance of high strength Q460 steel beams considering un-uniform temperature distribution. Fire Technology 54 (2), 437–460.

Wang, W.Y., Li, G.Q., 2014. An approach for evaluating fire resistance of high strength Q460 steel columns. Front. Struct. Civ. Eng. 8 (1), 26–35.

Wang, W.Y., Liu, T.Z., Liu, J.P., 2015. Experimental study on post fire mechanical properties of high strength Q460 steel. J. Constr. Steel Res. 114, 100–109.

Zhou, H.T., Wang, W.Y., Wang, K., Xu, L., 2019. Mechanical properties deterioration of high strength steels after high temperature exposure. Constr. Build. Mater. 199, 664–675.

CHAPTER 2

Thermal properties of steel at elevated temperature

Contents

2.1 Introduction

For structural fire engineering, material thermal properties of steels are needed for thermal analysis, including the thermal conductivity, specific heat, expansion coefficient, and density. The book focuses on the material properties of new types of steel, mainly on high strength steels.

Thermal conductivity and specific heat vary with temperature and are also influenced by the chemical composition (carbon content) of steel. Thermal conductivity and specific heat of steel can be measured using "steady-state" and "transient" test methods. Transient methods are preferred because the temperature-induced physicochemical changes that occur in steel can be captured. The thermal conductivity of conventional mild steel at room temperature is in the range 46–65 W/m°C. Carbon content has significant influence on the thermal conductivity of steel, with higher carbon content leading to lower thermal conductivity. The specific heat of steel is also influenced by the carbon content of steel when heated beyond the "Curie point" (also called Curie temperature, refers to the temperature beyond that at which a ferromagnetic substance exhibits

Material Properties of Steel in Fire Conditions
ISBN 978-0-12-813302-6
https://doi.org/10.1016/B978-0-12-813302-6.00002-3

paramagnetism). However, a review of the literature shows that no specific studies have been carried to evaluate the thermal properties of high strength steels, except the authors of these book carried out thermal tests on steels A325 and A490 used in bolts (Kodur et al., 2012). The coefficient of thermal expansion, defined as the percentage change in length of specimen per degree rise in temperature, is generally influenced by the temperature rise. Many researchers conducted tests on mild and nonstructural steels to determine thermal expansion at elevated temperatures. These tests did not consider the effect of high strength (of steel) on thermal expansion of steel. Comparison of various constitutive models indicated that the thermal properties of steel vary with the physicochemical changes that occur in steel at approximately 700°C. However, measurement of these thermal properties becomes complex beyond 700°C, when steel undergoes a phase change. Hence, the majority of test programs evaluate thermal properties (in the heating phase only) to 700°C, thus not capturing the influence of phase changes in steel.

This chapter describes the test on thermal properties of high strength steel A325 and A490, as well as thermal properties of mild steel suggested in current design codes, such as EC3 (2005), EC4 (2005), BS5950 (2000), and GB50017 (2017), etc.

2.2 Thermal properties of high strength bolt steel

2.2.1 Test procedures

Thermal conductivity and specific heat of A36, A325, and A490 steel specimens were measured during both the heating and cooling phases at nine temperatures: 20, 100, 200, 300, 400, 500, 600, 700, and 735°C. Kapton sensor was used for tests at room temperature (20°C), and mica sensor was used at the remaining temperatures. The sensor is sandwiched between two halves of the specimen, and the assembly is exposed to high temperatures generated in the electric furnace connected to the Hot Disk apparatus. The target temperature of the specimen was increased from 100 to 700°C in 100°C steps and, then, to 735°C to capture the effect of phase change on the thermal properties of steel. Following this, the specimen was cooled from 735 to 20°C; during the cool-down phase, readings were taken at 100°C intervals in the range 700–100°C and also at 20°C.

Test parameters, namely, probing depth, sensor type, initial thermal coefficient of resistance (TCR), hold time at steady state, measurement power, and measurement times, are programmed (as input) into the Hot

Disk apparatus by the user. For these tests, an initial TCR of 0.004693/K, probing depth of 15 mm, sensor measurement power of 1.0 W, hold time of 30 min, and measurement time of 10 s were used. For high-temperature measurements, a TCR value at each temperature was preset as per sensor type. At each target temperature, once thermal equilibrium is attained, thermal conductivity and specific heat are recorded. The temperature then increases for new measurements.

For thermal expansion measurements, test specimens were prepared as per E831-06 specifications (ASTM 2006). The steel specimen was placed on a pedestal in the moveable furnace of the TMA apparatus with a microexpansion probe set on the specimen. A small force of 0.02 N was applied to the probe on the specimen to maintain static contact between the specimen and the probe. A temperature ramp of 5°C/min was applied to raise the temperature to 1000°C and then decrease it back to room temperature at the same rate. At each temperature, the TMA apparatus recorded the changes in dimensions of the steel specimens in both heating and cooling stages.

2.2.2 Test results

Thermal conductivity of A36, A325, and A490 steel in the heating and cooling phases of fire is plotted in Fig. 2.1A. The thermal conductivity of these three steel specimens at room temperature is in the range 47–49 W/m°C and gradually decreases with temperature. Thermal conductivity of A490 steel is slightly lower than those of A36 and A325 steel throughout the temperature range. This could be attributed to the comparatively high carbon content of A490 steel. However, as temperature increases, the effect of carbon content on thermal conductivity diminishes and is almost negligible after 700°C. At 735°C, thermal conductivity drops to 28 W/m°C for all three types of steel. In the cooling phase, as temperature decreases, thermal conductivity increases and returns to its original value. Thus, the thermal conductivity of all three types of steel in the cooling phase is in close proximity to that in the heating phase. Irrespective of the irreversible microstructural changes that occur as a result of temperature variation, the thermal conductivity of steel is quite similar in the heating and cooling phases, as inferred by Kodur and Harmathy (2008) on the basis of the behavior of conventional steel.

Fig. 2.1B illustrates the specific heat values of A36, A325, and A490 steel recorded in the heating and cooling phases of fire. The specific heats of A36, A325, and A490 steel at room temperature were recorded as 490, 506, and 493 J/kg°C, respectively; specific heat increased slowly with

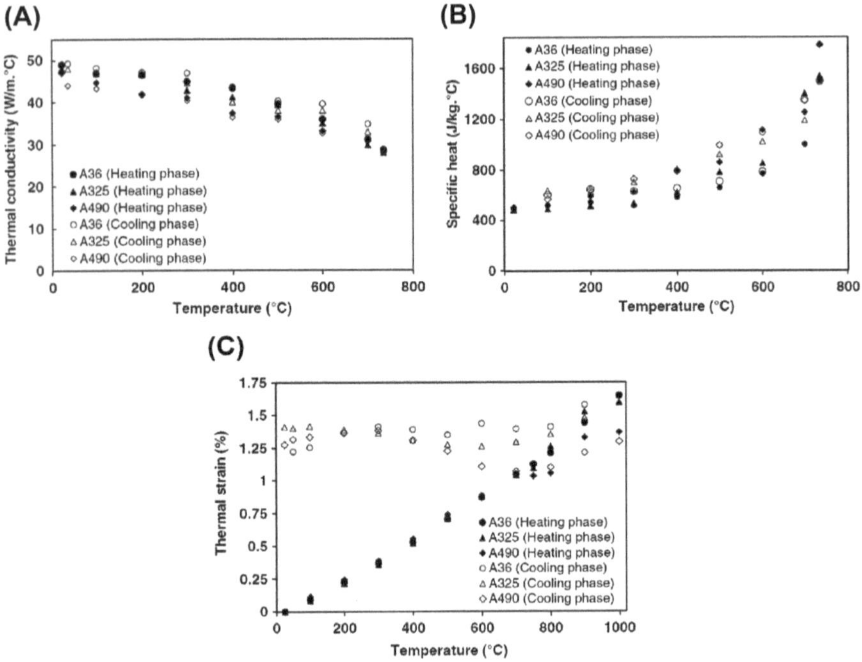

Figure 2.1 Thermal properties of A36, A325, and A490 steel in heating and cooling phases. (A) Thermal conductivity, (B) specific heat, and (C) thermal strain.

temperature up to 600°C. Specific heat increased at a faster rate in the range 600–700°C, and a sudden jump occurred at 735°C. At this temperature, the values recorded were 1538, 1487, and 1798 J/kg°C, respectively. When steel is heated beyond 700°C, its microstructure undergoes a phase transition from a body-centered cubic crystal structure, "α-ferrite," to a face-centered cubic crystal structure, "γ austenite." During this phase change, the steel absorbs considerable thermal energy (latent heat), resulting in a sudden increase in specific heat. As the temperature of steel decreases (in cooling phase), the specific heat of all three types of steel also decreases, and the values are in close proximity to that noted in the heating phase.

Thermal expansion of the three types of steel was evaluated as the percentage change in length over original length of the specimen and is plotted in Fig. 2.1C as a function of temperature, in both the heating and cooling phases. The thermal strain of A36 and A325 steel increases linearly with temperature in the range 20–1000°C. However, A490 experiences a linear increase in thermal strain up to 700°C, a constant plateau in the range 700–800°C, and then another linear increase in the range 800–1000°C.

The constant thermal strain for A490 steel at 700–800°C can be attributed to the phase change that occurs in A490 steel (having higher carbon content) at about 700°C. During this phase change, energy absorbed by the steel is used for the microstructural changes; hence, thermal expansion remains constant. As its crystalline structure transforms completely at about 800°C, steel starts to expand again. This phenomenon is less pronounced in A325 (medium carbon) steel and totally absent in A36 (low carbon) steel because of the lower carbon content of these steel types.

In the cooling phase, thermal strain of all three types of steel follows a trend different (constant) from that in the heating phase basically because of the irreversible changes that occur in the microstructure of steel when it is subjected to a thermal cycle of heating and cooling. This difference in thermal strain between A325 and A490 steel, in the heating and cooling phases, can be explained by the fact that, on heating beyond 700°C, their microstructure changes to γ-austenite, which on cooling turns into pearlite, which has a coarse microstructure, with alternate layers of α-ferrite and cementite. This microstructure possesses different expansion properties compared with the original tempered martensite steel of A325 and A490 bolts.

2.2.3 Comparison of measured properties with published data

The thermal properties of A36, A325 and A490 steel are now compared with those of conventional steel.

Thermal conductivity and specific heat of A36, A325, and A490 steel, measured in the current test program, are compared with specified ASCE and EC3 relationships and other published data in Fig. 2.2A,B, respectively. It should be noted that the specified thermal property values from published literature are derived for conventional steel (A36). It can be seen in Fig. 2.2A that the measured thermal conductivity of A36 and A325 steel lies within the range of ASCE and EC3 provisions. However, measured thermal conductivity of A490 steel is slightly lower than that specified in codes and standards, and this can be attributed to the higher carbon content of A490 steel. Measured specific heats of A36, A325, and A490 steel in both heating and cooling phases (of fire) agree with ASCE and Eurocode provisions up to 700°C. Beyond 700°C, as steel undergoes phase transition, specific heats of different types of steel follow different trends, which could be attributed to the varying chemical composition of steel.

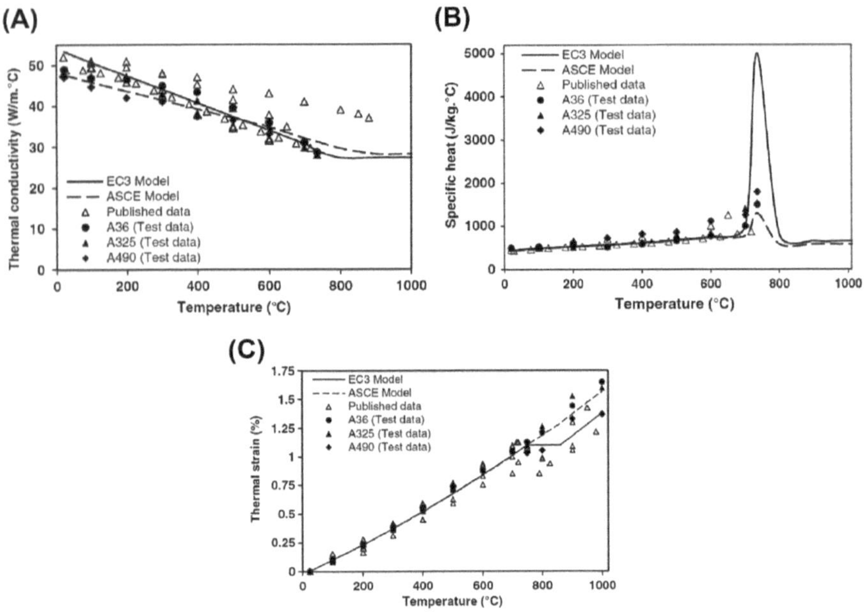

Figure 2.2 Comparison of test data with ASCE and EC3 provisions and published test data. (A) Thermal conductivity, (B) specific heat, and (C) thermal expansion.

Thermal expansion of all types of steel follows trends similar to the ASCE and EC3 provisions until 700°C (see Fig. 2.2C). However, beyond 700°C, thermal expansion of A36 and A325 steel in the heating phase of fire follows trends closer to ASCE provisions, whereas that of A490 steel follows trends closer to EC3 provisions in the heating phase of fire.

2.2.4 High-temperature property relations

Relationships for thermal conductivity, specific heat, and thermal expansion of A36, A325, and A490 steel are presented as a function of temperature in the heating and cooling phases of fire. These high-temperature property relationships are derived by regression analysis of measured property data. The coefficient of determination (R^2), which represents the proportion of the sum of squares of deviations of the response values about their predictor, defines the accuracy of the statistical model. The R^2 values for the proposed relationships are in the range 0.92–1, indicating the good fit of the test data.

Data from the thermal property measurements are utilized to develop thermal conductivity (k_t), specific heat (c_p), and thermal strain (ε_{th}) relationships for A36, A325, and A490 steel. The proposed relations for thermal conductivity, specific heat, and thermal expansion for A36, A325, and A490 steel in the heating and cooling phases of fire are given in Table 2.1. Thermal conductivity and specific heat of steel were not affected by the direction of the heat flow; hence, the proposed relations for thermal conductivity and specific heat are applicable in both the heating and cooling phases of fire. On the other hand, thermal expansion of steel in the cooling phase follows a different pattern than that in the heating phase; hence, separate relations are proposed.

2.3 Thermal expansion of mild steel

For carbon steels, the thermal strain is linear with the increase of temperature up to 700°C, as shown in Fig. 2.3. When the temperature exceeds 700°C, the thermal strain decreases with the elevation of temperature and the value of thermal strain is dependent of carbon content in steel, as shown in Fig. 2.4. This phenomenon is called "phase transformation" effect in medium and lower carbon steels. As far as we know, the test data of thermal expansion for new types steel are rare, and the thermal expansion of the new type steel can be determined according to specifications in current codes, such as ECCS (1989), EC3 (2005), EC4 (2005), BS5950 (2000), and GB50017 (2017), etc.

In EC3 (2005) and EC4 (2005), the thermal expansion of structural steels can be calculated by

$$\Delta l / l = 0.4 \times 10^{-8} T_s^2 + 1.2 \times 10^{-5} T_s - 2.416 \times 10^{-4}$$

$$20^\circ\text{C} \leq T_s \leq 750^\circ\text{C} \tag{2.1a}$$

$$\Delta l / l = 11 \times 10^{-3} \quad 750^\circ\text{C} < T_s \leq 860^\circ\text{C} \tag{2.1b}$$

$$\Delta l / l = 2.0 \times 10^{-5} T_s - 6.2 \times 10^{-3} \quad 860^\circ\text{C} < T_s \leq 1200^\circ\text{C} \tag{2.1c}$$

where Δl is expansion due to temperature elevation, l is the member length before expansion, and T_s is the temperature of steel.

The values of thermal expansion coefficient can be determined by EC3 (2005) and EC4 (2005) with the following expressions:

$$\alpha_s = 0.8 \times 10^{-8}(T_s - 20) + 1.2 \times 10^{-5} \quad 20^\circ\text{C} \leq T_s \leq 750^\circ\text{C} \tag{2.2a}$$

$$\alpha_s = 0 \quad 750^\circ\text{C} < T_s \leq 860^\circ\text{C} \tag{2.2b}$$

$$\alpha_s = 2.0 \times 10^{-5} \quad 860^\circ\text{C} < T_s \leq 1200^\circ\text{C} \tag{2.2c}$$

Table 2.1 Thermal property relationships for different types of steel.

Property	Steel type	Relationship
Thermal conductivity (in both heating and cooling phases), W/m °C	A36	$k_t = 49.85 - 0.021T$, $20°C \le T < 600°C$
		$k_t = 67.063 - 0.052T$, $600°C \le T \le 735°C$
	A325	$k_t = 50 - 0.027T$, $20°C \le T \le 735°C$
	A490	$k_t = 47 - 0.023T$, $20°C \le T \le 735°C$
Specific heat (in both heating and cooling phases), MJ/m^3 °C	A36	$c_p = 495 - 0.21T + 0.0014T^2$, $20°C \le T < 600°C$
		$c_p = 20.94T - 0.012T^2 - 7432$, $600°C < T \le 735°C$
	A325	$c_p = 490 - 0.19T + 0.0014T^2$, $20°C \le T \le 600°C$
		$c_p = 20.79T - 0.0118T^2 - 7350$, $600°C \le T \le 735°C$
	A490	$c_p = 490 - 0.39T + 0.001T^2$, $20°C \le T \le 600°C$
		$c_p = 15.274T - 9438.8$, $600°C < T \le 735°C$
Thermal expansion (heating phase), %	A36	$\varepsilon_{th} = 0.0017T - 0.1$, $20°C \le T \le 1000°C$
	A325	$\varepsilon_{th} = 0.0017T - 0.1$, $20°C \le T \le 1000°C$
	A490	$\varepsilon_{th} = 0.0015T - 0.1$, $20°C \le T \le 1000°C$
Thermal expansion (cooling phase), %	A36	$\varepsilon_{th} = 1.45$, $20°C \le T \le 1000°C$
	A325	$\varepsilon_{th} = 1.38$, $20°C \le T \le 1000°C$
	A490	$\varepsilon_{th} = 1.25$, $20°C \le T \le 1000°C$

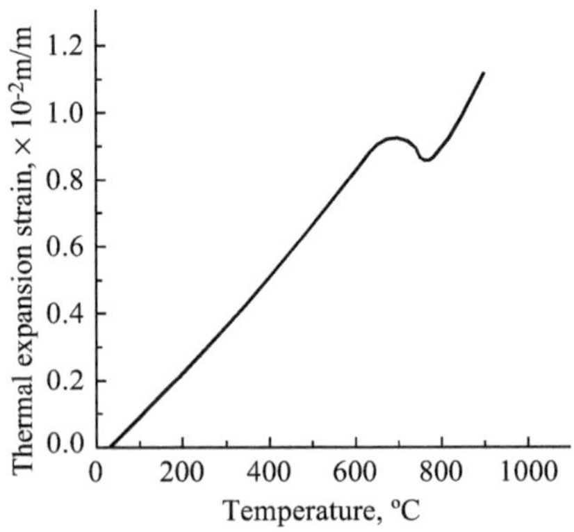

Figure 2.3 Thermal strain curve of carbon steel.

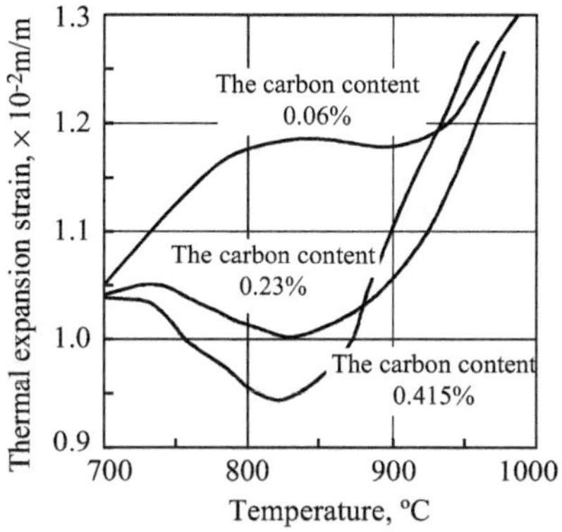

Figure 2.4 "Phase transformation" effect.

American Institute of Steel Construction provides a simple equation to predict the thermal expansion coefficient as a function of temperature.

$$\alpha_s = (11 + 0.062T_s) \times 10^{-6} \quad 20°C \le T_s \le 600°C \tag{2.3}$$

In Australian standard AS4100 (1990), the thermal expansion coefficient can be obtained by

$$\alpha_s = (11.4 + 0.01T_s) \times 10^{-6} \quad 20°C \le T_s \le 600°C \tag{2.4}$$

In Japan, the thermal expansion coefficient can be obtained by The Japan Building Synthetic Fire Prevention Design (1994).

$$\alpha_s = (11.0 + 5.75 \times 10^{-3} T_s) \times 10^{-6} \tag{2.5}$$

In China GB 50017(2017), the thermal expansion coefficient is very simple and a value of $\alpha_s = 1.2 \times 10^{-5}$ is used. The comparison of the thermal expansion coefficient obtained by various codes are plotted in Fig.2.5.

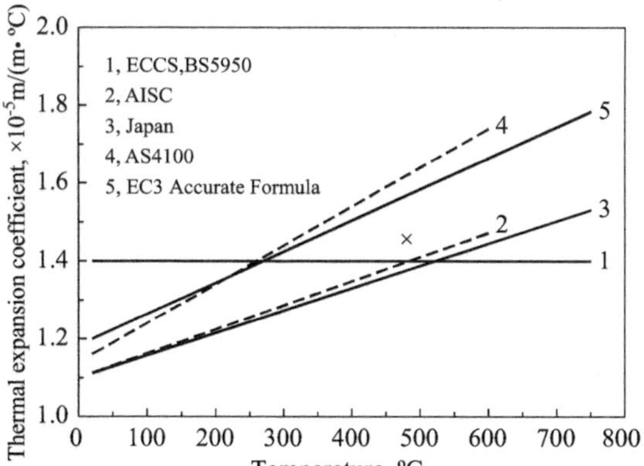

Figure 2.5 Comparison of thermal expansion coefficient of steel in various standards.

2.4 Specific heat of mild steel

Specific heat Cs in J/(kg·°C) or J/(kg·K) is the heat required or released to increase or decrease by 1°C by a unit quantity of material.

EC3 (2005) and EC4 (2005) suggested that the specific heat of steel is taken as follow:

For 20°C ≤ T ≤ 600°C,

$$C_s = 425 + 7.73 \times 10^{-1} T_s - 1.69 \times 10^{-3} T_s^2 + 2.22 \times 10^{-6} T_s^3 \quad (2.6a)$$

For 600°C ≤ T ≤ 735°C,

$$C_s = 666 - \frac{13302}{T_s - 738} \quad (2.6b)$$

For 735°C ≤ T ≤ 900°C,

$$C_s = 545 - \frac{17820}{T_s - 731} \quad (2.6c)$$

For 900°C ≤ T ≤ 1200°C,

$$C_s = 650 \quad (2.6d)$$

For simplifying the calculation, the specific heat of steel can take a constant value of 600.

BS5950 (2000) gives the following value for the specific heat as C520.

The Annex of ECCS (1989) gives the following temperature-dependent equation for the specific heat as

$$C_s = 470 + 20 \times 10^{-2} T_s + 38 \times 10^{-5} T_s^2 \quad (2.7)$$

The Japan Building Synthetic Fire Prevention Design (1994) gives the following equation for the specific heat as

$$C_s = 483 + 8.02 \times 10^{-4} T_s^2 \quad (2.8)$$

The China code for fire safety of steel structures in buildings (GB51249-2017) gives the temperature-independent thermal conductivity as 600.

Specific heats calculated by above equations are shown in Fig. 2.6. It can be seen that the specific heat of steel changes greatly with the elevation in temperature. In the temperature range of 0–600°C, the average value of specific heat is about 600 J/(kg·°C). Taking the specific heat as a temperature-independent value does not agree with the facts. However, it can greatly simplify the calculation and can be easily used by the engineer.

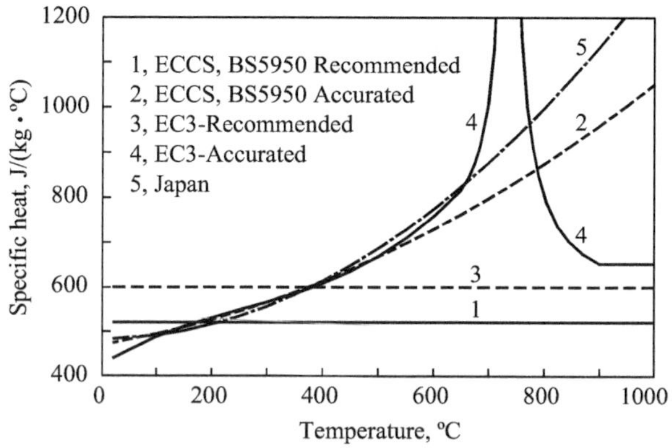

Figure 2.6 Comparison of specific heat of steel in various standards.

2.5 Thermal conductivity of mild steel

The conductivity in W/(m·°C) or W(m·K) is the property of a material representing its ability to conduct heat. EC3 (2005) and EC4 (2005) give the following equation:

For $20°C \leq T \leq 800°C$,

$$\lambda_s = 54 - 3.33 \times 10^{-2} T_s \tag{2.9}$$

For $800°C \leq T \leq 1200°C$,

$$\lambda_s = 27.3 \tag{2.10}$$

BS5950 Part 8 (2000) gives the following temperature-independent equation for λs as 37.5.

The temperature-dependent thermal conductivity described by BS5950 (2000) is

$$\lambda_s = 52.27 - 1.541 \times 10^{-2} T_s - 2.155 \times 10^{-5} T_s^2 \tag{2.11}$$

The Japan Building Synthetic Fire Prevention Design (1994) gives the following equation:

$$\lambda_s = 52.08 - 5.05 \times 10^{-5} T_s^2 \tag{2.12}$$

The China Code for Fire Safety of Steel Structures in Buildings (GB51249-2017) also gives a temperature-independent thermal conductivity of steel as 45.0.

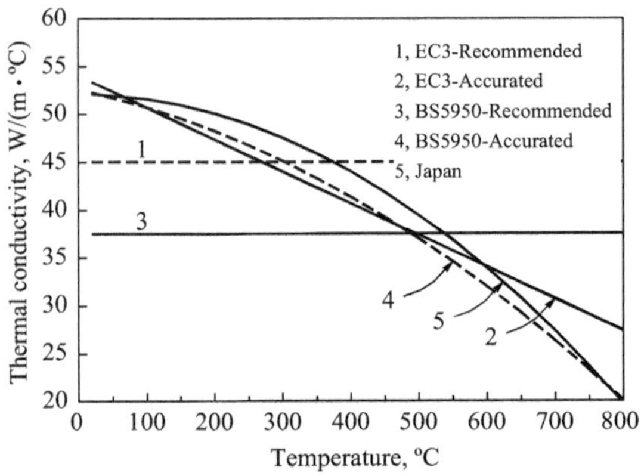

Figure 2.7 Comparison of thermal conductivity of steel in various standards.

The comparison of above equations is shown in Fig. 2.7. In the temperature range of 0–600°C, the average value of λs is very close to 45 W/(m·°C), and in the temperature range of 200–800°C, the average value of λs is very close to 37.5 W/(m·°C). Usually, the critical temperature of the structural component is in the temperature range of 300–600°C; hence, the thermal conductivity taking a value of 45 W/(m·°C) is reasonable.

2.6 Density

The density of steel may be taken as its ambient value of 7850 kg/m^3 over the normally experienced temperature range in a building fire.

References

AS4100, 1990. Australian Standard. Steel Structures, Standards, Australia.

ASTM, 2006. Standard test method for linear thermal expansion of solid materials by thermomechanical Analysis. ASTM E831, West Conshohocken, PA.

British Standard BS 5950, 2000. British Standard Institution. Part 8, Code of Practice for Fire Resistance Design.

ECCS, 1989. European Convention for Construction Steelwork. Calculation for Fire Resistance of Composite Structures. Technical Note No. 55, ECCS Technical Committee 3.

EN1993-1-2, 2005. European Committee for Standardization. Eurocode 3: Design of Steel Structures, Part 1.2, General Rules, Structural Fire Design.

EN1994-1-2, 2005. European Committee for Standardization. Eurocode 4: Design of Composite Steel and Concrete Structures, Part 1-2: General Rules, Structural Fire Design.

GB50017, 2017. State Standard of the P.R. China. Code for Design of Steel Structures. China Planning Press, Beijing, China.
GB51249, 2017. State Standard of the P.R. China. Code for Fire Safety of Steel Structures in Buildings. China Planning Press, Beijing, China.
Japan Construction Department, 1994. The Building Synthetic Fire Prevention Design.
Kodur, V.K.R., Kand, S., Khaliq, W., 2012. Effect of temperature on thermal and mechanical properties of steel bolts. J. Mater. Civ. Eng. 24 (6), 765–774.
Kodur, V.K.R., Harmathy, T.Z., 2008. Section 10: Properties of building materials. Chapter 1-10. In: SFPE handbook of fire protectionengineering. Society of Fire Protection Engineers, Quincy, MA, pp. 167–195.

CHAPTER 3

Tensile test on steels at elevated temperatures

Contents

Material Properties of Steel in Fire Conditions
ISBN 978-0-12-813302-6
https://doi.org/10.1016/B978-0-12-813302-6.00003-5

3.1 Introduction

Mechanical properties of steel are significant, not only for evaluating the behavior of individual steel members but also for predicting the performance of a whole structure. Under fire condition, the mechanical properties of steel deteriorate as temperature increases. The deterioration of mechanical properties is dependent on the temperatures of steel material.

In practice, the deterioration is represented by reduction factor of mechanical properties as recommended by some current design standards. This chapter presents test data of mechanical properties of high strength steels (HSSs) including Q460 steel, Q690 steel, S460 steel, S690 steel, S960 steel, BISPLATE 80 steel, and ASTM A572 steel.

3.2 Mechanical properties of high strength Q460 steel

To develop high-temperature strength and stiffness properties of high strength Q460 steel, Wang et al. (2013) carried out a comprehensive test. The test program consisted of two types of tests, one for strength and the other for elastic modulus. These tests were conducted at various temperature points in the range of 20 ~ 800°C.

3.2.1 Test setup

The test setup for tension strength tests is composed of an electrohydraulic servo universal testing machine, an electric furnace, and a data collection system. The universal testing machine has a loading capacity of 1000 kN, and the axial displacement of the specimen in tension is measured by a displacement transducer. An electric furnace mounted around the universal testing machine can generate temperatures up to 1200°C, with an accuracy of ±0.5°C through a temperature control system. Three thermocouples are mounted on the furnace to measure temperatures at the upper, middle, and lower zones. The average value of these three thermocouples is assumed as the representative furnace temperature. The furnace has an outer diameter of 380 mm and a length of 500 mm, and hence the heating length of a specimen is 500 mm. The universal testing machine and the furnace are connected to a computer and a SANS DCS-300 closed-loop control system and temperature; displacement and load during the test are recorded through software package, SANS Power Test V3.3 for DCS. The test setup for tension strength is shown in Fig. 3.1A.

The test setup for elastic modulus is composed of a power signal generator, an oscilloscope, an electric furnace, and transducers, as shown in Fig. 3.1B. The power signal generator can produce a sine wave, triangle wave, or square wave with a frequency range of 5 Hz through 500 kHz. The oscilloscope receives the electric signal and displays it on the screen. The electric furnace, similar to that in the tension setup, can generate temperatures up to 1200°C. The transducer transforms the electric signal to vibration signal and vice versa. When the test is conducted at

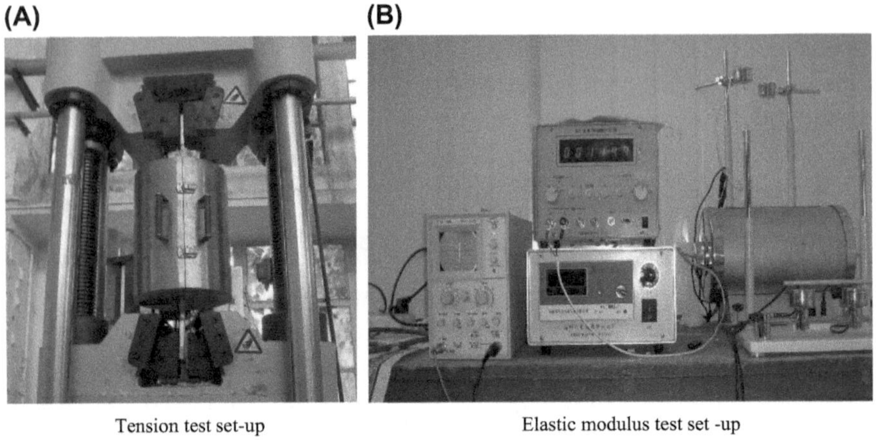

Figure 3.1 Test setup for high-temperature tension test and elastic modulus test.

room temperature, the specimen is put on the transducer outside the furnace. When the experiment is conducted at elevated temperatures, the specimen is put into the furnace and anchored through two copper wires.

3.2.2 Test specimen

The specimen for tension tests are 760 mm long, and their configuration is shown in Fig. 3.2. For tension tests, the recommended configuration is as shown in Fig. 3.2A. Preliminary tests reveal that the failure of HSS specimen in the range of 200–450°C always occurs at the ends of the section (cooler region) because the tension strength of the heated portion of steel increases with temperature, resulting from the blue brittleness effect. This effect forms a blue oxidation film on the surface of steels when they

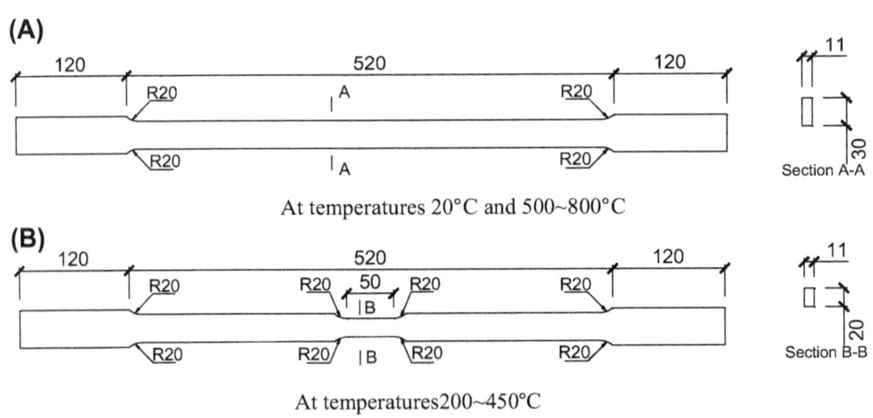

Figure 3.2 Configuration of specimen for tension strength tests (units: mm).

are subjected to approximately 300°C. This phenomenon is called blue brittleness, and this effect leads to a small increase in strength and decrease in the ductility property of steel. To capture this effect, the test specimen used for temperatures ranging from 200 to 450°C were reconfigured, as shown in Fig. 3.2B. For these specimen, the heating portion is 500 mm, so it is difficult to calculate the strain through measured axial displacements.

For these specimen, only stress—axial displacement curves are described in the test result section. All test specimen are prepared by cutting samples from a structural steel sheet with a nominal plate thickness of 11 mm. A total of 22 specimen are used in this experiment. At each temperature, two specimen (S-1 and S-2) are tested, and the average of the two values is the result. The actual dimensions of the specimen are measured using vernier calipers. The middle part of the specimen is laid in the furnace, and the two ends are connected by the grips of the electrohydraulic servo universal testing machine. For the elastic modulus test, a total of six specimen (S-1 to S-6) are prepared by cutting a Q460 steel plate. The specimen are of cylindrical shape with length of 160 mm and diameter of 7 mm. Special attention was paid in preparing the specimen to precise dimensions because the dimensions can have a large influence on the results. The measured sizes of the test specimen are shown in Table 3.1. Each specimen is placed in the furnace, and the temperature is elevated to the desired level. Six sets of data are generated at each predefined temperature, and the average of the six values is the resulting elastic modulus.

3.2.3 Test procedure

In the tension tests, the specimen is placed in the furnace, which is turned on to heat up to a predefined target temperature. The rate of heating is 10°C/min. When the target temperature is attained, the specimen is maintained at this temperature (in the furnace) for 20 min to reach a steady-state condition throughout the specimen. After that, a tensile load is

Table 3.1 Specimen sizes of Q460 steel used in elastic modulus test.

Specimen	Mass (g)	Length (mm)	Diameter (mm)
1	47.874	159.96	6.997
2	48.126	160.00	6.930
3	48.038	160.04	6.912
4	48.115	160.02	6.922
5	48.032	160.03	6.932
6	47.899	160.04	6.900

applied in increments on the specimen while keeping the temperature constant. A constant tensile loading at a rate of 0.5 kN/s is applied on the specimen, as specified by GB/T 228 (2002) and GB/T 4338 (2006b). The tension tests are conducted at target temperatures of 20, 100, 200, 300, 400, 450, 500, 550, 600, 700, and 800°C. For the high-temperature elastic modulus tests, the specimen is placed in the furnace and connected with two copper wires. The furnace is then turned on, and the specimen is heated to the predefined target temperature. After attaining a stable temperature throughout the specimen, a power signal is generated and adjusted until the resonance arises on the computer screen. At this point, the fundamental frequency of the specimen is noted.

The elastic modulus tests are conducted at 13 temperature points, namely, 20, 100, 200, 250, 300, 350, 400, 450, 500, 550, 600, 700, and 800°C. The room temperature elastic modulus was evaluated using the following equation, as specified by GB/T 2105 (1991):

$$E = 1.6067 \times 10^{-9} m l^3 f_1^2 T_1 / d^4 \quad (3.1)$$

where m is mass of the specimen (in g); l and d are length and diameter of the specimen (in mm); f_1 is fundamental frequency of the specimen (in Hz); T_1 is modification factor of the fundamental frequency of the specimen, which is shown by GB/T 2105 (1991) with the values of $\bar{r}/l$ and μ, in which $\bar{r}/d$ equals to 4 and d is diameter of the specimen; μ is Poisson's ratio of steel, which is measured as 0.3. The measured frequency at a given temperature and the various parameters from Table 3.1 were substituted into Eq. (3.1) to obtain the elastic modulus at a target temperature.

3.2.4 Test results

Data generated from tension and vibration frequency tests were utilized to obtain the failure characteristics, tensile strength, and elastic modulus of Q460 steel at various temperature points. The typical failure modes of specimen at various temperatures are shown in Fig. 3.3. As shown in the figure, the damage zones (failure surfaces) become smooth and flat with increasing temperatures.

3.2.4.1 Yield strength and ultimate strength

The recorded load and displacement at various temperatures were used to calculate stress and strain values. These stress—strain points are plotted in Fig. 3.4A for 20 and 100°C and also for various temperatures in the range of 500—800°C. In the temperature range of 200—450°C, the evaluation of

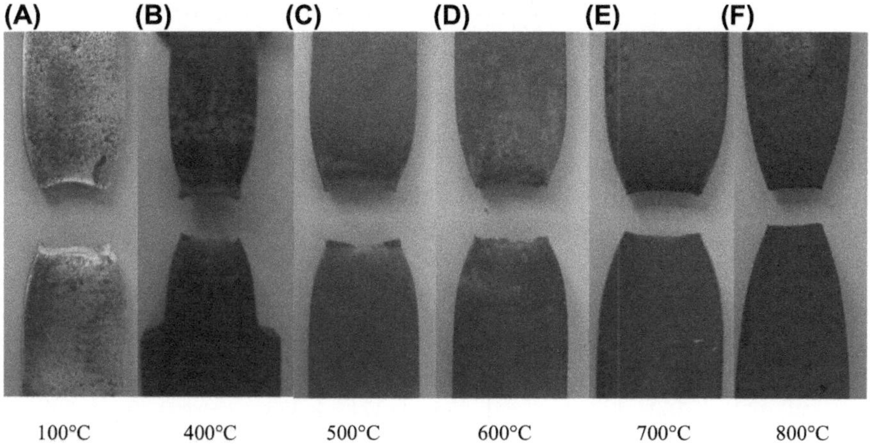

Figure 3.3 Photos of the tension specimen after failure at different temperatures.

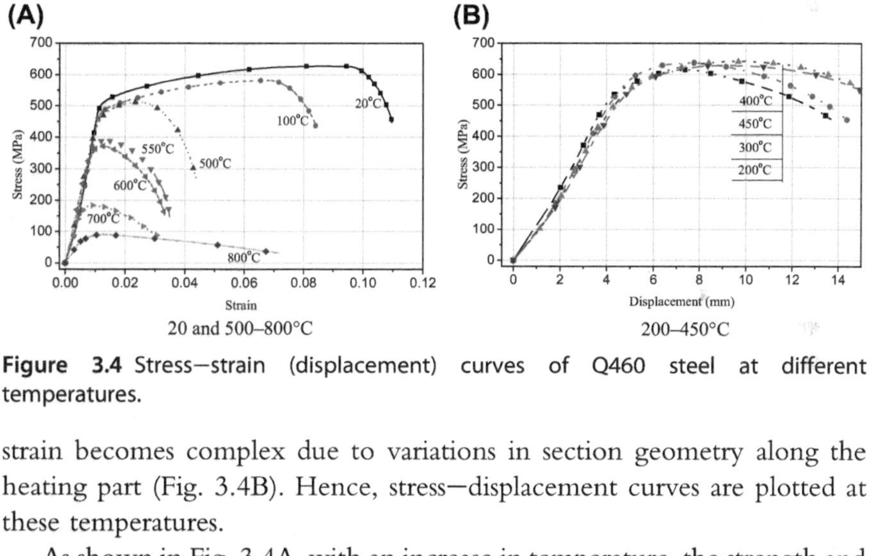

Figure 3.4 Stress—strain (displacement) curves of Q460 steel at different temperatures.

strain becomes complex due to variations in section geometry along the heating part (Fig. 3.4B). Hence, stress—displacement curves are plotted at these temperatures.

As shown in Fig. 3.4A, with an increase in temperature, the strength and ductility of Q460 steel generally decreases, except at 800°C, when the ductility increases. Results plotted in Fig. 3.4B show that in the temperature range of 200 ~ 450°C, the tensile strength is higher than at room temperature. This is due to the deformation aging of carbon and nitrogen interstitial atoms in the temperature range of 200—450°C. This phenomenon is called blue brittleness, and the temperature range at which this occurs varies with the composition of steel. From the data generated by the tension tests, the yield and ultimate strengths were evaluated and are given

Table 3.2 Measured yield and ultimate strength of Q460 steel at various temperatures.

	$f_{y,T}$ (MPa)			$f_{u,T}$ (MPa)				
T (°C)	S-1	S-2	Average	S-1	S-2	Average	$f_{y,T}/f_y$	$f_{u,T}/f_u$
20	500	506	503	620	632	626	1	1
100	444	440	442	587	573	580	0.88	0.93
200	530	548	539	610	618	614	1.07	0.98
300	574	576	575	635	639	637	1.14	1.02
400	526	510	518	664	622	643	1.03	1.03
450	532	—	532	628	—	628	1.06	1
500	421	439	430	501	523	512	0.85	0.82
550	374	—	374	392	—	392	0.74	0.63
600	372	362	367	375	371	373	0.73	0.6
700	180	184	182	180	188	184	0.36	0.29
800	88	90	89	88	94	91	0.18	0.15

At temperatures of 450 and 550°C, only one specimen was tested.

in Table 3.2. The yield strength at temperatures of 20 and 100°C, and in the range of 500–800°C, corresponds to a strain level of 1%, whereas the yield strength at temperatures of 200, 300, 400, and 450°C was obtained by identifying the inflexion of the stress–displacement curves. The ultimate strength at all temperature points was obtained, corresponding to maximum stress points on the curves. In addition, the yield (ultimate) strength reduction factor of Q460 steel was obtained as the ratio of yield (ultimate) stress at temperature T to that at room temperature and is given in Table 3.2. The test results show that in the temperature range of 200–450°C, the yield strength of Q460 steel is higher than at room temperature, but there is no obvious increase in ultimate strength. At 600°C, Q460 steel retains 70% of its room temperature strength, which is higher than for conventional mild steel. However, the yield strength and ultimate strength decrease at a faster rate beyond 600°C.

The yield strength of Q460 steel is compared with that reported in literature for other HSSs, such as AS/NZS 1554.4 (2004) and BISPLATE 80 (Chen et al., 2006), and mild steel, such as Japanese Industrial Standard G3106-SM41 (1999) and SM41 (Li and Zhang, 2001), and GB/T 700 (2006) and Q235 (Tan, 2000). The comparison of strength ratios plotted in Fig. 3.5 indicates that Q460 follows slightly different trends than other steels. In the temperature range of 150–500°C, Q460 steel has a higher yield strength reduction factor than BISPLATE 80 steel, although both are HSSs. The reduction factor of the yield strength of Q460 is much higher

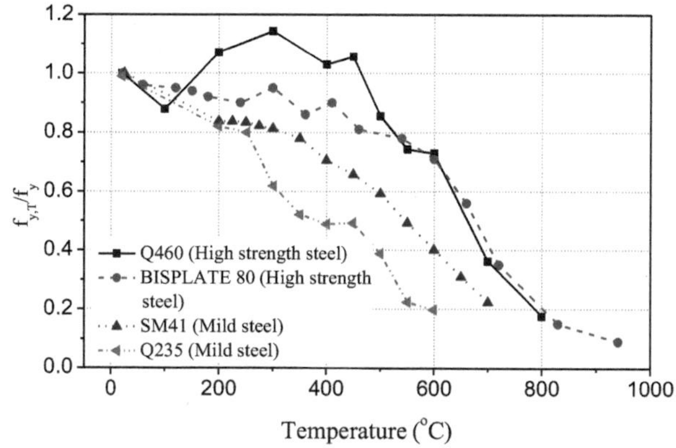

Figure 3.5 Comparison of the measured yield strength of Q460 steel with other steels.

Table 3.3 Chemical composition of Q460 and other steels.

Chemical composition (%)	High strength steel		Mild steel	
	Q460	**BISPLATE 80**	**Q235**	**SM41**
Carbon	≤0.2	0.16	0.12 ~ 0.2	≤0.18
Manganese	1.0 ~ 1.7	1.1	0.3 ~ 0.7	≤1.4
Silicon	≤0.55	0.2	≤0.3	≤0.35
Sulphur	≤0.035	0.003	≤0.045	≤0.035
Phosphor	≤0.035	0.01	≤0.045	≤0.035
Nickel	≤0.7	—	≤0.3	—
Titanium	0.02 ~ 0.2	—	—	—
Boron	—	0.001	—	—
Chromium	≤0.7	—	≤0.3	—
Aluminum	≤0.015	—	—	—
Niobium	0.015 ~ 0.06	—	—	—
Rhenium	0.02 ~ 0.2	—	—	—
Vanadium	0.02 ~ 0.2	—	—	—
Molybdenum	—	0.2	—	—
Cerium	—	0.2	—	—

A dash indicates that no data or insignificant amounts of data are available.

than those of SM41and Q235 at temperatures over 150°C. Additionally, Q460 steel has a higher strength retention than mild steel. These variations in strength trends and the better performance of Q460 steel can be attributed to the chemical composition of steel. The chemical compositions of these different steels are tabulated in Table 3.3 and show that Q460 steel contains more alloys (such as chromium and niobium) that are beneficial to fire-resistance properties.

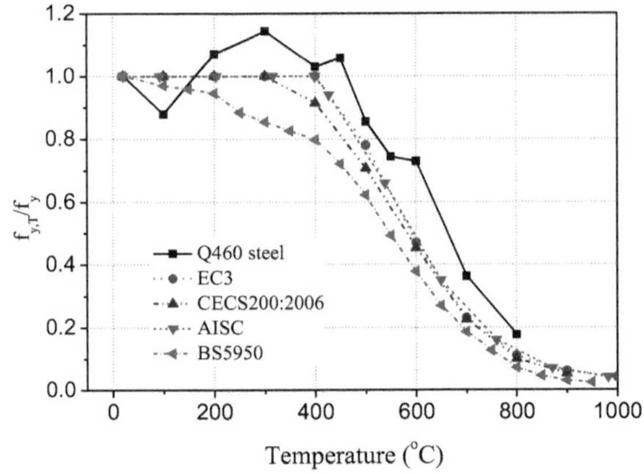

Figure 3.6 Comparison of the yield strength of Q460 steel with provisions in codes and standards.

The yield strength reduction factor from the tests for Q460 steel is also compared with current provisions in EN1993-1-2 (2005), CECS200 (2006), AISC (2005), and BS5950-8 (2003) in Fig. 3.6. These code provisions are based on measured data for mild steel. The predictions of the four provisions are similar and lower than the obtained strength results for Q460 steel for temperatures greater than 150°C. Hence, the use of code provisions for the fire-resistance design of high strength Q460 leads to conservative predictions. The ultimate strength of Q460 steel is compared with reported experimental results on other HSS and mild steels (Chen et al., 2006; Li and Zhang, 2001; Tan, 2000) in Fig. 3.7. Fig. 3.7 shows that strength trends for Q460 steel and BISPLATE 80 steel match

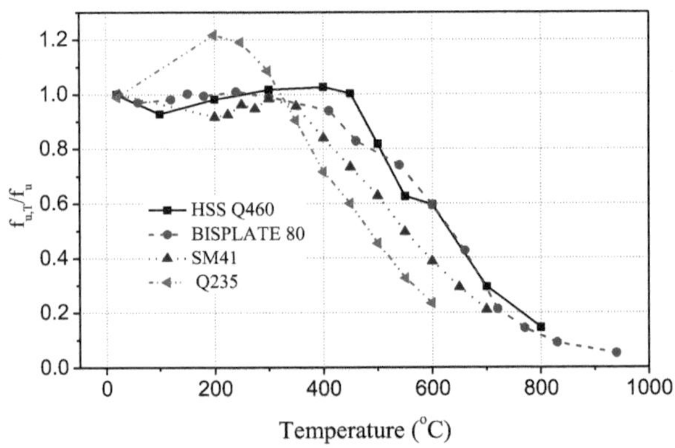

Figure 3.7 Comparison of the measured ultimate strength of Q460 steel with other steels.

well, and the ultimate strength reduction factor of these two steels is generally higher than SM41and Q235 steels. In the temperature range of 20–300°C, Q235 steel exhibits a more pronounced increase in strength due to an acute blue brittleness effect that has been observed in this particular grade of steel by other researchers (Chen et al., 2006; Li and Zhang, 2001; Tan, 2000).

3.2.4.2 Elastic modulus

The elastic modulus of Q460 steel, obtained through the frequency approach, is tabulated in Table 3.4 for various temperatures. In addition, the elastic modulus reduction factor (E_T/E) of Q460 steel is compared in Fig. 3.8 to other steels, BISPLATE 80 (Chen et al., 2006), SM41 (Li and Zhang, 2001), and Q235 (Tan, 2000). Fig. 3.8 shows that the elastic modulus reduction factor of Q460 steel compares well with BISPLATE 80 steel but is higher than those of SM41 and Q235, especially in the post-450°C range.

Fig. 3.9 shows a comparison of elastic modulus reduction factor (defined as the ratio of elastic modulus at a given temperature to that at room temperature) for Q460 steel with the predicted values in CEN, CECS, AISC, and BSI provisions. As shown in the figure, the elastic modulus reduction factor for Q460 steel is not as severe as that given in the current specifications throughout the temperature range of 20–800°C. This can be attributed to the different alloy composition in the high strength Q460 steel, especially its higher chromium and niobium content. Hence, the high strength Q460 steel retains higher stiffness at elevated temperatures than mild steel.

3.2.5 Prediction equations

Data generated from the tests are utilized to develop yield strength and elastic modulus relationships as a function of temperature for HSS Q460 steel. Because the reduction factor for yield strength (1% nominal stress) is more frequently used than ultimate strength in fire-resistance design, only yield strength formula is proposed. Both strength and elastic modulus are expressed in the form of exponential relations in the temperature range of 20–800°C. In fire engineering literature, both mean and lower bound fit are used. Researchers justify lower bound fit based on "conservative" argument, while others justify use of average value based on cost consideration and also a fact that fire is a rare event. Here, the exponential

Table 3.4 Measured elastic modulus of Q460 steel at various temperatures.

Temperature (°C)	Elastic modulus (GPa)						
	S-1	S-2	S-3	S-4	S-5	S-6	Average
20	209.16	210.62	209.19	209.66	208.32	211.05	209.67
100	205.71	206.99	205.83	205.91	204.55	207.04	206.01
200	200.77	201.38	201.22	201.3	200.29	202.09	201.18
250	198.01	197.8	198.29	198.04	197.36	199.15	198.11
300	194.62	194.24	195.05	194.8	193.17	195.59	194.58
350	191.41	190.09	191.52	190.96	190.62	191.09	190.95
400	187.92	185.34	187.39	185.89	180.88	185.39	185.47
450	182.89	180.04	182.37	180.58	177.18	181.02	180.68
500	178.25	174.82	178.03	174.74	170.19	176.09	175.35
550	173.96	169.37	172.53	169.6	163.05	169.71	169.7
600	168.23	162.83	156.25	159.24	156.93	157.9	160.23
700	141.84	141.22	131.1	130.41	127.04	128.66	133.38
800	111.53	101.68	108.39	99.09	86.5	97.11	100.72

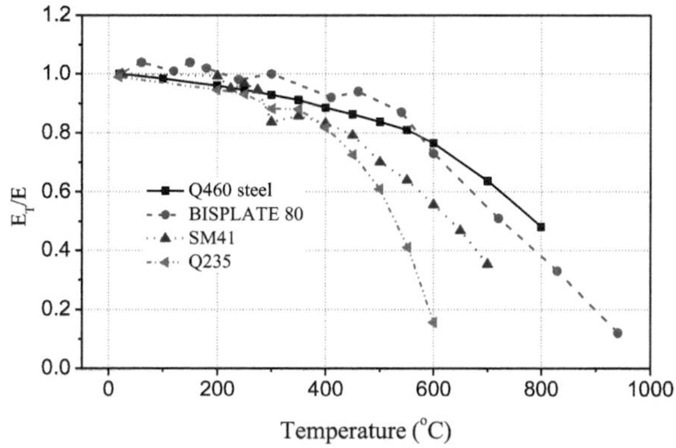

Figure 3.8 Comparison of the elastic modulus of Q460 steel with other steels.

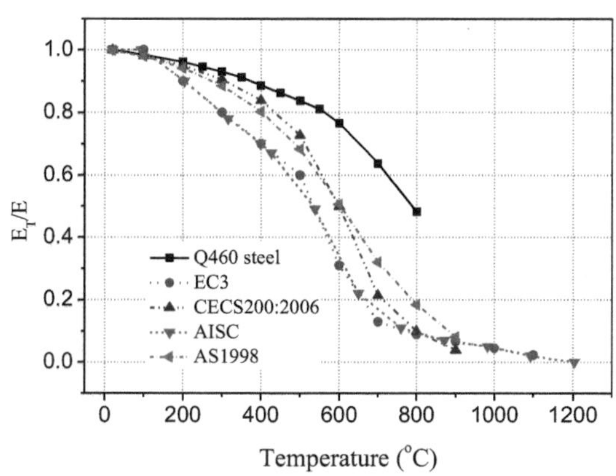

Figure 3.9 Comparison of the elastic modulus of Q460 steel with provisions in codes and standards.

relations were arrived at based on the "lower bound fit" to the test data and are listed in Eq. (3.2) and Eq. (3.3).

$$\frac{f_{y,T}}{f_y} = \begin{cases} 1 & 20°C \le T \le 450°C \\ 4.32e^{-T/880} - 1.6 & 450°C \le T \le 800°C \end{cases} \tag{3.2}$$

$$E_T \,/\, E = 1.02 - 0.035e^{T/280} \quad 20°C \le T \le 800°C \tag{3.3}$$

A comparison of reduction factors predicted by the proposed equations with the measured results and comparison with EC3 (2005b) models are shown in Fig. 3.10. As can be seen from Fig. 3.10A, the yield strength up to 450°C is taken to be same as that at room temperature. For temperatures exceeding 450°C, the predicted results from the proposed equations agree

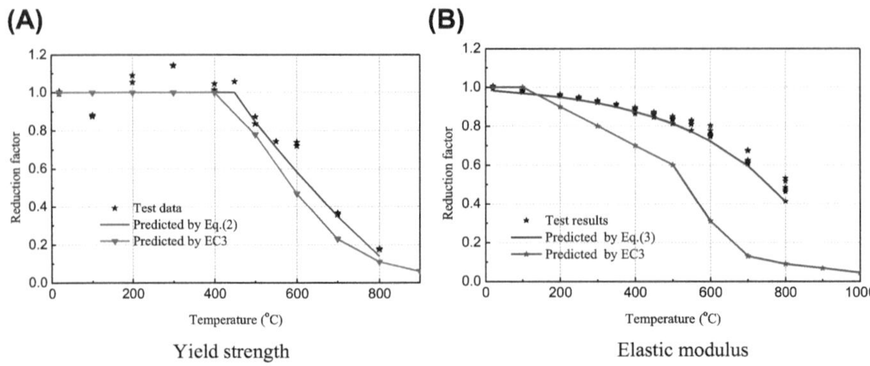

Figure 3.10 Comparison of reduction factors as predicted with measured results and EC3 models.

well with measured results. As to the elastic modulus, shown in Fig. 3.10B, the results from the proposed equation compare well with the lower bound of the measured test data throughout the temperature range. For the yield strength, the predicted results have similar trend, but a little higher than that from EC3 model. As to the elastic modulus, the predicted results are much higher than that from EC3 model. Therefore, EC3 models for strength and stiffness degradation are overly conservative and therefore not applicable to the high strength Q460 steel.

3.3 Mechanical properties of high strength Q690 steel

The authors carried out a standard tensile test to determine the temperature-dependent mechanical properties of Q690 steel over a temperature range of 20–900°C. From these tests, strength and elastic modulus of Q690 steel at elevated temperatures were derived. Based on the limited test data, the equations of reduction factors can be fitted as a function of temperature.

3.3.1 Test setup

The tensile testing machine used in this study was a SANS-CMT5305 electromechanical universal testing machine of 300 kN capacity. The high-temperature chamber with a temperature controller, which can reach the maximum temperature of 1200°C, was used to heat the specimen. The extensometer was placed on the displacement collection instrument to measure the strain of the specimen. The photo of the test setup is shown in Fig. 3.11.

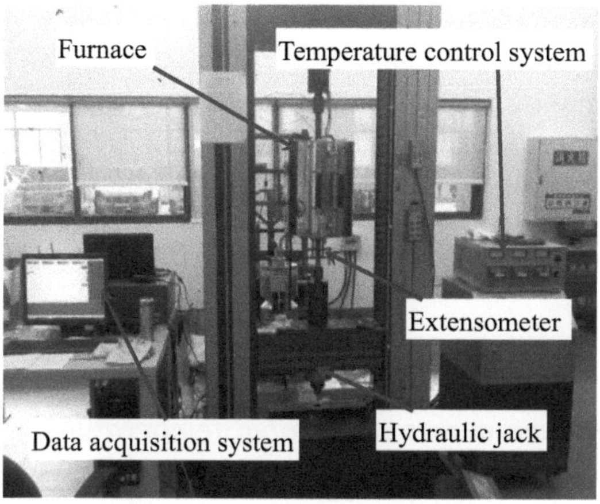

Figure 3.11 Tensile test setup.

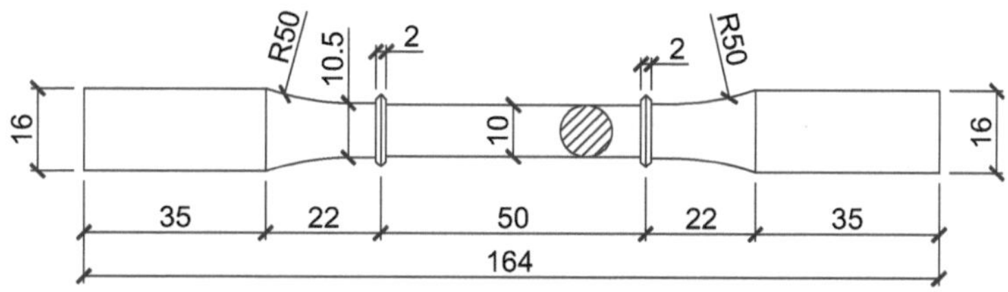

Figure 3.12 Dimensions of tension specimen (unit: mm).

3.3.2 Test specimen

The specimen were obtained from the same Q690 steel sheet with a nominal thickness of 20 mm. Q690 steel is a high strength structural steel, with nominal yield strength of 690 MPa. The shape and dimensions of the specimen were prepared in accordance with GB/T228.1 (2010), as presented in Fig. 3.12. Two round lips were manufactured at the end of gauge length to measure tensile strain. To improve the reliability of test results, at every temperature, two specimen were prepared.

3.3.3 Test procedures

In the tension tests, the specimen was heated up to a predefined target temperature and then loaded until it failed while maintaining the same temperatures. In this investigation, the tension tests are conducted at target temperatures of 300, 400, 500, 550, 600, 700, 800, and 900°C, and the heating rate was about 20°C/min. Two groups of tests were carried out for each temperature. When the target temperature is attained, the specimen is maintained at this temperature for 20 min to ensure the total specimen achieves uniform temperature. And then tensile load was applied to the specimen until failure. Strain control was used in this investigation. When strain ε is lower than 1%, the tensile rate was 0.003/min, and when strain ε is higher than 1% but lower than 2%, the tensile rate uniformly increased to 0.02/min. When strain ε is higher than 2%, the tensile rate maintains the rate of 0.02/min. This load system meets the requirements of GB/T4338 (2006).

3.3.4 Test results

3.3.4.1 Stress—strain curves at elevated temperatures

The tested stress—strain curves at different temperature levels are presented in Fig. 3.13, respectively. In these figures, stress—strain curves are plotted up to 30% strain, which is the maximum range of extensometer. The stress—strain curves at 300—550°C include strains from the start of loading to the fracture of the coupons. As can be seen in Fig. 3.13, the yield plateau at elevated temperatures disappears. At elevated temperatures, gradually loss in the tensile strength can be clearly observed. As can be seen in Fig. 3.13A—D, ductility, as measured by the final elongation of the coupons, changes little up to 500°C, then increases in the range of 550—900°C.The stress—strain curves increases rapidly when the strain was 1% in the temperatures range of 550—900°C. The reason is that the load rate increases when the strain was 1%, and the stress—strain curves were influenced significantly by the load rate at the temperatures range of 550—900°C.

3.3.4.2 Yield stress

Because the yield plateau disappears at elevated temperatures, yield strength of steel was determined according to different strain levels. The elevated temperature yield strengths of steel at strain levels of 0.2%, 1%, or 2% offset were all recommended (BS5950-8, 2003). Hence, the yield strengths at above three strain levels were obtained from this experimental investigation for comparison. The definitions of yield strengths at corresponding strain levels are plotted in Fig. 3.14. The 0.2% yield strength (noted as $f_{0.2}$) is the intersection point of the stress—strain curve and the proportional line offset by 0.2% strain. The yield strengths of $f_{1.0}$ and $f_{2.0}$ at the strain levels of 1.0% and 2.0%, respectively, are those values corresponding to the intersection points of the stress—strain curve and the vertical lines specified at given strain levels. The average of yield strength at the same temperature as well as reduction factors were shown in Table 3.5.

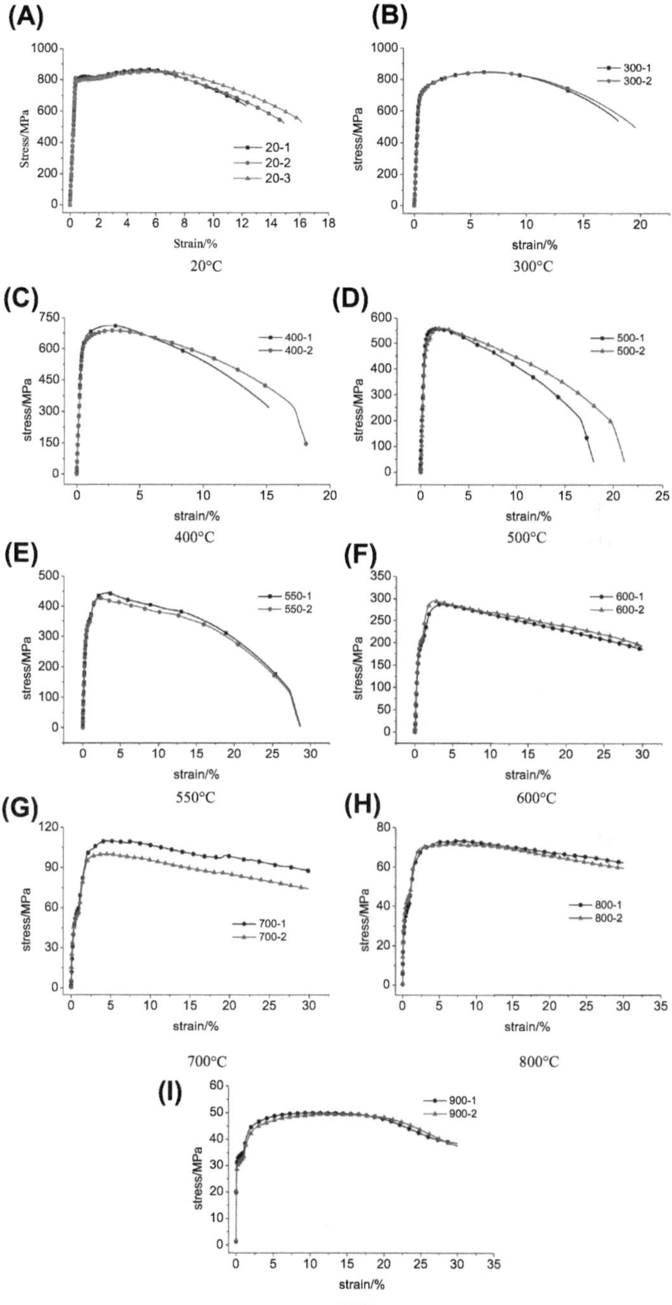

Figure 3.13 Stress–strain curves for Q690 steel at various temperatures.

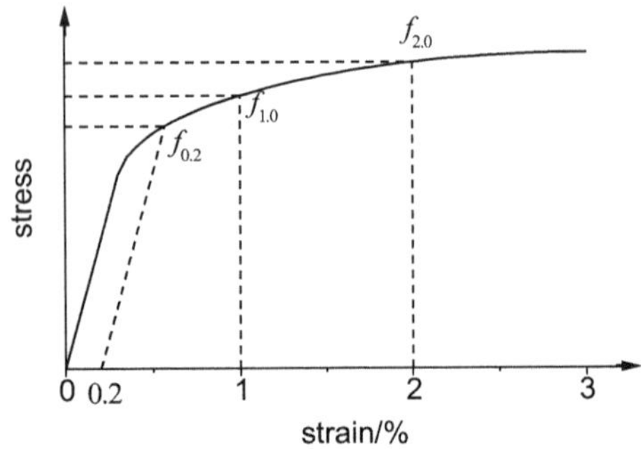

Figure 3.14 Definition of yield strength.

The yield strength $f_{2.0}$ was obtained when the tensile rate was changed. At some temperatures, $f_{2.0}$ is much bigger than $f_{1.0}$ and $f_{0.2}$. For example, at 700°C, the reduction factor of $f_{2.0}$ is 0.12, and the reduction factor of $f_{1.0}$ is 0.076. Yield strength decreases notably at temperatures ranged from 500 to 700°C. The reduction factor of $f_{0.2}$ is 0.66 at 500°C, which is 0.061 at 700°C.

3.3.4.3 Ultimate stress

To obtain the ultimate strength $f_{u,T}$ at elevated temperatures, the maximum value of stress in the stress—strain curves is used. The average of ultimate strengths at the same temperature and reduction factors were shown in Table 3.5. The results show the trend of reduction in ultimate strength is similar to that observed for yield strengths at elevated temperatures.

3.3.4.4 Elastic modulus

The elastic modulus is an important factor influencing the load-bearing capacity of steel structures. The elastic modulus at elevated temperature is defined as the slope of the stress—strain curve in the linear-elastic range, and reduction factors are shown in Table 3.5. As can be seen from Table 3.5, the elastic modulus E_T decreases considerably in the temperature range of 300—800°C. At 800°C, the reduction factor of elastic modulus E_T is just 0.066. However, the reduction factor of elastic modulus at 900°C increases to 0.13, which is different from the other kinds of steels (Wang et al., 2013).

3.3.4.5 Specimen failure

Fig. 3.15 shows the failed steel specimen after the tensile tests. Obvious necking phenomenon was observed. The ultimate elongation increases with temperatures. As we all know, when the load rate is lower, the

Table 3.5 Strength and elastic modulus and reduction factor of Q690 steel at elevated temperatures.

Temperature/°C	$f_{0.2}$/MPa	$f_{1.0}$/MPa	$f_{2.0}$/MPa	$f_{u,T}$/MPa	E_T/GPa	$f_{0.2}/f_y$	$f_{1.0}/f_y$	$f_{2.0}/f_y$	$f_{u,T}/f_u$	E_T/E
20	797.5	797.5	797.5	859.2	210	0.88	0.94	0.99	0.98	0.91
300	706.4	746.4	790	845.2	191	0.79	0.84	0.87	0.81	0.77
400	627.7	666.8	695	699.7	162.5	0.66	0.67	0.7	0.65	0.57
500	506.4	535.9	556.5	558.1	120.5	0.4	0.45	0.54	0.51	0.5
550	318.9	355.8	429.5	436.6	104	0.23	0.26	0.35	0.34	0.29
600	179.9	208.6	279.5	291.3	61	0.061	0.073	0.12	0.12	0.11
700	47.9	58.3	97.5	105	22.5	0.046	0.055	0.083	0.085	0.066
800	36.8	44	66.1	72.7	13.95	0.04	0.043	0.054	0.058	0.13
900	31.7	34.05	43.35	49.7	26.75	0.88	0.94	0.99	0.98	0.91

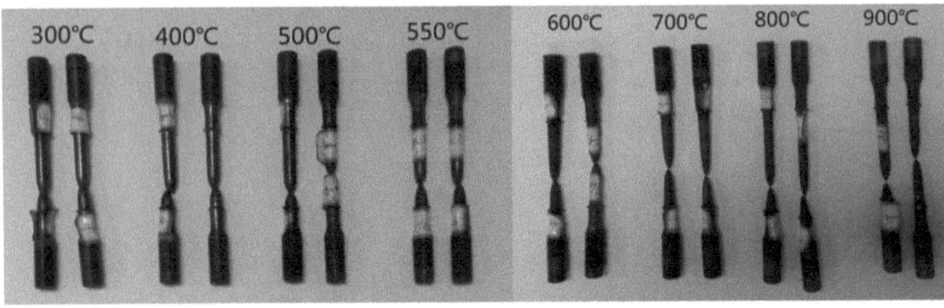

Figure 3.15 Photos of test specimen after fracture.

ultimate elongation is bigger. Ductility at 800–900°C has become very large. If the load rate is slow, the ultimate elongation will exceed the range of the acquisition system for displacement. The load rate increased properly when the strain exceeds 30%. This is the reason why the ultimate elongation of specimen at the same temperature range of 800–900°C was quite different.

3.3.5 Proposed stress–strain curves

Ma et al. (2015) proposed an implicit expression (Eq. 3.4) for expressing stress–strain relations of steel with gradual transition from elastic stage to plastic stage based on Ramberg–Osgood model. The stress–strain relations ensure the response follows four points, e.g., (ε_p, f_p), $(\varepsilon_{0.01}, f_{0.01})$, $(\varepsilon_{0.2}, f_{0.2})$, and (ε_y, f_y). Hence, the same relation was used to fit stress–strain relations of Q690 steel at elevated temperatures. The stress–strain relationship was determined by a set of material parameters, E, $\sigma_{0.01}$, $\sigma_{0.2}$, σ_u, ε_u, n_0, and m. The exponent m is determined by fitting the model to the tested stress–strain curve data, as shown in Table 3.6. The comparisons of the predictions from the constitutive model and the test results from the study are presented in Fig. 3.16. Generally, the agreement is acceptable for the above-generated results on two sets of specimen.

$$\varepsilon_p = \varepsilon - \frac{\sigma}{E} = 0.002\left(\frac{\sigma}{\sigma_{0.2}}\right)^n \tag{3.4a}$$

Table 3.6 Fitted values m.

Temperature/C	300	400	500	600	700	800	900
m	0.40	1.0	1.0	1.0	1.0	0.55	0.48

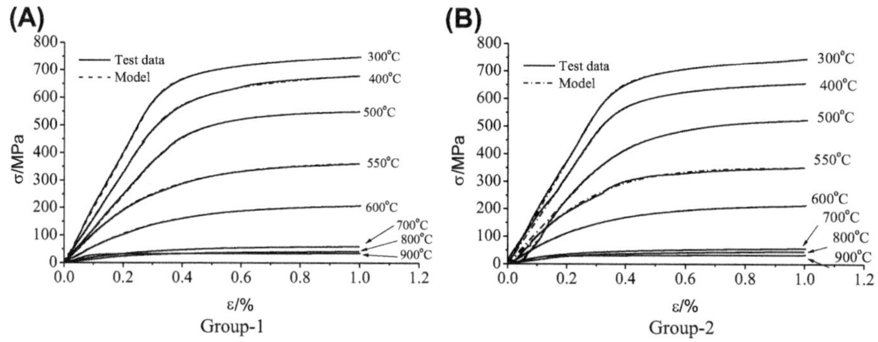

Figure 3.16 Comparison of fitted stress–strain curves with test results.

$$n = f(\varepsilon_p) = n_0 + K(\varepsilon_p)^m \tag{3.4b}$$

$$n_0 = \lg\frac{0.2}{0.01}/\lg\left(\frac{f_{0.2}}{\sigma_{0.01}}\right) \tag{3.4c}$$

$$K = \left[\lg\frac{\varepsilon_{pu}}{0.002}/\lg\left(\frac{f_y}{f_{0.2}}\right) - n_0\right]/(\varepsilon_{pu})^m \tag{3.4d}$$

where ε is the total strain of steel; ε_p is the plastic strain of steel; σ is the stress of steel; $\sigma_{0.2}$ is the static 0.2% tensile proof stress; n is strain-hardening exponent; n_0 is original strain-hardening exponent; K is modular coefficient in Ramberg–Osgood expression; $\sigma_{0.01}$ is the static 0.01% tensile proof stress; m is exponent for the modified Ramberg–Osgood model; ε_{pu} is the plastic strain at ultimate strength.

3.3.6 Prediction equations

Current relation in Eurocode 3 cannot predict the mechanical properties of high strength Q690 steel at elevated temperatures. Therefore, new predictive equations (Eqs. 3.5–3.6) are proposed for temperature-dependent properties reduction factors of high strength Q690 steel.

$$E_T / E = \frac{1}{1 + (T/534)^{5.5}} \tag{3.5}$$

$$f_{y,T} / f_y = \frac{1}{1 + (T/538)^{10}} \tag{3.6}$$

where E_T is the elastic modulus at elevated temperature; E is the elastic modulus at room temperature; $f_{y,T}$ is the yield strength at elevated

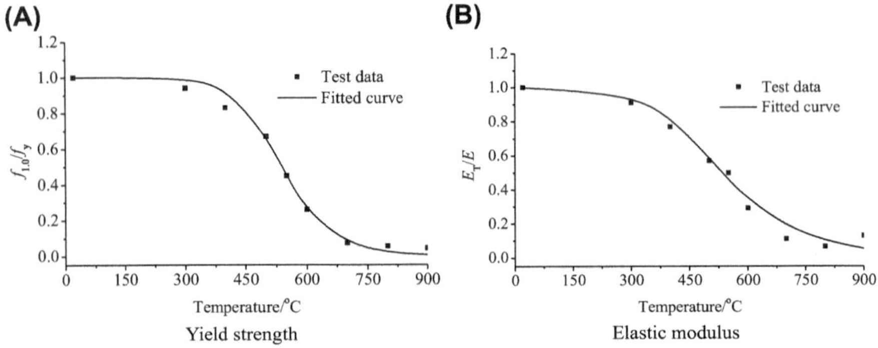

Figure 3.17 Comparison between test results and fitting values. (A) Yield strength and (B) Elastic modulus.

temperature; f_y is the yield strength at room temperature; and T is the temperature in steel.

The comparisons of test data with fitted curve, calculated by applying Eqs. (3.5 and 3.6), are shown in Fig. 3.17. It is obvious that the fitted curves closely match the measured properties values in tests.

3.4 Mechanical properties of high strength S460 steel

To reveal the deterioration of mechanical properties of the HSS S460N under steady-state and transient-state fire condition, tensile tests were conducted by Qiang et al. (2012a, 2012b) at various elevated temperatures ranged from 20 to 700°C.

3.4.1 Test device

The tensile tests were conducted using a Gleeble 3800 System, which is a fully integrated digital closed-loop control thermal and mechanical testing system. The direct resistance heating system of the Gleeble 3800 can heat specimen at rates of up to 10,000°C/s or can hold steady-state equilibrium temperatures. Gleeble 3800 has a digital control system, which provides all signals necessary to control thermal and mechanical test variables simultaneously through the digital closed-loop thermal and mechanical servo systems. In this study, 3 pairs of thermocouples provided signals for accurate feedback control of specimen temperatures. The air temperature inside the furnace was obtained by Gleeble 3800 System. The Gleeble 3800 mechanical system is an integrated hydraulic servo system capable of exerting as much as 200 kN of static force in compression or 100 kN intension. A noncontact laser extensometer was used to obtain the strain of specimen. The experimental process was controlled by some predefined programming options using QuikSim software. All the data obtained from the tests were recorded and could be monitored during testing.

3.4.2 Test material and specimen

All test specimen were cut from the same S460NL steel sheet ordered for this experimental study with a nominal thickness of 5 mm S460NL is weldable normalized steel for steel constructions, with high yield strength and high toughness, produced in compliance with EN10025−3 (2004). S460NL is the grade designation abbreviation of this steel, where S means structural steel; 460 is its minimum yield strength; N means normalized rolled delivery condition; and L means low notch toughness testing temperature. The chemical composition of the tested HSSS460N was shown in Table 3.7. The shapes and dimensions of the specimen were prepared in accordance with EN 10002−5 and ASTM E21-09 (2009). As shown in Fig. 3.18, one hole was provided at each end of the specimen to fix it to the grips of Gleeble 3800 by using two steel bars.

Table 3.7 Chemical composition of HSS S460N material (%).

C	Si	Mn	P	S	Cr	Cu
0.172	0.483	1.5	0.012	0.005	0.02	0.025
N	Nb	Ni	Ti	V	Al	Mo
0.0051	0.046	0.018	0.002	0.087	0.037	0.002

(A) (B)

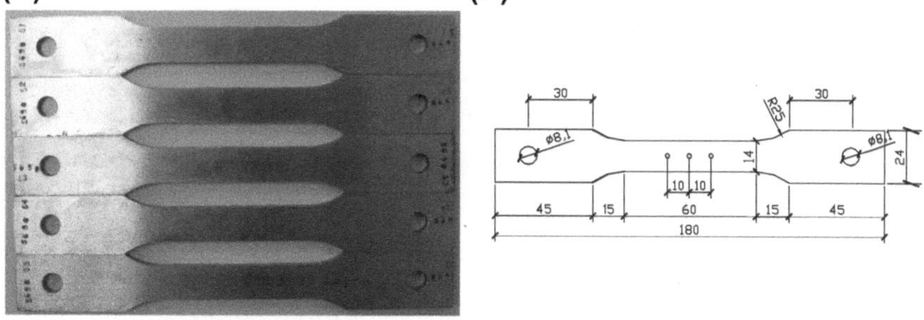

Test coupons Dimensions

Figure 3.18 (A) Tensile test coupon and (B) dimensions.

3.4.3 Test method

To assess the mechanical properties of steels, the commonly used method is to conduct tensile tests either in steady state or transient state. Comparing with transient-state test method, the steady-state test method is used more commonly in research as it is easier to perform and provides the stress–strain curves directly. However, the transient-state test method is considered to be more realistic in simulating the performance of steel components in a natural fire. To compare the difference between these two methods on HSS, both steady-state test method and transient-state test method were conducted in this experimental study.

3.4.4 Test procedure

Tensile tests at elevated temperatures were carried out in the furnace of Gleeble 3800 system. Three thermocouples were welded on the surface of the specimen to measure its temperature of the specimen, as shown in Fig. 3.18B.

In steady-state test, the specimen was heated up to a specified temperature and then loaded until it failed while maintaining the same temperature. The heating rate used herein was uniform 50 °C/min, and the preselected temperatures were 100, 200, 300, 350, 400, 450, 500, 550, 600, 650, and 700°C. After reaching the preselected temperature, approximately 10 min was required for the temperature to stabilize, to ensure a uniform temperature within the central part of the specimen. And then tensile load was applied to the specimen until failure. In the steady-state tests, strain control was used according to the strain obtained from the extensometer. The strain rate used herein was 0.005/min, which satisfied the requirement of ASTM E21-09 (2009). To compare the elevated temperature effect on mechanical properties of S460N, tensile tests were conducted at ambient temperature as well.

In transient-state tests, the specimen was under some constant tensile load, while the temperature rose until failure occurred. The used stress levels were preselected; they were 100, 150, 200, 250, 300, 350, 400, 450, 500, 600, 650, 700, 750, and 800 MPa. The heating rate in all transient-state tests was 10°C/min, which corresponded to a natural fire. The total strain and temperature of the specimen during the whole test procedure were recorded. The thermal elongation of specimen should be subtracted from the total strain, and then the strains of specimen at various temperatures could be obtained. Under some constant stress level, the strain–temperature curves obtained from transient-state tests can be transferred to stress–strain curves at various elevated temperatures.

3.4.5 Experimental results

The reduction factors of elastic modulus $E_{T/}E_{20}$ for S460N at various elevated temperatures were obtained from this experimental study and presented in Table 3.8.

In current design standards, the reduction factors of yield strength recommended by EC3 (2005) are based on the strain levels of 0.2% and 2.0%, and in BS5950-8 (2003), different reduction factors are given based on three strain levels: 0.5%, 1.5%, and 2.0%. In AISC (2005), American Society of Civil Engineers (ASCE) (1992), or AS 4100 (1998), no specification on strain level accompanies the given reduction factors for yield strength. The yield strength at elevated temperatures used herein was defined based on the total strain of 2.0%, to agree with the normally used method in literature. The yield strengths and reduction factors $f_{y\theta}/f_{y20}$ at elevated temperatures from both steady-state test and transient-state test were shown in Table 3.9.

The ultimate strengths reduction factors at elevated temperatures f_{uy}/f_{u20} obtained under steady state and transient state were given in Table 3.10.

Ductility of steel is defined based on the deformation that steel can undergo before fracture. The stress—strain curves at different temperatures obtained from this steady-state test were plotted in Fig. 3.19. The effect of temperature on the ductility of HSS S460N was studied by comparing the strain values at fracture. It is interesting to note that its ductility at 200°C is much lower compared with that at ambient temperature, while at 300 and 350°C, it increases gradually and from 400 to 600°C it reduces gradually. From 600 to 700°C, the ductility of S460N is in the same level.

Typical failure modes of HSS S460N tensile specimen at various elevated temperatures obtained from steady test in this experimental study were shown in Fig. 3.20. It can be seen that for all specimen, necking appeared before failure. No brittle failure was observed for all temperatures from 20 up to 700°C, which indicated that the lack of ductility was not a big concern for HSS S460N at elevated temperatures.

3.5 Mechanical properties of high strength S690 steel

To understand elevated temperature dependence of mechanical properties of HSS S690, an experimental investigation has been carried out by Qiang et al. (2012), using both steady-state and transient-state methods at temperatures ranged from 20 to 700°C.

Table 3.8 Elastic modulus and reduction factors of S460N at elevated temperatures.

Steady-state test			Transient-state test		
Temperature (°C)	Elastic modulus (MPa)	Reduction factor	Temperature (°C)	Elastic modulus (MPa)	Reduction factor
20	202,812	1	20	201,925	1
100	199,770	0.985	100	199,704	0.989
200	178,677	0.881	200	175,675	0.87
300	161,974	0.799	300	159,925	0.792
350	144,372	0.712	350	141,751	0.702
400	135,621	0.669	400	134,482	0.666
450	117,225	0.578	450	118,126	0.585
500	103,205	0.509	500	97,328	0.482
550	75,852	0.374	550	72,491	0.359
600	59,019	0.291	600	54,924	0.272
650	50,390	0.248	650	44,827	0.222
700	31,008	0.153	700	26,654	0.132

Table 3.9 Yield strength and reduction factors of S460N at elevated temperatures.

Steady-state test			Transient-state test		
Temperature (°C)	**Yield strength (MPa)**	**Reduction factor**	**Temperature (°C)**	**Yield strength (MPa)**	**Reduction factor**
20	504	1	20	523	1
100	497	0.987	100	517	0.989
200	501	0.994	150	510	0.975
300	504	1.001	200	507	0.97
350	496	0.984	250	505	0.966
400	478	0.949	300	503	0.962
450	442	0.877	350	501	0.958
500	372	0.739	400	493	0.942
550	282	0.559	450	470	0.899
600	209	0.415	500	403	0.771
650	158	0.313	550	334	0.639
700	94	0.187	600	259	0.495
			650	199	0.381
			700	129	0.247

Table 3.10 Ultimate strength reduction factors of S460N at elevated temperatures.

Steady-state test		Transient-state test	
Temperature (°C)	**Reduction factor**	**Temperature (°C)**	**Reduction factor**
20	1	20	1
100	0.945	100	0.998
200	0.969	150	0.969
300	1.023	200	0.968
350	1.024	250	0.968
400	0.88	300	0.968
450	0.75	350	0.968
500	0.601	400	0.968
550	0.443	450	0.897
600	0.328	500	0.693
650	0.249	550	0.556
700	0.157	600	0.421
		650	0.278
		700	0.206

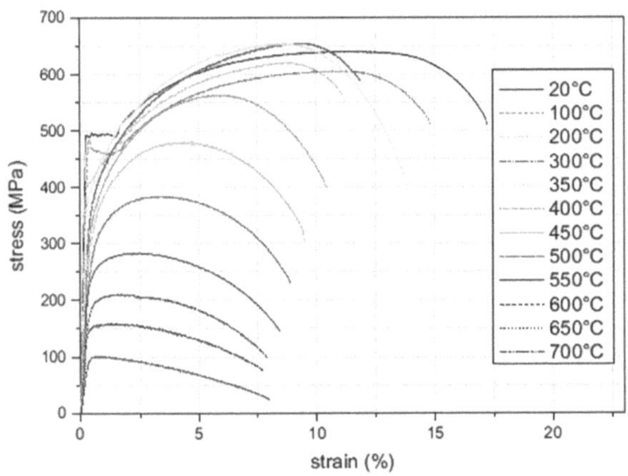

Figure 3.19 Stress—strain curves of S460N at elevated temperatures.

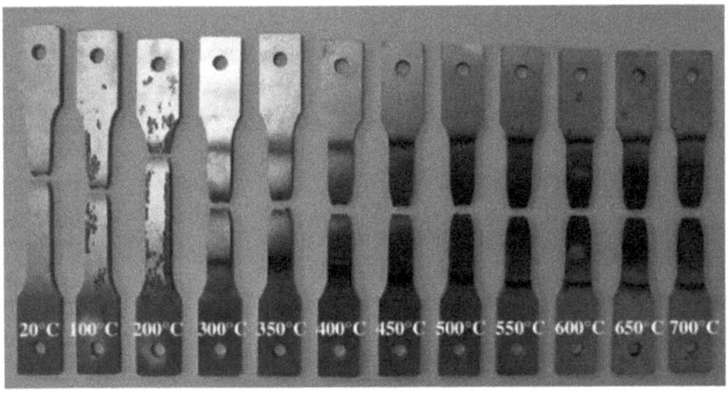

Figure 3.20 Failure modes of S460N specimen at various elevated temperatures.

3.5.1 Test device

The tensile setup on S690QL steel at elevated temperature is similar to the test setup used for S460 steel, and the detail of the test setup is introduced in Section 3.4.1.

3.5.2 Test material and specimen

All test specimen were cut from the same S690QL steel sheet ordered for this investigation with a nominal thickness of 5 mm. S690QL is a high strength structural steel produced in compliance with EN 10025-6(2009). The material is heat-treated using the quenched and tempered process and has good bending and welding properties. S690QL is the grade designation abbreviation of this steel, where S means structural steel, 690 is its minimum

Table 3.11 Chemical composition of high strength steel S690QL material (%).

C	Si	Mn	P	S	Cr	Cu	Mo
0.16	0.21	0.85	0.012	0.001	0.35	0.03	0.2
N	Nb	Ni	Ti	V	Al-g	B-g	Zr
0.0026	0.025	0.05	0.006	0	0.093	0.0024	0

yield strength, Q means quenching and tempering, and L means low notch toughness testing temperature. The chemical composition of the tested HSS S690 was shown in Table 3.11. The shapes and dimensions of the specimen were similar to the specimen in Section 3.4.2.

3.5.3 Test method

In steady-state test, the specimen was heated up to a specified temperature and then loaded until it failed while maintaining the same temperature. In this investigation, the heating rate was 50°C/min, and the preselected temperatures were 100, 200, 300, 400, 500, 550, 600, and 700°C, respectively. At least two tests were carried out for each temperature. After reaching the preselected temperature, approximately 10 min was required for the temperature to stabilize, to ensure a uniform temperature within the central part of the specimen. And then tensile load was applied to the specimen until failure. Strain control was used in this investigation and the strain rate used herein was 0.005/min, which satisfied the requirement of ASTM Standard E21-09 (2009). To compare the elevated temperature effect on mechanical properties of S690, tensile tests were also conducted at ambient temperature.

In transient-state tests, the specimen was under some constant tensile load while the temperature in the furnace rose until failure occurred. The stress levels used herein were preselected; they were 100, 200, 250, 300, 400, 500, 600, 650, 700, 800, 850, and 900 MPa. The heating rate in all transient-state tests was 10°C/min. The total strain and temperature of the specimen during the whole test procedure were recorded. And the thermal elongation of specimen should be subtracted from the total strain, and then the strains of specimen at various temperatures could be obtained. Under some constant stress level, the strain—temperature curves obtained from transient-state tests can be transferred into stress—strain curves at various elevated temperatures, as shown in Fig. 3.21.

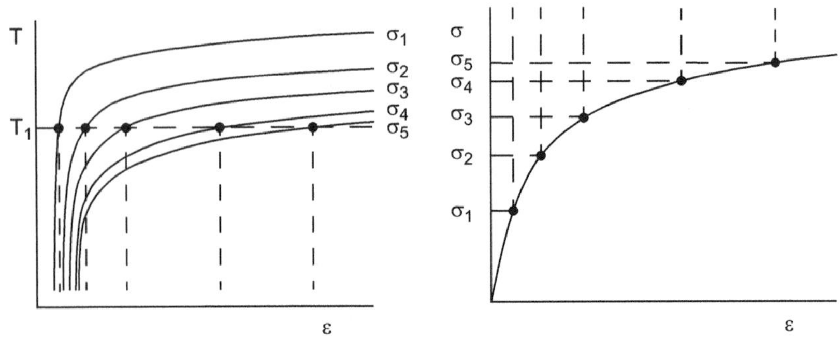

Figure 3.21 Transient-state stress—strain derivation.

3.5.4 Test results

The elastic modulus and its reduction factor at elevated temperatures of S690 were calculated based on the results from both steady- and transient-state test and presented in Table 3.12.

Because of the absence of a well-defined yield point, the elevated-temperature yield strengths at strain levels of 0.2%, 0.5%, 1.5%, and 2.0% were used by researchers. Hence, the yield strengths at above four strain levels were obtained from this experimental investigation for comparison. The reduction factors of yield strength at elevated temperatures at various

Table 3.12 Elastic modulus and reduction factors.

Steady-state test			Transient-state test		
Temperature (°C)	Elastic modulus (MPa)	Reduction factor	Temperature (°C)	Elastic modulus (MPa)	Reduction factor
20	204,690	1	20	205,890	1
100	204,592	1	100	202,267	0.982
200	179,146	0.875	200	178,874	0.869
300	171,819	0.839	250	176,515	0.857
400	158,608	0.775	300	173,242	0.841
500	140,127	0.685	350	160,829	0.781
550	111,788	0.546	400	151,457	0.736
600	76,105	0.372	450	142,459	0.692
700	28,848	0.141	500	133,146	0.647
			550	110,642	0.537
			600	76,238	0.37
			650	42,027	0.204
			700	20,480	0.099

Table 3.13 Reduction factors of yield strengths at various strain levels from steady-state test.

	Reduction factors			
Temperature (°C)	**0.20%**	**0.50%**	**1.50%**	**2.00%**
20	1	1	1	1
100	0.947	0.874	0.958	0.968
200	0.884	0.854	0.956	0.982
300	0.879	0.751	0.952	0.975
400	0.794	0.794	0.864	0.85
500	0.628	0.605	0.655	0.624
550	0.554	0.438	0.557	0.533
600	0.38	0.345	0.382	0.371
700	0.1	0.114	0.133	0.133

Table 3.14 Reduction factors of yield strengths at various strain levels from transient-state test.

	Reduction factors			
Temperature (°C)	**0.20%**	**0.50%**	**1.50%**	**2.00%**
20	1	1	1	1
100	1	0.989	0.91	0.923
200	0.9	0.878	0.836	0.868
250	0.9	0.875	0.831	0.861
300	0.8	0.872	0.826	0.855
350	0.8	0.839	0.813	0.839
400	0.8	0.812	0.786	0.798
450	0.8	0.763	0.73	0.738
500	0.6	0.631	0.716	0.716
550	0.5	0.542	0.554	0.554
600	0.4	0.397	0.445	0.445
650	0.3	0.213	0.278	0.278
700	0.2	0.228	0.203	0.203

strain levels from steady-state test were presented in Table 3.13, while that from transient-state test were shown in Table 3.14.

The ultimate strengths and their reduction factors at elevated temperatures obtained under steady state and transient state were given in Table 3.15.

Ductility of steel is defined based on the deformation that steel can undergo before fracture. The stress—strain curves at different temperatures obtained from this steady-state test were plotted in Fig. 3.22. The dependence on temperature of the ductility of HSS S690 was studied by comparing the strain values at fracture. It is interesting to note that its

Table 3.15 Ultimate strength and reduction factors at elevated temperatures.

Steady-state test			Transient-state test		
Temperature (°C)	**Ultimate strength (MPa)**	**Reduction factor**	**Temperature (°C)**	**Ultimate strength (MPa)**	**Reduction factor**
20	821	1	20	899	1
100	796	0.97	100	850	0.946
200	814	0.991	200	850	0.946
300	789	0.961	250	850	0.946
400	680	0.828	300	850	0.946
500	548	0.668	350	850	0.946
550	458	0.558	400	800	0.89
600	310	0.377	450	700	0.779
700	107	0.13	500	650	0.723
			550	501	0.558
			600	399	0.444
			650	252	0.281
			700	201	0.223

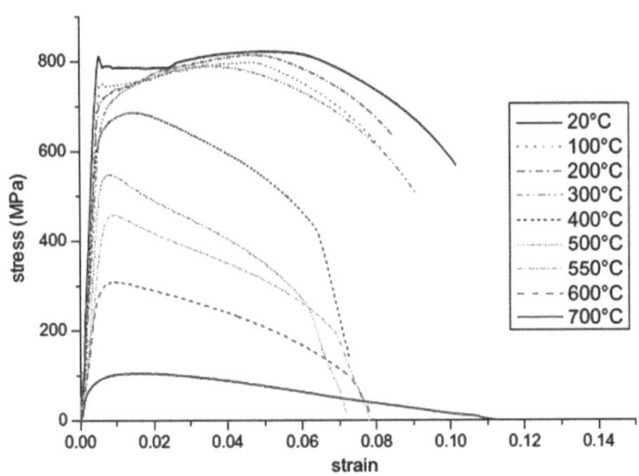

Figure 3.22 Stress–strain curves at various temperatures from steady-state test.

ductility at 100°C is lower compared with that at ambient temperature, while at 200 and 300°C, it increases gradually and at 400°C it reduces again and thereafter it starts to improve again. At 700°C, the ductility of S690 is higher than that at ambient temperature. The ductility changes at temperatures in the range of 100–400°C can be attributed to chemical transformations taking place in the steel base. With increasing temperature, these chemical transformations retard and ductility of steel increases as long as temperature becomes dominant.

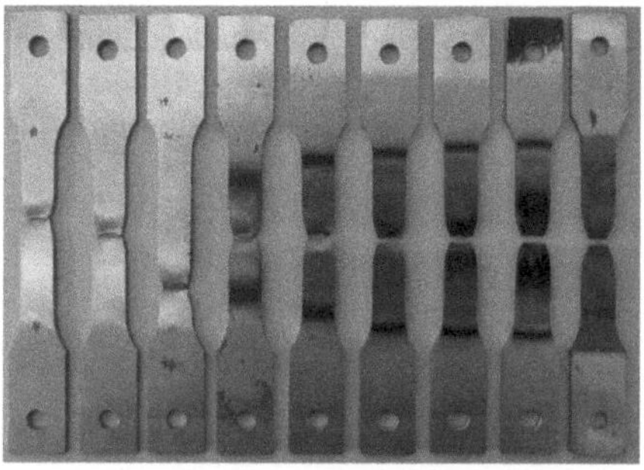

Figure 3.23 Failure modes of S690 specimen at elevated temperatures under steady state.

Typical failure modes of HSS S690 tensile specimen at different elevated temperatures obtained from steady-state test in this investigation were shown in Fig. 3.23. It can be seen for all specimen necking appeared before failure. No brittle failure was observed for all temperatures from 20°C up to 700°C, which is promising for fire safety of steel structures with HSS S690.

3.5.5 Comparison and analysis

The reduction factors of yield strengths at different strain values obtained from the above two methods were compared with European, American, Australian, and British design standards, as described in Figs. 3.24–3.27.

In Fig. 3.24, it can be seen that the prediction of AISC is always nonconservative for yield strength of S690 at strain level of 0.2%, while the

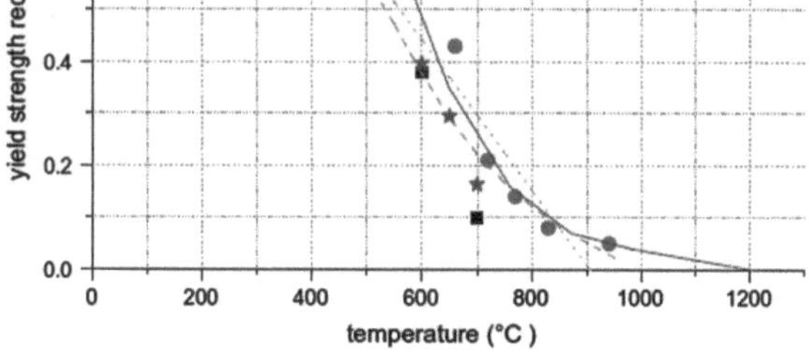

Figure 3.24 Yield strength reduction factors versus temperatures at strain level 0.2%.

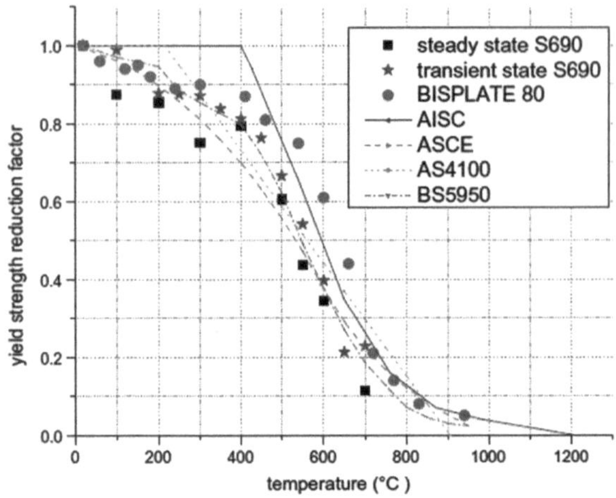

Figure 3.25 Yield strength reduction factors versus temperatures at strain level 0.5%.

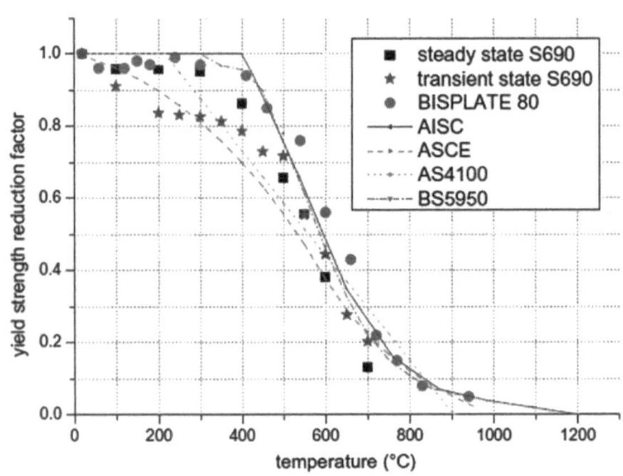

Figure 3.26 Yield strength reduction factors versus temperatures at strain level 1.5%.

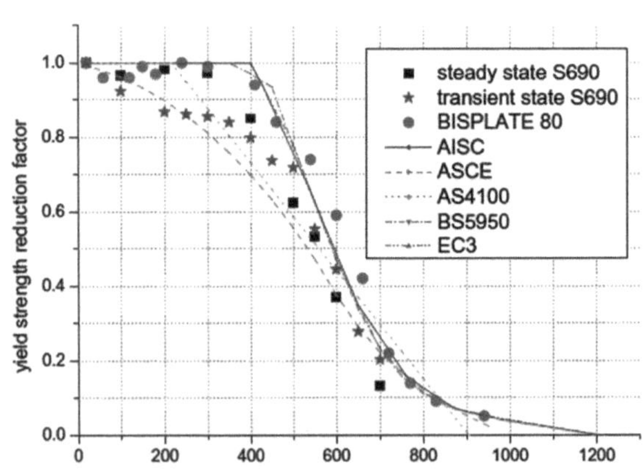

Figure 3.27 Yield strength reduction factors versus temperatures at strain level 2.0%.

prediction of AS 4100 is conservative from 400 to 550°C but nonconservative for other temperatures, and similarly, ASCE is conservative from 300 to 550°C. Fig. 3.25 demonstrates that at strain level 0.5%, the prediction of AISC is generally nonconservative, and the prediction of BS5950 is generally similar to the results under transient-state condition but nonconservative for that under steady-state condition except for temperatures from 400 to 600°C, and AS 4100 is conservative from 350 to 500°C but nonconservative for other temperatures. As shown in Fig. 3.26, at strain level of 1.5%, the predictions of AISC and BS5950 are nonconservative, and AS 4100 is conservative from 400 to 550°C but not the case for other temperatures, while ASCE is conservative from 300°C to 550°C but not the case for other temperatures. Fig. 3.27 shows that at strain level of 2.0%, the prediction of EC3 is nonconservative, and the predictions of other standards are similar with the conclusion obtained at strain level of 1.5%.

At the above four different strain levels, the yield strength reduction factors of BISPLATE 80 obtained by Chen and Young are generally nonconservative in comparison with that of S690 from this experimental investigation. Hence, using the yield strength reduction factors of BISPLATE 80 to predict that of S690 is not safe.

From the comparison of this investigation with current design standards and available research on HSS BISPLATE 80, it can be concluded that the reduction of yield strength at elevated temperatures depends on steel grades. Thus, it is necessary to propose separate reduction factors of yield strength for different steel grades.

The ultimate strength reduction factors of HSS S690 obtained from this experimental investigation were compared with that of HSS BISPLATE 80 reported by Chen et al. (2006) and also with the prediction of American standard AISC (2005). The comparison was presented in Fig. 3.28, from which it can be found that the prediction of AISC is not conservative for S690. Moreover, the ultimate strength reduction factors of BISPLATE 80 are generally higher than that of S690 for both transient state and steady state. It means neither the recommendation of AISC nor the reduction factor for BISPLATE 80 is safe when using to predict the ultimate strength of S690 at elevated temperatures.

3.6 Mechanical properties of high strength S960 steel

A detailed experimental study on S960, under both steady-state fire condition and transient-state fire condition were carried out by

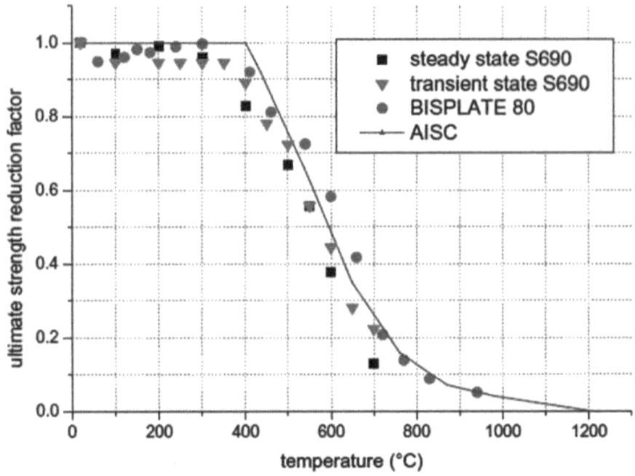

Figure 3.28 Comparison of ultimate strength reduction factors.

Qiang et al. (2016). The elastic modulus, yield strength, and ultimate strength of S960 in fire are obtained and compared with the recommendations of leading design standards for steel structures (i.e., EC3, ASCE, AISC, AS 4100, and BS 5950), as well as with HSS S460 and S690. The comparison with design standards reveals the necessity of including specified recommendations on the deterioration of mechanical properties in fire for very high strength structural steels in current leading design standards. Further, some unique predictive equations are proposed for the elevated-temperature mechanical properties of S960.

In the tensile test, the test setup, specimen, test method, and test procedure are similar to the tension test on S460 and S690 steel, introduced in Section 3.4 and 3.5.

3.6.1 Test material

All test specimen were cut via water cutting from a S960QL steel sheet ordered for this study with a nominal thickness of 5 mm S960QL is a very high strength structural steel produced in compliance with EN 10025-6 (2009). The material is heat-treated using the quenched and tempered process and has good bending and welding properties. S960QL is the grade designation abbreviation of this steel, where S means structural steel, 960 is its minimum yield strength, Q represents quenching and tempering, and L means low notch toughness testing temperature. The chemical composition of the tested very HSS S960 is shown in Table 3.16.

Table 3.16 Chemical composition of HSS S690QL material provided by the supplier (%).

C	Si	Mn	P	S	Cr	Cu	Mo
0.19	0.39	1.43	0.008	0.0009	0.58	0.02	0.45
N	Nb	Ni	Ti	V	Al-g	B-g	
0.0047	0.022	0.05	0.004	0.03	0.044	0.0001	

3.6.2 Test results

3.6.2.1 Stress–strain curves

The stress–strain curves of S960 at various fire temperatures obtained from this steady-state test are plotted in Fig. 3.29. The stress–strain curves of S960 under various transient-state fire conditions transferred from the experimentally obtained strain–temperature curves under various constant stress levels are presented in Fig. 3.30.

3.6.2.2 Failure mode

The typical failure mode of very HSS S960 tensile specimen at various elevated temperatures obtained from the steady-state test in this study is shown in Fig. 3.31. It can be seen that for all specimen necking appeared before failure. No brittle failure can be observed for all temperatures from 20°C up to 700°C, which is promising for fire safety of steel structures with members made of VHSS S960.

The typical failure mode of VHSS S960 tensile specimen under transient-state fire conditions at various stress levels obtained from this

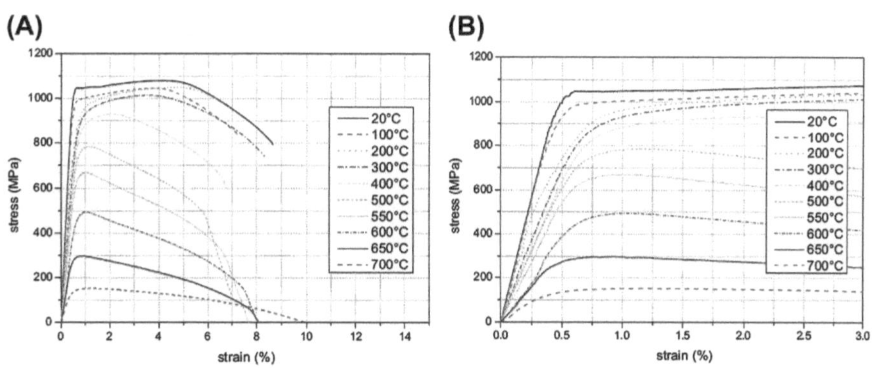

Figure 3.29 Stress–strain curves of S960 at various temperatures under steady state. (A) Full range of strain-stress curve and (B) With strain range of 3% of strain-stress curve.

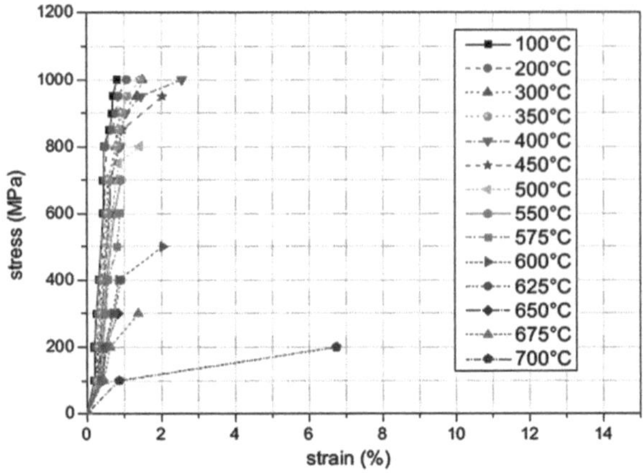

Figure 3.30 Stress–strain curves of S960 at various temperatures under transient state.

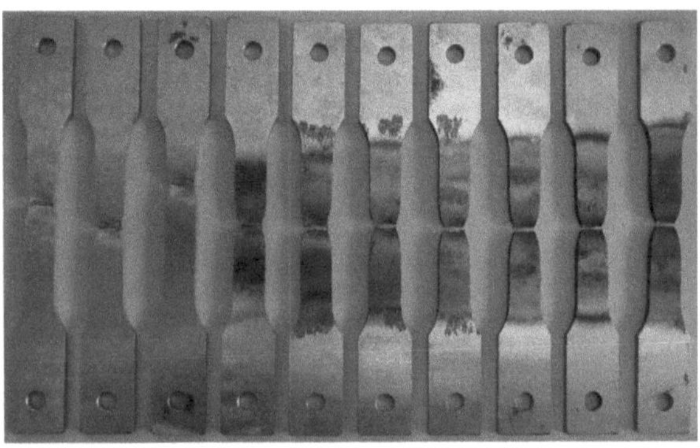

Figure 3.31 Failure mode of S960 specimen at elevated temperatures under steady-state condition.

experimental study is presented in Fig. 3.32. It can be seen that for all specimen necking appeared before failure. No brittle failure can be observed for any stress levels up to 1000 MPa. The fire temperature at which the specimen fails under a constant tensile stress level is called failure temperature at this specified stress level under transient-state fire condition. The failure temperatures of S960 at various stress levels up to 1000 MPa obtained in this study are presented in Table 3.17. It can be found that within its nominal yield stress level under transient state, S960 fails at fire temperatures in excess of 400°C.

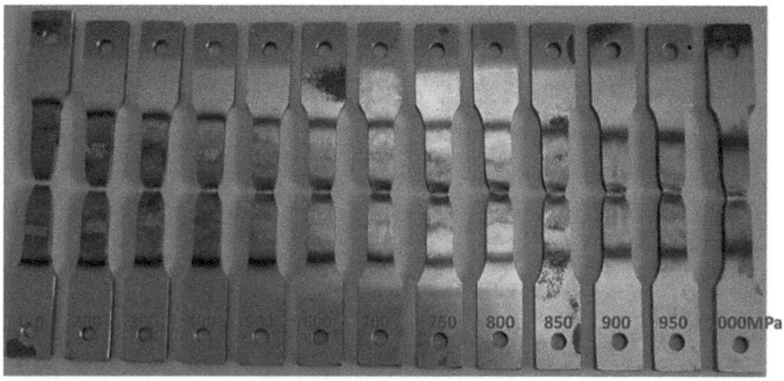

Figure 3.32 Failure modes of S960 specimen at various stress levels under transient-state condition.

Table 3.17 Failure temperatures of S960 at various stress levels under transient-state fire condition.

Stress level (MPa)	Failure temperature (°C)
100	800
200	700
300	690
400	665
500	613
600	602
700	594
750	584
800	580
850	545
900	540
950	430
1000	402

3.6.2.3 Elastic modulus

In fire, the elastic modulus of structural steel is determined from the stress—strain curve at the corresponding elevated temperature, as shown in Fig. 3.33. The elastic modulus and reduction factors E_θ/E_{20} of S960 at various elevated temperatures were obtained from this experimental study and presented in Table 3.18. Furthermore, the results from this experimental study are compared with some current leading design standards for steel structures (i.e., EC3, AISC, and AS 4100), as shown in Fig. 3.34. It appears that the reduction factors of the elastic modulus given in EC3 and

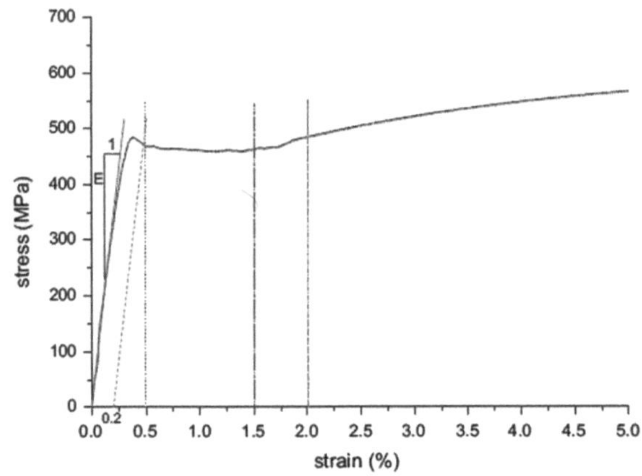

Figure 3.33 Mechanical properties determination of structural steel.

Table 3.18 Elastic modulus and reduction factors of S960 at elevated temperatures.

Steady-state test			Transient-state test		
Temperature (°C)	**Elastic modulus (MPa)**	**Reduction factor**	**Temperature (°C)**	**Elastic modulus (MPa)**	**Reduction factor**
20	213,586	1	20	213,586	1
100	211,141	0.989	100	213,764	1.001
200	185,292	0.868	200	184,325	0.863
300	172,992	0.81	300	173,432	0.812
—	—	—	350	161,055	0.754
400	154,812	0.725	400	150,819	0.706
—	—	—	450	149,254	0.699
500	115,089	0.539	500	126,152	0.591
550	106,959	0.501	550	108,696	0.509
600	70,059	0.328	600	62,893	0.294
—	—	—	625	59,121	0.277
650	51,096	0.239	650	48,900	0.229
700	36,084	0.169	700	21,601	0.101

AISC are generally appropriate when used to predict the elastic modulus of VHSS S960; the prediction of AS 4100 is nonconservative for VHSS S960.

3.6.2.4 *Yield strength*

Because of the absence of a well-defined yield point, the elevated-temperature yield strengths at strain levels 0.2%, 0.5%, 1.5%, and 2.0% were used by researchers. Hence, the yield strengths of VHSS S960 at

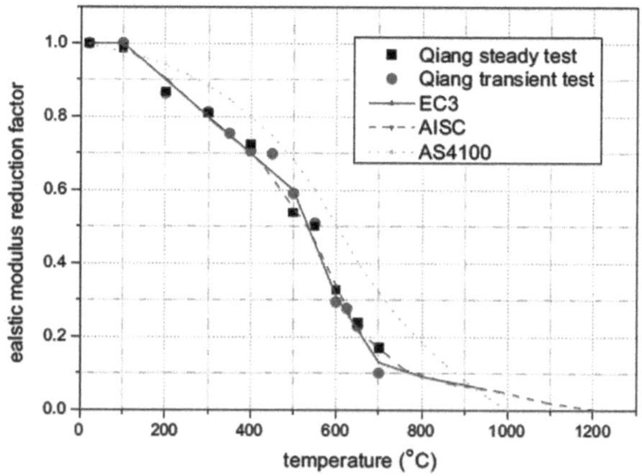

Figure 3.34 Comparison on elastic modulus of S960 at elevated temperatures.

above four strain levels were obtained from this experimental study for comparison. The yield strength reduction factors $f_{y\theta}/f_{y20}$ of VHSS S960 at elevated temperatures from both the steady-state test and the transient-state test are shown in Table 3.19. The yield strength reduction factors of S960 at different strain values obtained from the above two methods are compared with European, American, Australian, and British design standards, as described in Figs. 3.35–3.38.

In Fig. 3.35, it can be seen that the prediction of AISC is always nonconservative for the yield strength of S960 at strain level 0.2%, while the prediction of AS 4100 is conservative from 350 to 600°C but nonconservative for other temperatures, and similarly ASCE is conservative from 300 to 600°C. Fig. 3.36 demonstrates that at strain level 0.5%, the prediction of AISC is generally nonconservative, and the prediction of BS5950 is generally nonconservative when the temperature is below 500°C, while ASCE is generally nonconservative except at fire temperatures ranging between 450 and 550°C; AS 4100 is conservative from 450 to 550°C but nonconservative for other temperatures. As shown in Fig. 3.37, at strain level 1.5%, the predictions of AISC and BS5950 are nonconservative when the fire temperature is below 400°C, and AS 4100 is conservative from 300 to 600°C but not the case for other temperatures, while ASCE is generally conservative, except at 700°C. Fig. 3.38 shows that at strain level 2.0%, the prediction of EC3 is nonconservative when the fire temperature is below 400°C, and the predictions of other standards are similar to that obtained at strain level of 1.5%.

Table 3.19 Reduction factors of yield strengths at various strain levels from steady-state test.

Steady-state test					Transient-state test				
	Reduction factors					Reduction factors			
Temperature (°C)	0.20%	0.50%	1.50%	2.00%	Temperature (°C)	0.20%	0.50%	1.50%	2.00%
20	1.000	1.000	1.000	1.000	20	1.000	1.000	1.000	1.000
100	0.947	0.899	0.964	0.968	100	0.953	0.807	0.951	0.945
200	0.904	0.714	0.945	0.956	200	0.925	0.800	0.951	0.945
300	0.859	0.673	0.923	0.936	300	0.863	0.732	0.949	0.943
—	—	—	—	—	350	0.857	0.715	0.946	0.940
400	0.819	0.622	0.874	0.880	400	0.813	0.701	0.913	0.928
—	—	—	—	—	450	0.806	0.697	0.874	0.909
500	0.734	0.590	0.738	0.710	500	0.740	0.605	0.763	0.788
550	0.631	0.513	0.618	0.589	550	0.679	0.576	0.718	0.721
600	0.472	0.364	0.454	0.431	600	0.462	0.251	0.513	0.569
650	0.280	0.259	0.271	0.259	650	0.345	0.263	0.391	0.401
700	0.138	0.130	0.144	0.141	700	0.102	0.192	0.206	0.210

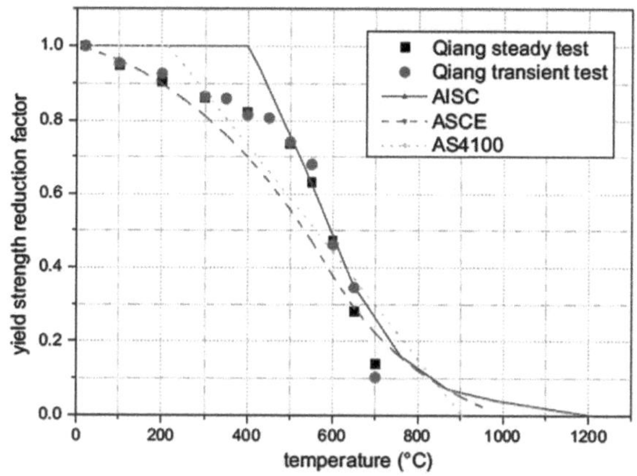

Figure 3.35 Yield strength reduction factors of S960 versus temperatures at strain level 0.2%.

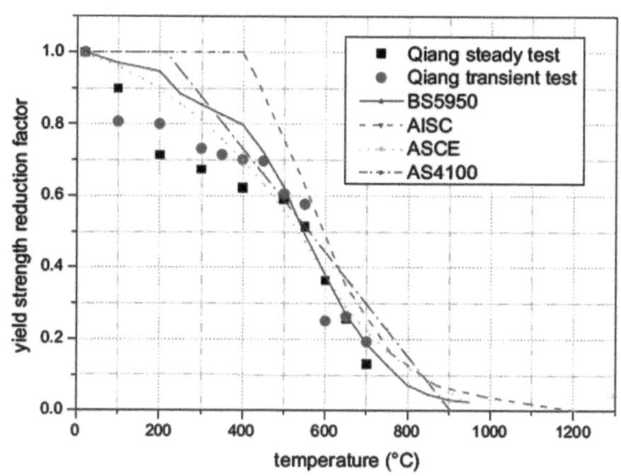

Figure 3.36 Yield strength reduction factors of S960 versus temperatures at strain level 0.5%.

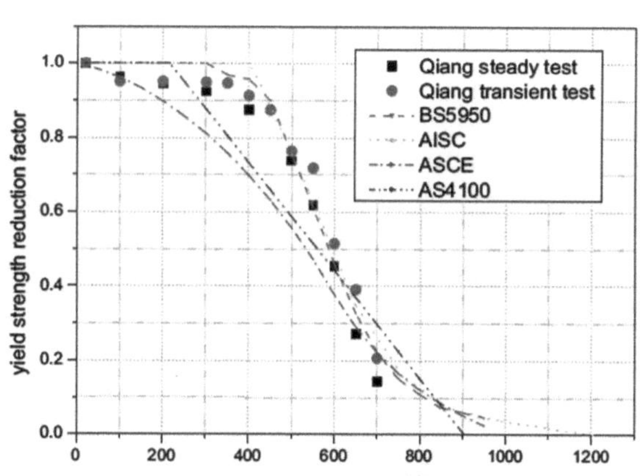

Figure 3.37 Yield strength reduction factors of S960 versus temperatures at strain level 1.5%.

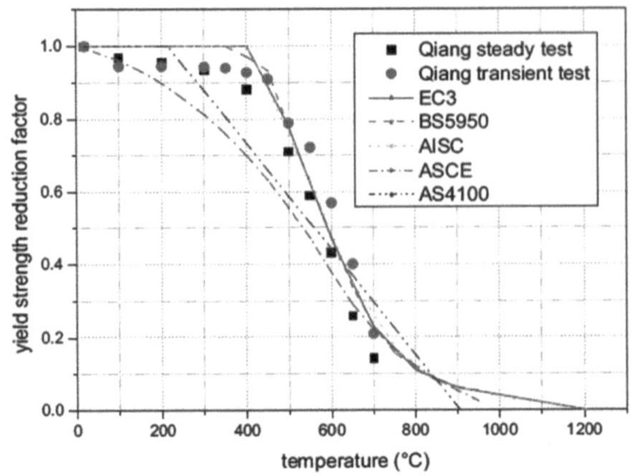

Figure 3.38 Yield strength reduction factors of S960 versus temperatures at strain level 2.0%.

From the comparison of test results with leading design standards which were mainly obtained from mild steels, it can be concluded that the reduction of yield strength at elevated temperatures depends on the steel grades and manufacturing method. It means that the recommendations on strength deterioration of structural steels under fire conditions in the leading design standards are not appropriate to very high strength steels. Thus, it is necessary to propose unique reduction factors of yield strength for very high strength steel grades.

3.6.2.5 Ultimate strength

The ultimate strengths of S960 and their reduction factors at various elevated temperatures obtained under steady state and transient state are given in Table 3.20. The ultimate strength reduction factors of VHSS S960 obtained from this experimental study are compared with the recommendation of American standard AISC (2005). The comparison is presented in Fig. 3.39, from which it can be found that the prediction of AISC is not conservative for S960 when the temperature is below 400°C or beyond 600°C, which means the recommendation of AISC is not always conservative when used to predict the ultimate strength of S960 exposed to fire.

3.6.3 Predictive equations

Some specified predictive equations were developed herein based on the experimental results to describe the deteriorations of mechanical properties of VHSS S960 exposed to various fire temperatures, to offer an accurate

Table 3.20 Ultimate strength and reduction factors of S960 at elevated temperatures.

Steady-state test			Transient-state test		
Temperature (°C)	Ultimate strength (MPa)	Reduction factor	Temperature (°C)	Ultimate strength (MPa)	Reduction factor
20	1079	1	20	1079	1
100	1048	0.971	100	1000	0.927
200	1052	0.975	200	1000	0.927
300	1016	0.942	300	1000	0.927
—	—	—	350	1000	0.927
400	934	0.866	400	1000	0.927
—	—	—	450	950	0.88
500	786	0.728	500	800	0.741
550	670	0.621	550	700	0.649
600	494	0.458	600	500	0.463
650	296	0.274	650	300	0.278
700	153	0.142	700	200	0.185

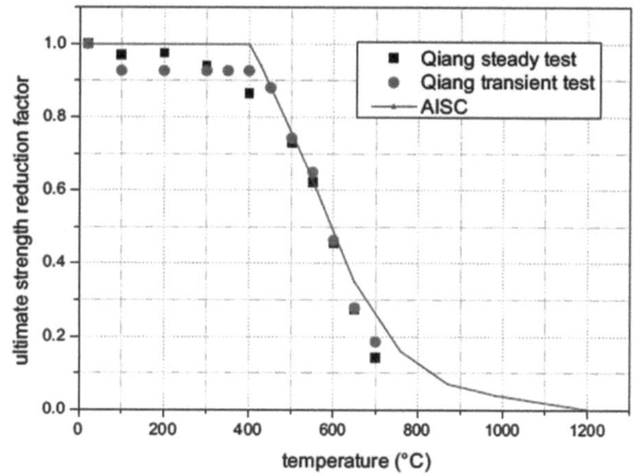

Figure 3.39 Comparison on ultimate strength reduction factors of S960 at elevated temperatures.

choice for structural engineers to conduct safe fire-resistance design of steel structures with S960. As the elevated temperature is the main reason causing the deterioration of material properties of steel, the equations are developed as a function of the elevated temperature Θ, which specimen have been exposed to.

3.6.3.1 Elastic modulus

Eq. (3.7) is developed based on the experimental results to present the deterioration of the elastic modulus of S960 exposed to fire temperatures in the form of reduction factors, as shown below. The predictions of Eq. (3.7) are compared with experimental results and shown in Fig. 3.40. It can be observed that good agreement between Eq. (3.7) and experimental results exists.

$$\frac{E_\theta}{E_{20}} = -1.8 \times 10^{-9}\theta^3 + 2.319 \times 10^{-7}\theta^2 - 5.845 \times 10^{-4}\theta + 1.021 \qquad 20 \le \theta \le 700 \tag{3.7}$$

$$\frac{f_{y\theta}}{f_{y20}} = 1.006 - 2.834 \times 10^{-4}\theta \quad 20 \le \theta \le 400 \tag{3.8a}$$

$$\frac{f_{y\theta}}{f_{y20}} = -3.21 \times 10^{-6}\theta^2 + 9.75 \times 10^{-4}\theta + 1.011 \quad 400 < \theta \le 700 \tag{3.8b}$$

$$\frac{f_{y\theta}}{f_{y20}} = 1.004 - 1.818 \times 10^{-4}\theta \quad 20 \le \theta \le 400 \tag{3.9a}$$

$$\frac{f_{y\theta}}{f_{y20}} = -6.024 \times 10^{-6}\theta^2 + 4.205 \times 10^{-3}\theta + 0.216 \quad 400 < \theta \le 700 \tag{3.9b}$$

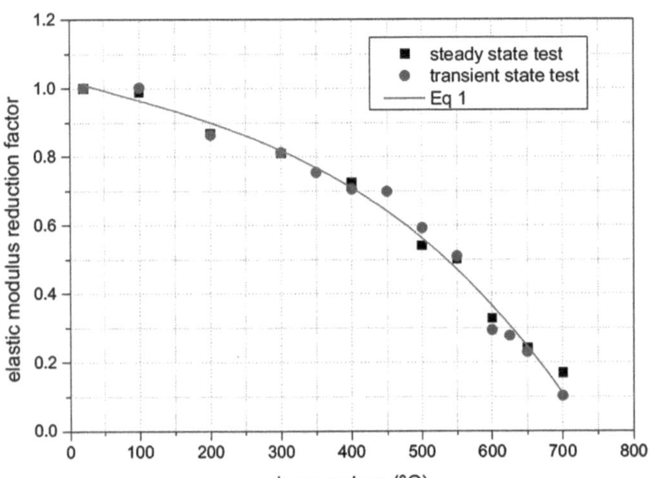

Figure 3.40 Validation of predictive equation on elastic modulus reduction factors of S960 in fire.

$$\frac{f_{u\theta}}{f_{u20}} = 1.005 - 2.607 \times 10^{-4}\theta \quad 20 \leq \theta \leq 400 \tag{3.10a}$$

$$\frac{f_{u\theta}}{f_{u20}} = -5.055 \times 10^{-6}\theta^2 + 3.01 \times 10^{-3}\theta + 0.497 \quad 400 < \theta \leq 700 \tag{3.10b}$$

3.6.3.2 Yield strength

Because the deviations between yield strength reduction factors of S960 under steady-state fire conditions and those under transient-state fire conditions are not small, two sets of equations were developed based on the experimental results at strain level of 2.0%, to predict the deterioration of yield strength of S960 exposed to fire. Eq. (3.8) is proposed to present the yield strength reduction factors of S960 under steady-state fire condition, while Eq. (3.9) is worked out to present those of S960 under transient-state fire condition.

Their predictions are compared with test results and described in Fig. 3.41. The comparison shows that there is good agreement between Eq. (3.8) and test results under steady-state fire condition, and Eq. (3.9) is in good agreement with the test results under transient-state fire condition. For conservative consideration, Eq. (3.8) is recommended for presenting the deterioration of the yield strength reduction factor of S960 in fire.

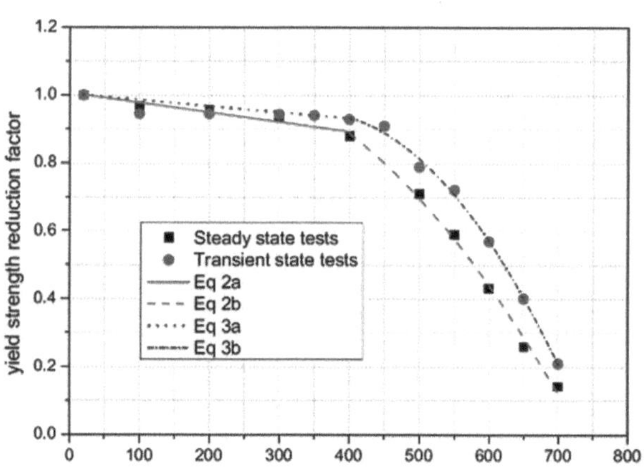

Figure 3.41 Validation of predictive equations on yield strength reduction factors of S960 in fire.

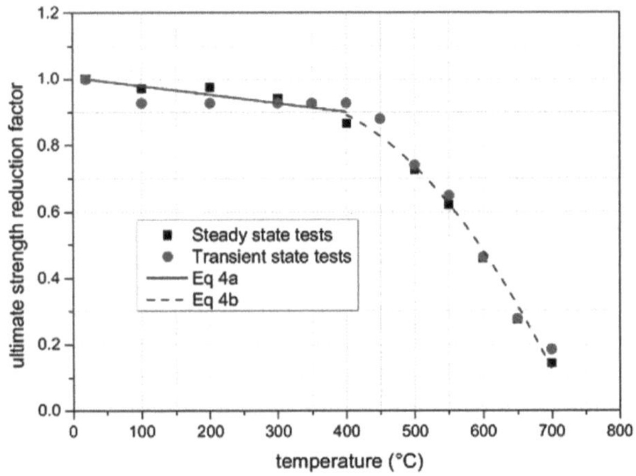

Figure 3.42 Validation of predictive equation on ultimate strength reduction factors of S960 in fire.

3.6.3.3 Ultimate strength

Eq. (3.10) is worked out to present the ultimate strength reduction factors of VHSS S960 exposed to various fire temperatures up to 700 °C. Its predictions are compared with test results and shown in Fig. 3.42. The comparison shows that there is good agreement between Eq. (3.10) and the experimental results.

3.7 Mechanical properties of BISPLATE 80 steel

3.7.1 Experimental investigation

3.7.1.1 Testing device

The tensile testing machine used in this study was an MTS 810 universal testing machine of 100 kN capacity. The heating device used was an MTS Model 653 high-temperature furnace with a maximum temperature of 1400°C, as shown in Fig. 3.43. The furnace was controlled by an MTS model 409.83 temperature controller. An MTS Model 632.53F-11 axial extensometer was used to measure the strain of the central region of the coupon specimen. The test devices are detailed in Chen and Young (2004).

3.7.1.2 Test specimen

The coupon test specimen were obtained from structural steel sheets. The structural steel sheets are HSS BISPLATE 80 (approximately equivalent to ASTM A 514, EN 10137-2 Grade S690Q and JIS G 3128) with a nominal plate thickness of 5.0 mm. A total of 34 tests (18 steady-state tests and

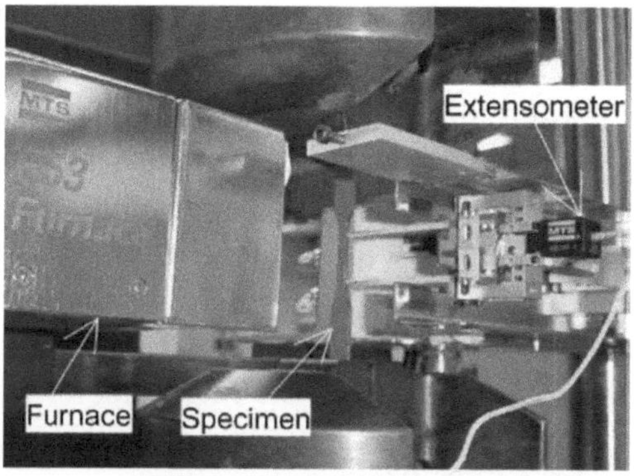

Figure 3.43 Testing device.

Table 3.21 Arrangement of steady- and transient-state tests.

Test method	High strength steel	Mild steel	Total
Steady state	18	23	41
Transient state	16	—	16

Table 3.22 Chemical composition of high strength steel and mild steel used in tests.

Chemical composition (Typical)	C	P	Mn	Si	S	Cr	Mo	B	CEΣIIWΠ	PCM
BISPLATE80	0.16	0.01	1.1	0.2	0.003	—	0.2	0.001	0.4	0.25

16 transient-state tests) were conducted in this study, and the arrangement of the tests is shown in Table 3.21. The chemical composition of the tested HSS is shown in Table 3.22. The test specimen were prepared in accordance with the ASTM Standard E 21-92 (1997) and Australian Standard AS 2291 (1979). The actual dimensions of the cross-sectional areas of the coupon specimen within the gauge length were measured using a micrometer. The measured dimensions were used to determine the cross-sectional area of each coupon.

3.7.2 Testing procedure

3.7.2.1 Steady-state test

In the steady—state tests, the specimen was heated up to a specified temperature and then loaded until it failed while maintaining the same temperature. In this study, thermal expansion of the specimen was allowed by maintaining zero tension load during the heating process. After reaching the preselected temperature, approximately 2 min was required for the temperature to stabilize, and after another 15 min, the tensile load was applied to the specimen. This would allow the heat to transfer into the specimen. The external thermocouple indicated that the variation of the specimen temperature within the gauge length was less than 6°C (±3°C) during the tests. In the steady-state tests, strain control was used in the tensile testing machine. A constant tensile loading rate of 0.2 mm/min was used, and the strain rate obtained from the extensometer was approximately 0.006/min, which is within the range 0.005 ± 0.002/min as specified by the ASTM Standard E 21-92 (1997).

3.7.2.2 Transient-state test

In the transient-state tests, the specimen was under constant tensile load while the temperature was raised. The stress levels selected in the tests were 1, 50, 100, 150, 200, 250, 300, 350, 400, 450, 500, 550, 600, 650, and 700 MPa. The temperatures specified in the temperature controller ranged from 100 to 1000°C at an interval of 100°C. The strain of the specimen at a given temperature was recorded using the extensometer 10 min after the temperature reached the specified value. The ultimate strength of the specimen was defined as the point at which strain kept increasing at a given value of temperature. In the tests, there are two reasons for the temperature to rise step by step. First, there is a rapid loss of strength for the loaded specimen, and the loading machine could not follow the sudden load drop under load control. Second, the strain data for different specified temperatures should be obtained because the results of the transient-state tests need to be converted to stress—strain curves.

3.7.3 Thermal elongation in transient-state test

Thermal elongation of the specimen was determined at a tensile stress level of 1 MPa (0.15 ksi), which is close to free thermal expansion, and compared with the thermal elongation calculated according to the ASCE Manual (1992), BS5950-8 (2003), and European Code 3 Part 1.2 (2001) in Fig. 3.44. The thermal elongation of the strain in percentage (%) terms was plotted on the vertical axis of the graph, and the horizontal axis represents varying temperatures. The comparison indicates that the test values of thermal elongation of HSS are less than the values predicted by the ASCE (1992), BS5950-8 (2003), and EN1993-1-2(2005). Although the 1 MPa (0.15 ksi) tensile stress was almost negligible at ambient temperature for

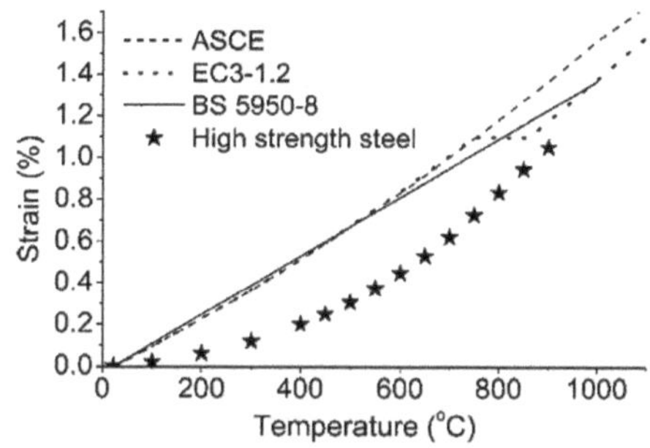

Figure 3.44 Comparison of thermal elongation predicted by ASCE, BS5950-8 and EN1993-1-2(2005) (EC3-1.2) with test results of high strength steel.

determining the thermal elongation, it, however, slightly affected the elongation when the temperature increased. As the thermal elongation was determined for specimen loaded at a stress level of 1 MPa (0.15 ksi), the elastic modulus obtained from the transient-state tests was slightly underestimated.

3.7.4 Determination of strength and elastic modulus

The yield strengths at strain levels of 0.2%, 0.5%, 1.5%, and 2.0% were obtained for the purposes of comparison with design standards prediction as these strain levels are widely accepted. The 0.2% yield strength ($f_{0.2}$) is the intersection point of the stress–strain curve and the proportional line offset by 0.2% strain. The yield strengths of $f_{0.5}$, $f_{1.5}$, and $f_{2.0}$ at the strain levels of 0.5, 1.5, and 2.0, respectively, are those values corresponding to the intersection points of the stress–strain curve and the vertical lines specified at given strain levels, as shown in Fig. 3.45. Serration of the stress–strain

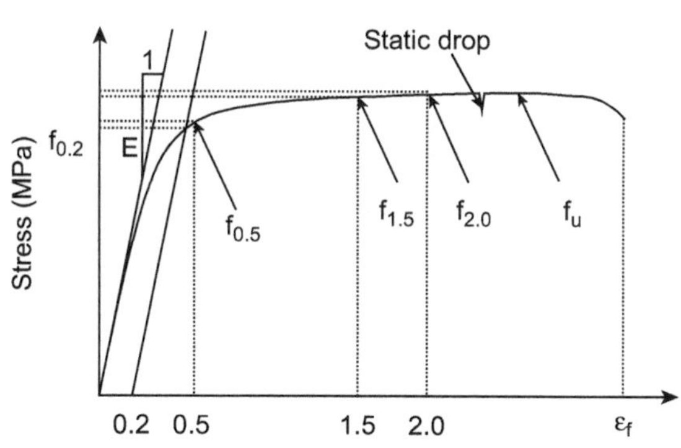

Figure 3.45 Definition of symbols.

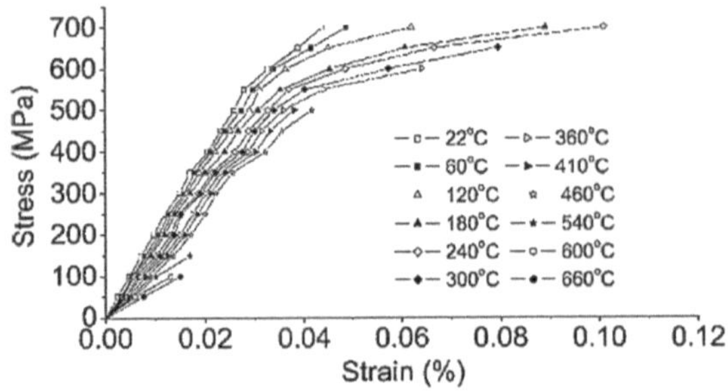

Figure 3.46 Stress—strain curve of high strength steel obtained from transient-state test results.

curve was observed at high temperatures and the intersection point was the mean value determined from the serration. The elastic modulus was determined from the stress—strain curve based on the tangent modulus of the initial elastic linear curve.

For the transient-state tests, the results are first converted into stress—strain curves, as shown in Fig. 3.46. The specimen were loaded to a given stress level, and the elastic moduli at different temperatures of each specimen can be determined from the stress—strain curves obtained from the transient-state tests. The data for each specimen at varying temperatures were normalized with respect to the initial elastic modulus at ambient temperature (normal room temperature) of each specimen so that the influence of elastic modulus variation could be eliminated. Some repeat tests were conducted, and the deviations between these test results were quite small with a maximum difference of 4%.

The material properties obtained from the tests for HSS at normal room temperature are presented in Table 3.23. The reduction factors ($f_{0.2,T}/f_{0.2,normal}$, $f_{0.5,T}/f_{0.5,normal}$, $f_{1.5,T}/f_{1.5,normal}$, $f_{2.0,T}/f_{2.0,normal}$) determined from the ratio of different yield strengths at different temperatures to that at ambient temperature (22°C) for the four strain levels of 0.2%, 0.5%, 1.5%,

Table 3.23 Material properties of high strength steel and mild steel at normal room temperature.

Steel	$f_{0.2,normal}$ (MPa)	$f_{0.5,normal}$ (MPa)	$f_{1.5,normal}$ (MPa)	$f_{2.0,normal}$ (MPa)	$f_{u,normal}$ (MPa)	E_{normal} (GPa)	ε_f (%)
High strength steel	789	790	813	823	847	223	7

Table 3.24 Reduction factors of yield strength and elastic modulus of high strength steel.

T (°C)	E_T/E_{normal}	$f_{0.2,T}/f_{0.2,normal}$	$f_{0.5,T}/f_{0.5,normal}$	$f_{1.5,T}/f_{1.5,normal}$	$f_{2.0,T}/f_{2.0,normal}$
22	1	1	1	1	1
60	1.04	0.95	0.96	0.96	0.96
120	1.01	0.94	0.94	0.96	0.96
150	1.04	0.96	0.95	0.98	0.99
180	1.02	0.92	0.92	0.97	0.97
240	0.98	0.89	0.89	0.99	1
300	0.99, 1.00*	0.90, 0.88*	0.91, 0.89*	0.98, 0.97*	0.99, 0.98*
410	0.92	0.87	0.87	0.94	0.94
460	0.94	0.8	0.81	0.85	0.84
540	0.87	0.75	0.75	0.76	0.74
600	0.73	0.6	0.61	0.56	0.59
660	0.73	0.43	0.44	0.43	0.42
720	0.51	0.21	0.21	0.22	0.22
770	0.49	0.14	0.14	0.15	0.14
830	0.33	0.08	0.08	0.08	0.09
940	0.12	0.05	0.05	0.05	0.05

* = second test.

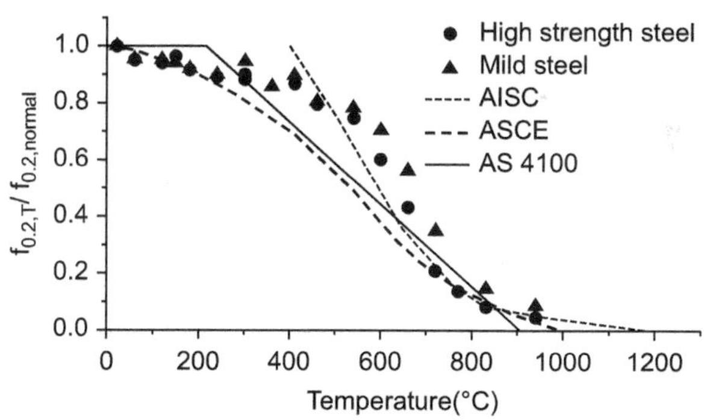

Figure 3.47 Comparison of reduction factors of 0.2% yield strength predicted by AISC, ASCE, and AS 4100 with test results.

and 2.0%, respectively, are presented in Table 3.24. The test results for 0.2% yield strength of HSS are plotted in Fig. 3.47. The vertical axis of the graph presents the reduction factor $f_{0.2,T}/f_{0.2,\text{normal}}$ and the horizontal axis represents the variation in temperature.

The reduction factors of 0.2% yield strength obtained from the tests were compared with the AISC Specification (2005), ASCE Manual (1992), and Australian Standard AS 4100 (1998) predictions, as shown in Fig. 3.47. The comparison indicates that the prediction from the AISC Specification is generally adequate but unconservative for the temperatures of 410 and 460°C. It can be seen that the prediction from the ASCE Manual (1992) is generally conservative for HSS, but slightly unconservative for temperatures of 60, 120, 770, and 830°C for HSS. It is also shown that the AS 4100 provides a conservative prediction for the test results of HSS for temperatures ranging from 300 to 660°C and greater than 830°C but is unconservative for temperatures ranging from 720 to 830°C and less than 300°C.

The reduction factors of yield strength for strain levels of 0.5%, 1.5%, and 2.0% are compared with the EN1993-1-2(2005) and BS5950-8 (2003), as shown in Figs. 3.48–3.50. The reduction factors of 0.5% and yield strength predicted by the BS5950-8 are conservative for the test results of

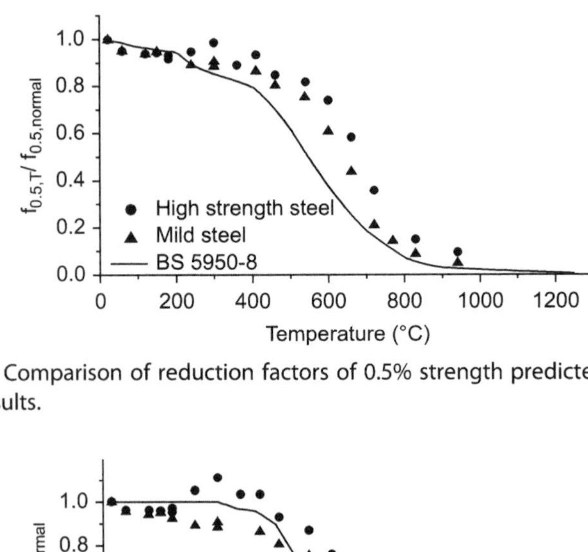

Figure 3.48 Comparison of reduction factors of 0.5% strength predicted by BS5950-8 with test results.

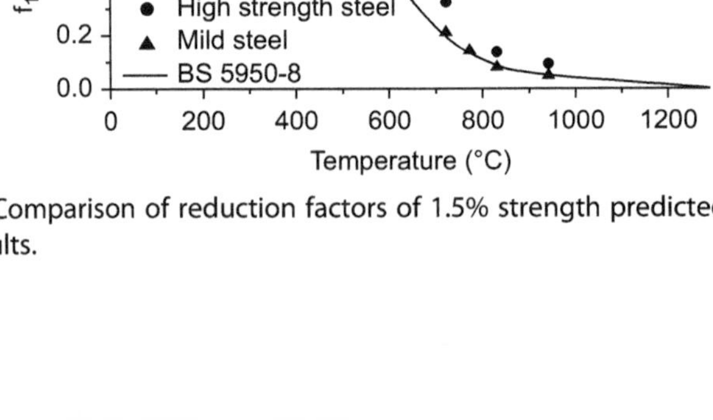

Figure 3.49 Comparison of reduction factors of 1.5% strength predicted by BS5950-8 with test results.

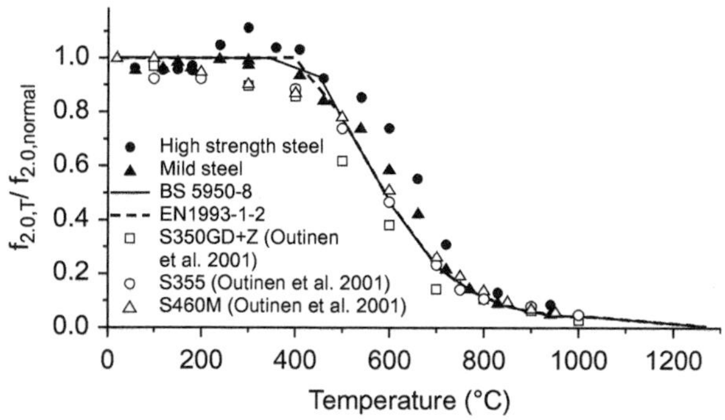

Figure 3.50 Comparison of reduction factors of 2.0% strength predicted by BS5950-8 and EN1993-1-2 with test results.

HSS for temperatures ranging from 240 to 940°C but unconservatively predicted for temperatures less than 240°C. The reduction factors of 1.5% and 2.0% yield strength predicted by the BS5950-8 are conservative for the test results of HSS for temperatures ranging from 240 to 940°C but unconservatively predicted for temperatures below 240°C. The reduction factors of 2.0% yield strength predicted by the EN1993-1-2(2005) is conservative for HSS for temperatures ranging from 240 to 940°C but unconservatively predicted for HSS for temperatures below 240°C. The reduction factors of yield strength for the strain level of 2.0% are compared with the transient-state test results of structural steel S350GD+Z, S355, and S460M conducted by Outinen et al. (2001). It is shown that the test results obtained in this study generally agree well with the test results obtained by Outinen et al. (2001), except that the test results by Outinen et al. are generally lower than the test results in this study for temperatures ranging from 240 to 720°C.

3.7.4.1 Elastic modulus

The reduction factors of elastic modulus of HSS obtained from steady- and transient-state tests were compared with AISC (2005), ASCE (1992), AS 4100 (1998), and EN1993-1-2(2005) predictions, as shown in Fig. 3.51. The reduction factors of elastic modulus of HSS obtained from the transient-state tests were also presented in Table 3.25. It can be seen that the reduction factors of elastic modulus of HSS obtained from the steady-state tests are higher than those obtained from the transient-state tests, as shown in Fig. 3.51. The prediction of the reduction factors of elastic modulus obtained from the AISC (2005) are close to those predicted by EN1993-1-2(2005). The prediction of the reduction factors of elastic modulus obtained from

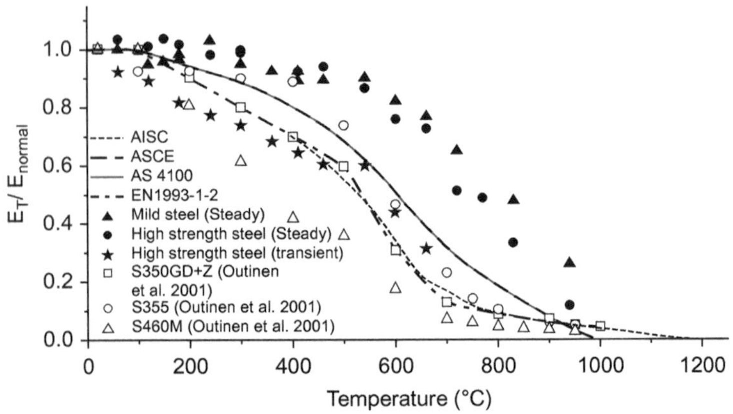

Figure 3.51 Comparison of elastic modulus predicted by AISC, ASCE, AS4100, and EN1993-1-2 with test results.

the ASCE (1992) is identical to those predicted by AS 4100 (1998). This is because of the ASCE (1992) and AS 4100 (1998) adopting the same elastic modulus relationships with varying temperatures. It can be seen that the reduction factors predicted by AISC (2005), ASCE (1992), AS 4100 (1998), and EN1993-1-2(2005) are generally conservative for the steady-state test results obtained from this study for HSS. The reduction factors of elastic modulus predicted by the AISC and EN1993-1-2 are unconservative for the HSS obtained from transient-state tests, except for temperatures ranging from 540 to 660°C. It is shown that the predictions of elastic modulus using the approach of ASCE (1992) and AS 4100 (1998) are unconservative for the HSS obtained from the transient-state tests. The transient-state test results of the elastic modulus of HSS obtained in this study are different to those transient-state test results conducted by Outinen et al. (2001), as shown in Fig. 3.51. It is shown that the reduction factors of elastic modulus obtained from steady-state tests are generally higher than those transient-state test results. The reduction factors of yield strength and elastic modulus of the HSS are compared in Table 3.26.

3.7.4.2 Ultimate strength

The ultimate strength of the specimen is defined as a specified load for the temperature reached, at which the specimen undergoes a continuous elongation at an appreciable rate. This specified load was considered as the ultimate strength of the specimen at that particular temperature in the transient-state tests. In Table 3.27, the ultimate strengths of the HSS obtained from the transient-state tests ($f_{t,u,T}$) are compared with the ultimate

Table 3.25 Elastic modulus of high strength steel obtained from transient-state tests.

Temperature (°C)	22	60	120	180	240	300	360	410	460	540	600	660
E_T (GPa)	209	192.9	186.5	170.7	161.7	154.3	142.9	134.9	126.6	125.8	92	66.3
E_T/E_{normal}	1	0.92	0.89	0.82	0.77	0.74	0.68	0.64	0.61	0.6	0.44	0.32

Table 3.26 Comparison of yield strength and elastic modulus between high strength steel and mild steel.

T (°C)	$f_{0.2,T}/f_{0.2,normal}$	E_T/E_{normal}	*T* (°C)	$f_{0.2,T}/f_{0.2,normal}$	E_T/E_{normal}
22	1	1	460	0.8	0.94
60	0.95	1.04	540	0.75	0.87
120	0.94	1.01	600	0.6	0.73
150	0.96	1.04	660	0.43	0.73
180	0.92	1.02	720	0.21	0.51
240	0.89	0.98	830	0.08	0.33
300	0.89	1	940	0.05	0.12
410	0.87	0.92			

Table 3.27 Comparison of ultimate strength obtained from transient- and steady-state tests for high strength steel.

Temperature (°C)	$f_{t,u,T}$ (MPa)	$f_{u,T}$ (MPa)	$f_{u,T}$-drop (MPa)
22	>700	856,838*	838,820*
60	>700	812	793
120	>700	822	811
150	>700	840	824
180	>700	833	812
240	>700	846	812
300	>700	853,830*	809,786*
410	600–650	787	735
460	500–550	694	642
540	350–400	620	550
600	150–200	498	407
660	100–150	357	250
720	50–100	178	109
770	50–100	119	57
830	1–50	75	34
940	<1	43	22

strengths obtained from the steady-state tests ($f_{u,T}$) with and without consideration of the static drop. A static drop of the stress–strain curve is obtained by pausing the applied strain for 1 min. This allowed the stress relaxation associated with plastic strain to take place; hence, the effect of loading rate can be eliminated, as shown in Fig. 3.44. The ultimate strength obtained from steady-state tests with the consideration of the static drop are closer to the results obtained from the transient-state tests compared with the results without the consideration of the static drop, as shown in Table 3.27.

3.8 Mechanical properties of high strength A572 steel

A comprehensive test program was designed to undertake mechanical property tests on ASTM A572 steel. The tensile strength tests were carried at various temperatures in 200–1000°C range. Rectangular steel coupons were cut from A572 Gr. 50 steel sheets. The total length of the test coupon is 609.6 mm, and the length of tapered section is 304.8 mm. The cross section of the reduced section is 9.53 × 12.7 mm, in which 9.53 mm is the width and 12.7 mm is the thickness of the steel coupon. The configuration of a typical test coupon used in tensile strength tests is shown in Fig. 3.52. The chemical composition of A572 steel is slightly different from that of conventional A36 steel. The specifications for chemical composition of both types of steel, together with actual composition of A572 steel used in current test program (as specified by the manufacturer), are tabulated in Table 3.28. The high strength and toughness in A572 steel is achieved by lowering the carbon content to about 0.06% and adding alloy elements such as vanadium (V = 0.08%). This is in comparison with a higher carbon content of 0.25% and alloy elements in A36 steel, as illustrated in Table 3.28. This combination of low carbon content and presence of alloy elements also enhances ductility and weldability properties.

3.8.1 Test equipment

The test setup for high-temperature tensile strength tests was composed of a steel frame that can apply tensile or compressive loading, an electric furnace, and a data acquisition system. The universal testing machine has a loading capacity of 200 kN, and the axial deformations in the test coupon subjected to tension is measured by a displacement transducer (±38 mm linear

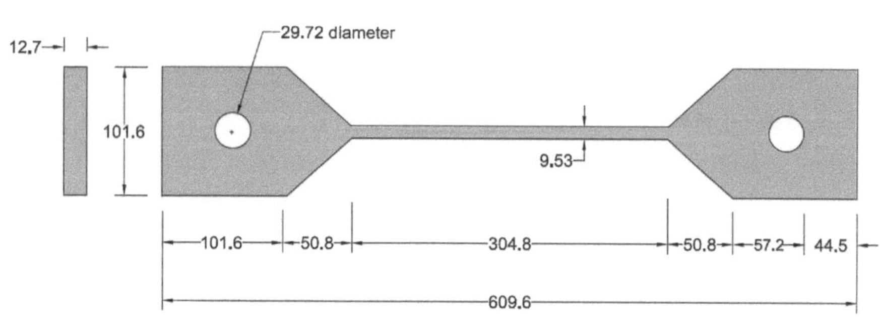

All dimensions are in millimeters

Figure 3.52 Steel coupon for high-temperature tensile strength tests.

Table 3.28 Chemical composition of A572 Gr. 50 steel as compared with A36 steel.

Chemical composition	A36 steel (ASTM limits)	A572 grade 50 steel (ASTM limits)	A572 grade 50 steel (actual)
Carbon (C) %	0.25–0.290	0.23 max	0.06
Manganese (Mn) %	1.03	1.35 max	0.91
Phosphorous (P) %	0.040	0.04 max	0.009
Silicon (Si) %	0.280	0.40 max	0.02
Sulfur (S) %	0.050	0.05 max	0.009
Copper (Cu) %	0.20	–	0.12
Vanadium (V) %	–	0.06 max	0.0818
Cobalt (Co) %	–	0.05 max	–
Iron (Fe) %	Remainder	Remainder	Remainder

variable differential transformer [LVDT]), with 0.0254 mm sensitivity, that is placed outside the furnace. The programmable furnace that surrounds the steel coupon can generate any target temperature up to 1000°C at any given rate of temperature rise and can keep the target temperature in steel coupon constant (inside the furnace) for the test duration. Three thermocouples are mounted on the furnace wall to measure temperatures at the upper, middle, and lower zones. The average values of these three thermocouples are taken to be the furnace temperature. The furnace has an inner diameter of 203 mm and a height of 305 mm. Therefore, the heating length of the specimen is to be 305 mm. The loading machine and the furnace are connected to a data acquisition system and computer wherein temperature, load, and displacement throughout the test are recorded. The test setup for high-temperature tensile test is illustrated in Fig. 3.53.

3.8.2 Test procedure

Tensile strength tests were conducted at temperatures of 20, 400, 500, 600, 700, and 800°C. For these tensile strength tests, each steel coupon was instrumented with a thermocouple at the midlength of the reduced area of the specimen and placed in the furnace. This thermocouple records the actual temperature progression on the steel coupon during the test because temperature inside the furnace might not be the same as the coupon temperature. Following the placement of coupon, the furnace is turned on

Figure 3.53 Test setup for high-temperature tensile strength tests.

to heat up the coupon to a predefined target temperature. A rate of heating of 10°C/min was adopted to heat the coupon to a target temperature (Wang et al., 2013). When the target temperature is attained, the specimen was kept at this temperature for 30 min to attain a steady-state condition throughout the specimen. Then, a tensile load was applied incrementally on the coupon while keeping the target temperature constant.

The applied loading is controlled manually at a loading rate of 0.5 kN/s. The tensile displacement during the test was measured using highly sensitive LVDT. Both LVDT and the loading cell were connected to the data acquisition system to record the readings during the test.

3.8.3 Tensile strength at elevated temperatures

The variation of stress—strain response of A572 steel at elevated temperatures is evaluated using data from uniaxial tensile strength tests carried out at various target temperatures. The displacements recorded at various load levels, during tension tests, are utilized to generate stress—strain response of A572 steel, and these are plotted in Fig. 3.54. It can be seen in this figure that the general trend of stress—strain curve is linear elastic up to yielding of steel, which varies with temperature level, followed by nonlinear response.

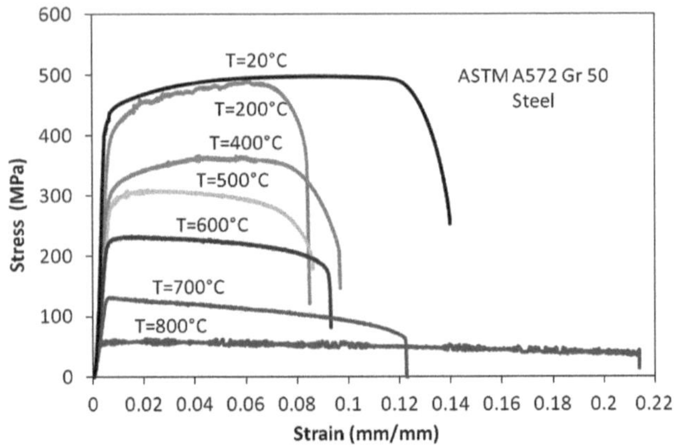

Figure 3.54 High-temperature stress–strain response of A572 Gr. 50 steel.

Therefore, the stress at which yielding occurs in steel coupon is different at different temperatures.

No well-defined yield plateau is recorded in the room temperature test for A572 steel. This can be attributed to very low levels of carbon content and presence of alloys such as vanadium, in the chemical composition of A572 steel as tabulated in Table 3.28. Past the yielding stress, steel undergoes plastic deformation with increasing stress up to reaching ultimate (stress) point. Once the ultimate stress is reached, plastic deformation continues through unloading phase (necking) up to rapture. In addition, A572 steel exhibits some level of strain hardening in the temperature range of 20–400°C, while this phenomenon disappears at temperatures beyond 400°C. This agrees with reported trends in the literature for conventional carbon steel (EC3, 2005).

From the tensile strength tests data, the yield stress is evaluated based on a strain level of 0.2%, while the ultimate stress is defined as the maximum stress recorded during the tensile strength test on the coupon. The significant points on stress–strain curve from high-temperature uniaxial tensile strength tests are summarized in Table 3.29. The yield and ultimate strength and modulus reduction factors given in this table are calculated as a ratio of yield and ultimate strength (or elastic modulus) at a given (elevated) temperature to that at room temperature. As illustrated in Table 3.29, with increasing temperature, modulus of elasticity (stiffness), yield strength, and ultimate strength of A572 steel decrease significantly. The extent to which steel undergoes plastic deformation, measured in terms of ductility, can be evaluated through percentage of elongation and reduction in area of the test

Table 3.29 Summary of high-temperature tensile strength tests on A572 Gr. 50 steel coupons.

Temperature (°C)	$f_{y,T,0.2\%}$ (MPa)	$f_{u,T}$ (MPa)	Rupture strain	Elongation (%)	Reduction in area (%)	$E_{,T}/E_{20}$	$f_{y,T}/f_{y20}$	$f_{u,T}/f_{u20}$
20	408.9	498.5	0.140	5.7	70.9	1.00	1.00	1.00
200	399.8	489.6	0.085	4.2	53.3	0.81	0.98	0.98
400	301.4	365.9	0.097	4.8	67.1	0.59	0.74	0.73
500	282.2	311.1	0.086	4.8	59.3	0.54	0.69	0.62
600	224.9	232.3	0.093	4.4	44.6	0.49	0.55	0.47
700	123.2	131.6	0.123	6.3	59.2	0.30	0.30	0.26
800	53.9	62.4	0.214	13.2	55.6	0.16	0.13	0.13

Figure 3.55 Tensile strength failure modes of A572 Gr. 50 steel at various temperatures.

coupon at rupture. These values at various temperatures are also tabulated in Table 3.29. Test data in this table indicate that, with increasing temperature, the elongation of steel decreases except for temperatures beyond 700°C. However, the reduction in cross-sectional area decreases with increasing temperature from 20 to 800°C. This can be attributed to necking phenomenon that is well defined for steel coupon at room temperature as compared with that at high temperature, which raptures with less necking (flat surface), and this can be seen in Fig. 3.55. Therefore, A572 steel exhibits higher expansion but lesser necking at temperatures beyond 700°C.

Results from tensile strength test on coupons are utilized to derive strength (yield strength) and stiffen (elastic modulus) reduction factors for A572 steel at high temperatures. These reduction factors are compared with reduction factors specified in ASCE manual, EC3, and those published data in the literature, in Fig. 3.56. The strength and stiffness reduction factors specified in EC3 and ASCE are for conventional carbon steel (EN1993-1-2, 2005; ASCE, 1992). These reduction factors in EC3 are derived based on data from transient tests, and yield strength at a given temperature is evaluated based on 2.0% strain level. In the case of ASCE, the high-temperature yield strength is based on stress corresponding to a strain level of 0.2%. However, no information is provided in associated references on specific test conditions that are used in mechanical property

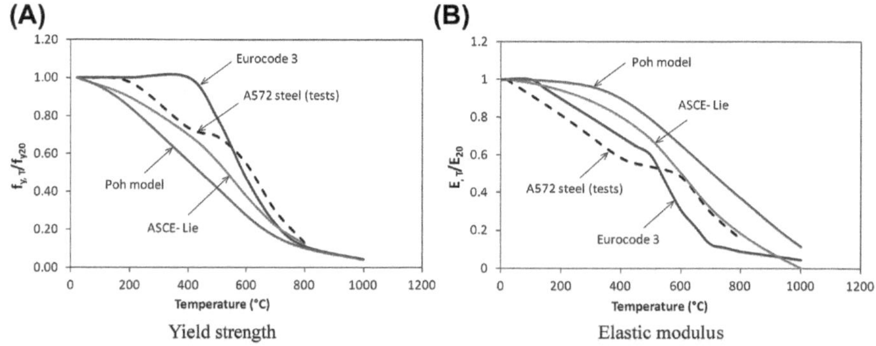

Figure 3.56 Comparison of measured strength and stiffness reduction factors of A572 steel with predicted values from codes. (A) Yield strenght and (B) Elastic modulus.

tests to derive ASCE strength reduction factors of steel (Kodur et al., 2010; Kodur and Harmathy, 2008).

It can be seen from Fig. 3.56A that the strength reduction curve for A572 steel lies between the strength reduction curves predicted from ASCE and EC3 relations. A closer review of this figure shows that strength reduction factors for A572 steel is more comparable with ASCE values for temperatures up to 400°C, while it compares well with EC3 for temperatures beyond 600°C. Accordingly, applying EC3 reduction factors to evaluate tensile strength of A572 steel in the temperature range 300−600°C can lead to overestimation, while applying ASCE factors beyond temperature of 400°C can lead to underestimation of tensile strength. These variations are attributed to chemical composition of A572 steel that is different as compared with that of conventional carbon steel as illustrated in Table 3.28. In addition, these variations can be attributed to different test conditions adopted in EC3 (transient tests) that account for partial creep strain. Fig. 3.56B shows comparison of measured elastic modulus reduction factors for A572 steel with corresponding predicted values from ASCE and EC3 relations. It can be seen that elastic modulus reduction factors for A572 follow the same trend as EC3 with an average of 10% variation in 20−500°C temperature range, while there is higher variation (average of 20%) with corresponding values from ASCE. For higher temperatures, in 600−800°C range, the elastic modulus reduction factors for A572 steel are comparable with predicted values from ASCE, while there are significant variations as compared with EC3 values. These variations are attributed to the chemical composition that is different for A572 steel as compared with conventional carbon steel as discussed previously.

Reduction factors for strength and elastic modulus of A572 steel are also compared with corresponding values from literature (Poh model) as shown in Fig. 3.56 (Poh, 2001). Reduction values from Poh model, developed for conventional steel, show significant variation in strength and elastic modulus of A572 steel. It can be seen that strength reduction curve predicted from Poh model lies below corresponding reduction curves from ASCE, EC3, and A572 steel, while elastic modulus reduction curve from Poh model lies above the elastic modulus curves from ASCE, EC3, and A572 steel. The substantial variation between Poh model and reduction factors of A572 is because Poh model is a general stress—stress model and does not take the chemical composition of steel into account.

3.8.4 Proposed reduction factors for strength and elastic modulus

For evaluating high-temperature tensile strength and elastic modulus of A572 Gr. 50 steel, the following relations are proposed:

For 20°C ≤ T_s ≤ 800°C,

$$\frac{f_{y,T}}{f_{y20}} = (-1.75e-11)T_s^3 - (1.38e-6)T_s^2 + (1.52e-5)T_s + 1.0 \quad (3.11)$$

$$\frac{f_{y,T}}{f_{y20}} = (-1.75e-11)T_s^3 - (1.38e-6)T_s^2 + (1.52e-5)T_s + 1.0 \quad (3.12)$$

$$\frac{E_T}{E_{20}} = (-1.98e-9)T_s^3 + (2.2e-6)T_s^2 - (1.59e-3)T_s + 1.0 \quad (3.13)$$

3.9 Mechanical properties of high strength steel RQT-S690

3.9.1 The test device

In the elevated temperature test, tensile tests were conducted by Chiew et al. (2014) on an ISTRON5584 servohydraulic universal testing machine with maximum pulling capacity of 60 kN at 800°C. The heating device attached to the tensile test machine is a split tube furnace that can heat the specimen at a rate of 30°C/min up to 1050°C. Inside the furnace, there were three temperature detectors monitoring the inner temperatures from top to bottom. In addition, a thermocouple was attached to the specimen directly to measure its real-time temperature profile. As for strain measurement, a high-temperature extensometer with gauge length of

25 mm and maximum strain capacity of 30% was used to capture the load–strain relationships. The configuration of the setup is shown in Fig. 3.57.

To ensure consistence in specimen dimension and test environment, in the postfire residual strength test, the postfire specimen were tested in the same tensile test machine with the furnace turned off. The laboratory heating furnace Nabertherm LH216 with a maximum heating capacity of 1200°C and robust refractory bricks inside were employed to simulate the heating and cooling in fire condition (Fig. 3.58). The internal dimensions (width × depth × height) of the oven were 500 × 500 × 700 mm. As the

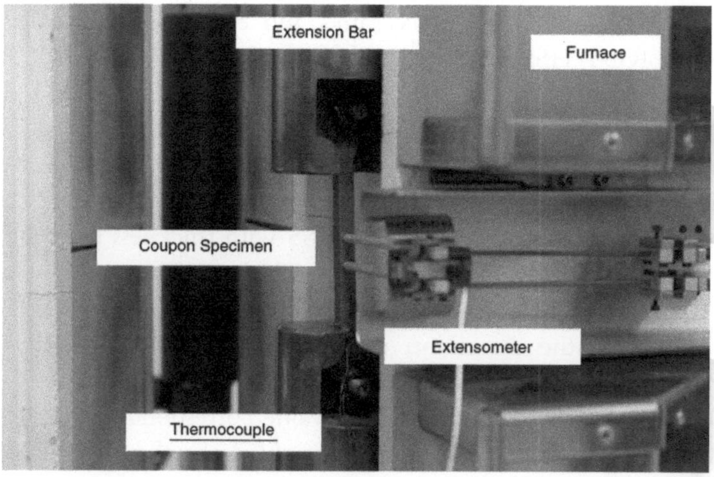

Figure 3.57 The high-temperature tensile test machine with heating chamber.

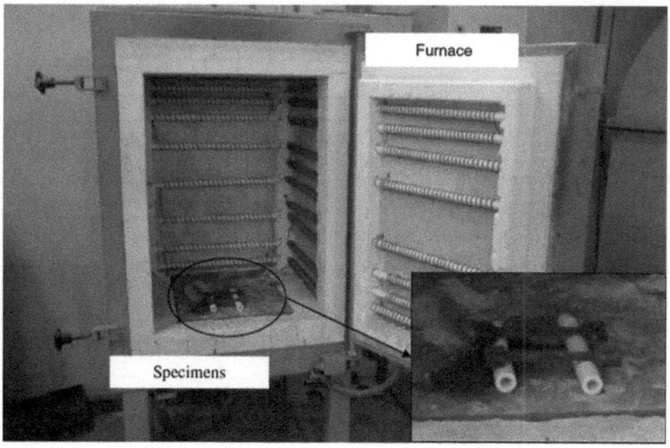

Figure 3.58 The furnace employed for heating the specimen.

thermal inertia of these specimen is very small compared with that of the furnace and they are located so as not to impede the airflow, a homogenous atmosphere has been achieved. To structurally isolate the members during heat treatment, specimen were simply supported on ceramic roller supports to allow flexibility of the specimen during thermal expansion as maintained.

3.9.2 Tested materials and specimen

RQT steel plates exhibit better homogeneity in through-thickness mechanical properties compared with traditional directly quenched and tempered steel plates. The materials and specimen studied in this chapter have a specified nominal yield strength of 690 N/mm^2, tensile strength between 790 N/mm^2 and 930 N/mm^2, and elongation capacity around 15%. These RQT HSS plates comply with the EN 10025-6 (2009) grade S690 specification, which is approximately equivalent to the ASTM A514 steel ASTM (2009).

The mechanical properties of RQT-S690 HSS at room temperature acquired from coupon tensile test are shown in Table 3.30. The chemical composition of the tested RQT-S690 HSS with thickness between 8 and 30 mm is shown in Table 3.31, in which both the maximum values and the typical values are provided. The calculated carbon equivalent (CE) value corresponding to the maximum element content state (second row of Table 3.31) is 0.43, while that corresponding to the typical values (last row of Table 3.31) is 0.4. Note that the CE value corresponding to the typical value is within the suggested welding limit of 0.40 by the AWS (2008), and this grants RQT-S690 HSS better weldability compared with other HSS alloy steels.

The specimen were cut from 8 mm thick RQT-S690 HSS plates in the longitudinal direction. The configuration was designed according to BS EN 10002-5, as shown in Fig. 3.59. Two holes were drilled at both ends to fix the specimen into the load cell during the elevated temperature test. Both the elevated temperature and the postfire residual strength tests adopted the

Table 3.30 Characteristic strengths of RQT-S690 high strength steel at room temperature (MPa).

Strengths	$f_{0.2,n}$	$f_{0.5,n}$	$f_{1.5,n}$	$f_{2,n}$	$f_{u,n}$	E_n	Elongation (%)
RQT-S690 HSS	769.0	772.6	810.2	817.1	849.8	201.3	14.7

Table 3.31 Chemical composition of RQT-S690 high strength steel (%).

Elements	C	Si	Mn	S	P	Cr	Mo	Nb	Ti	V	Ni	Cu	B	Al	Carbon equivalent
Maximum	0.20	0.50	1.60	0.010	0.025	0.25	0.20	0.060	0.040	0.08	0.70	0.20	0.004	0.060	0.43
Typical	0.14	0.40	1.35	0.003	0.012	0.01	0.12	0.035	0.025	0.05	0.01	0.01	0.002	0.035	0.40

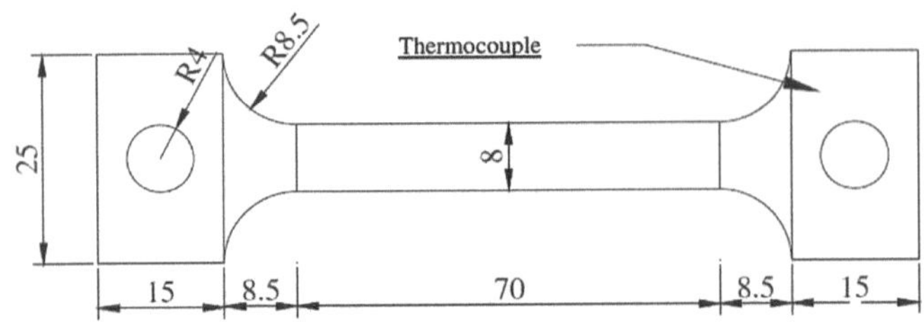

Figure 3.59 The dimensions of the specimen.

same specimen configuration. During the heat treatment for postfire test specimen, it was found that oxidized layer would form on the surface if the coupons were heated to temperatures above 600 °C. Hence, those specimen were polished and remeasured before the first postfire tensile test.

3.9.3 The test methods

This study adopted the steady-state method because continuous stress–strain curves are more favorable for potential complex structure fire-resistance analysis in the future. The heating rates were set as 20°C/min. Thermal expansion of the specimen was allowed by maintaining slight tension loads during heating. When the preselected temperature was reached, an additional of 10 min of heating was allowed to maintain and ensure the uniformity of the temperature distribution within the coupon. The tests were carried out at every 100°C increment from 100°C up to 800°C as well as at a supplementary temperature of 450°C. The test stopped at 800°C because strength at that temperature was already too low to carry any meaningful load (less than 10% of the nominal strengths at room temperature). At each prescribed temperature, two compulsory tests were always conducted, and another repeated test would be done if the difference between the first two compulsory test results was larger than 2%. The constant loading rate was set as 0.2 mm/min and is approximately equivalent to a strain rate of 0.003/min, which is within the required range of 0.001–0.005/min by the EN 10002-5 (1992).

3.9.4 Test results

3.9.4.1 General descriptions

The test results in terms of temperature-dependent stress–strain curves are shown in Fig. 3.60. It is found that the strength of the HSS was generally not affected if the maximum temperature attained was below 300°C. A rapid drop in strength could be observed from 400 to 600°C. In this temperature

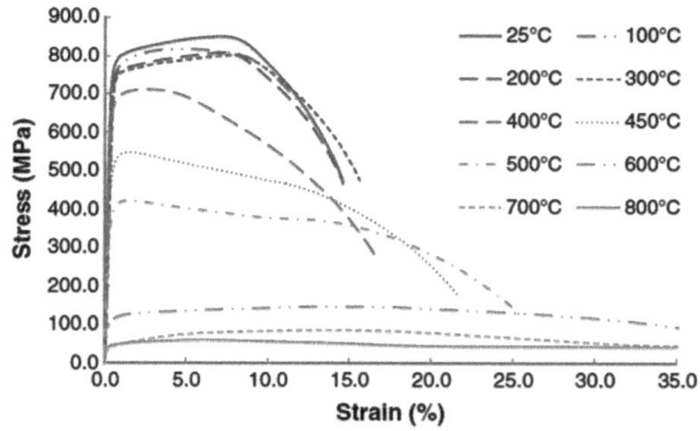

Figure 3.60 Stress–strain curves of RQT-S690 HSS at elevated temperatures.

range, the material lost most of its strength so that only about 10% of the nominal strengths were left at 700°C. On the other hand, the ductility increased remarkably whenever the maximum temperature exceeded 300°C. The elongations above 600°C were so large that they were not shown completely in Fig. 3.60. However, this conclusion happens to be on contrary to the findings about quenched and tempered steel S690QL by Qiang et al. (2012). According to their report, ductility of S690QL decreased from 10.5% at room temperature to less than 8% at 600°C.

3.9.4.2 Determination of characteristic strength and elastic modulus

As all the stress strain curves do not show any obvious yield stage, the yield strengths at different strain levels, i.e., 0.5%, 1.5%, and 2.0%, as well as the proportional limit strength and ultimate strength were captured as the characteristic strengths for further data analysis. The selection essentially followed the recommendation of characteristic strengths in fire-resistance design standards EN1993-1-2 (2005) and BS5950-8 (2003). Subsequently, the reduction factors $f_{p,t}/f_{p,n}$, $f_{0.5,t}/f_{0.5,n}$, $f_{1.5,t}/f_{1.5,n}$, $f_{2.0,t}/f_{2.0,n}$, and $f_{u,t}/f_{u,n}$ (see Fig. 3.61 for definition) are calculated as the ratio of yield strengths at elevated temperatures to the corresponding strengths at room temperature (25°C). The analysis for elastic modulus was carried out in this way too, i.e., studying the reduction factor of the tangent modulus at the linear-elastic part (E_t/E_n). The data obtained from the tests are listed in Table 3.32.

To verify the test results and obtain a better understanding about the RQT-S690 HSS at elevated temperatures, the characteristic strength reduction curves obtained from this study are compared with available data

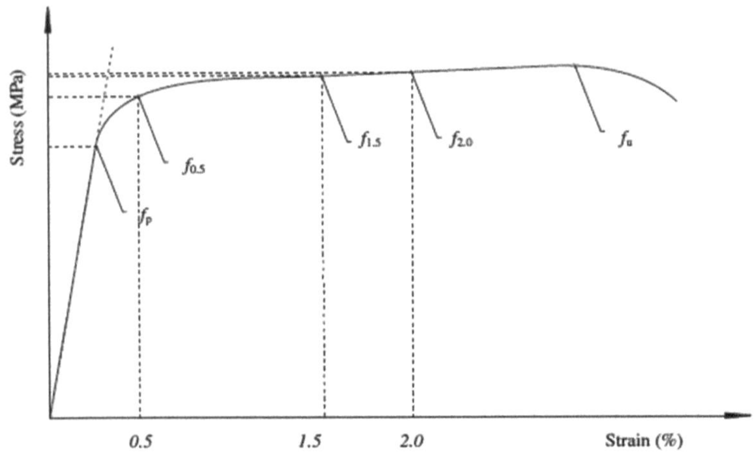

Figure 3.61 Definitions of characteristic strengths.

Table 3.32 Reduction factors of characteristic strengths and elastic modulus.

T (°C)	$f_{p,t}/f_{p,n}$	$f_{0.5,t}/f_{0.5,n}$	$f_{1.5,t}/f_{1.5,n}$	$f_{2.0,t}/f_{2.0,n}$	$f_{u,t}/f_{u,n}$	E_t/E_n
25	1.00	1.00	1.00	1.00	1.00	1.00
100	0.99	1.00	0.96	0.96	0.96	1.01
200	0.94	0.98	0.98	0.97	0.95	1.02
300	0.89	0.95	0.98	0.98	0.97	0.96
400	0.85	0.89	0.87	0.87	0.84	1.01
450	0.61	0.66	0.68	0.67	0.64	0.91
500	0.40	0.51	0.52	0.51	0.50	0.77
600	0.14	0.15	0.16	0.16	0.19	0.66
700	0.05	0.06	0.07	0.07	0.10	0.38
800	0.05	0.06	0.07	0.07	0.07	0.29

in the literature including similar researches on (i) the cold-formed steel (CFS) CF-S355J2H (reference (Outinen, 2007), yield strength 526 MPa, tensile ratio 1.09), (ii) the quenched and tempered steel plate BISPLATE-80 (reference (Chen et al., 2006), yield strength 789 MPa, tensile ratio 1.07), (iii) the S690QL HSS (reference (Qiang et al., 2012), yield strength 789 MPa, tensile ratio 1.04), and (iv) the reduction equations in fire design standards including BS5950-8 and EC3 (Table 3.33). In reference (Outinen, 2007), the CF-S355J2H CFS coupons were taken from the nonwelded faces of rectangular hollow section of 50 × 50 × 3 mm. This material showed no obvious yield stage at all temperatures, and the hardening effects by cold work vanished at 600–700°C. Both BISPLATE-80 plate and S690QL HSS are fabricated up to EN10215-5 grade S690 by

Table 3.33 Reduction factors of RQT-S690 high strength steel after being heated and cooled down.

Temperature (°C)	$f_{y,t}/f_{0.2,n}$	$f_{2.0,t}/f_{2.0,n}$	$f_{u,t}/f_{u,n}$	E_t/E_n
25	1.00	1.00	1.00	1.00
400	1.07	1.00	0.99	0.95
600	0.98	0.93	0.93	0.96
800	0.61	0.58	0.65	1.01
900	0.36	0.41	0.52	0.88
1000	0.35	0.39	0.50	1.02

quenching tempering technique. Although RQT-S690 HSS, BISPLATE-80 plate, and S690QL HSS exhibit similar elevated temperature strengths at room temperature, there is a significant difference between their ductility. At room temperature, RQT-S690 HSS had the best ductility of 14.7% (in terms of elongation on 50 mm gauge length), while the corresponding values for BISPLATE-80 plate and S690QL HSS are only 7% and about 10%, respectively.

3.9.4.3 Strengths

Based on the data shown in Table 3.32, the reduction curves of different characteristic yield strengths for RQT-S690 HSS are plotted in Fig. 3.62. It can be seen that all the reduction curves shared the same trends, except for slight fluctuation for the curve of fp. Generally, the reduction curves can be divided into three stages. In the first stage, which is from room temperature to 400°C, the deterioration of strength was less than 15%. In the second stage, which is from 400 to 600°C, the strength dropped rapidly and significantly (less than 15% of the nominal strength was left at 600°C).

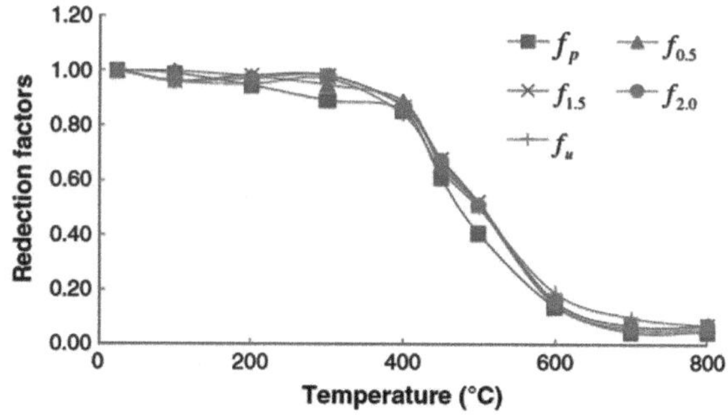

Figure 3.62 Reduction factor curves of different characteristic strengths of RQT-S690 high strength steel at elevated temperatures.

Finally, in the third stage, which is above 600°C, the strengths became too small to carry loads.

Comparisons between the proportional limit strength of the different material are shown in Fig. 3.63A. RQT-S690 HSS, CF-S355J2H CFS, and EC3 were taken into consideration. EC3 seemed to be overly optimistic about the strengths at elevated temperatures. It was much higher than the curves for RQT-S690 HSS and CF-S355J2H. Even at 600–800°C when strengths of RQT-S690 HSS and CF-S355J2H approached the lower bound, the EC3 strength curve is still descending. Besides, RQTS690 HSS showed better fire resistance over CF-S355J2H CFS below 500 °C but worse afterward. As for the 0.5% strain strength (Fig. 3.63B), the BS5950-8 standard was found to be generally conservative below 400°C for all the steels except for S690QL HSS. At above 400°C, however, it was only conservative for BISPLATE-80 plate with CF-S355J2H CFS and RQT-S690HSS deteriorated much faster than the others until 700°C when negligible strength remained. The reduction curves of 1.5% strain (Fig. 3.63C) and 2.0% strain (Fig. 3.63D) strengths were very similar to

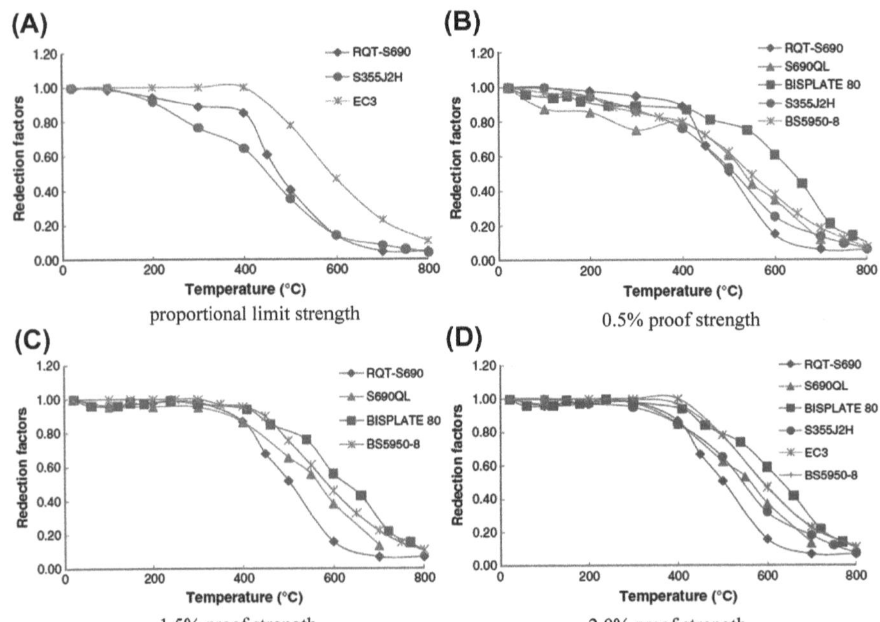

Figure 3.63 Reduction factor curves of RQT-S690 high strength steel in comparison with available data from literature. (A) Proportional limit strength (B) 0.5% proof strength (C) 1.5% proof strength and (D) 2.0% proof strength.

those of 0.5% strain curves, but the differences among the materials at low temperatures were much smaller. For 2.0% strain strength, EC3 adopted the same reduction curve as BS5950-8. However, the strength reduction curves of all the materials except for BISPLATE-80 plate were lower than that of BS5950-8 and EC3 above 400°C.

Based on the above results, it is found that RQT-S690 HSS, S690QL HSS, and CF-S355J2H CFS behave similarly at elevated temperatures. 400°C was the critical temperature above which significant strength deterioration could be observed. The strength reduction curves of both BS5950-8 and EC3 are predicted in a nonconservative manner at high temperatures (above 400°C). RQT-S690 HSS showed better resistance to temperatures below 400°C but worse behavior above 400°C compared with other materials. The mechanical properties of BISPLATE 80 plate were not conservative enough to represent heat or work hardened steels when the temperature was above 400°C, which confirmed Qiang et al.'s findings (Qiang et al., 2012).

3.9.4.4 Elastic modulus

The deterioration of elastic modulus at elevated temperatures for each material is presented in Fig. 3.64 in terms of reduction factors at specific temperatures and is compared with the EC3 predictions. Generally, these curves clustered in two groups. The lower group consists of EC3, CF-S355J2H CFS, and S690QL HSS. The curves of EC3 and CF-S355J2H are almost the same and are the lowest of all the curves. S690QL HSS shared the same reduction law below 300°C or above 700°C but was slightly higher than those two between 300 and 700°C. It should be noted that the reduction for this group started right after 100°C, which is rather different from the upper group, i.e., RQT-S690 HSS and BISPLATE plate.

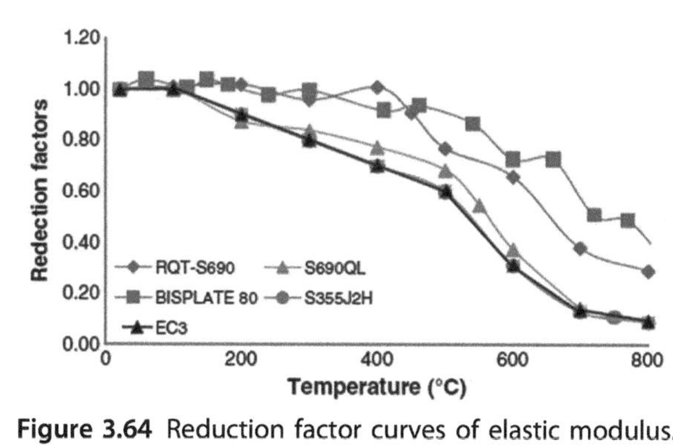

Figure 3.64 Reduction factor curves of elastic modulus.

Members of the upper group did not show obvious reduction in elastic modulus for temperatures below 450°C and decreased slowly afterward until 700°C. Overall, the reduction curves of strength were more predictable than those of elastic modulus, and there was no obvious relationship between these two properties. In other words, with sufficient understanding about a material's properties at room temperature and its fabrication techniques, it may be possible to roughly estimate the strength behavior at elevated temperatures but not for elastic modulus (Bhadeshia and Honeycombe, 2006). Nevertheless, the prediction curve of elastic modulus of EC3 has been proven to be conservative enough for all the four materials discussed.

References

AISC, 2005. American Institution of Steel Construction. Specification for structural steel buildings, Chicago.

AS 4100, 1998. Australia Standards. Steel Structures. Sydney, Australia.

AS/NZS 1554.4, 2004. Standards Australia. Structural Steel Welding Part 4: Welding of High Strength Quenched and Tempered Steels. Sydney, Australia.

ASCE, 1992. Structural Fire Protection. ASCE Manual and Reports on Engineering Practice No. 78, New York.

ASTM, 1997. E21−92: Standard Test Methods for Elevated Temperature Tension Tests of Metallic Materials. Annual Book of ASTM Standards, Vol. 03.01: Metals-Mechanical Testing; Elevated and Low-Temperature Tests; Metallography, West Conshohochken, PA.

ASTM, 2009. Standard Specification for High-Yield-Strength, Quenched and Tempered Alloy Steel Plate, Suitable for Welding. West Conshohocken, PA.

AWS, 2008. Structural Welding Code. Steel. American National Standards Institute, Miami.

BS5950-8, 2003. British Standard Institution. Structural Use of Steel Work in Building. 8: Code of Practice for Fire Resistant Design.

Bhadeshia, H.K.D.H., Honeycombe, R.W.K., 2006. Steels: Microstructure and Properties, 3th ed. Elsevier Science & Technology, Oxford, United Kingdom.

CECS 200, 2006. China Association for Engineering Construction Standardization CECS200. Chinese Technical Code for Fire Safety of Steel Structure in Buildings. China Plan Press, Beijing, China.

Chen, J., Young, B., 2004. Mechanical properties of cold-formed steel at elevated temperatures. In: Proc., 17th Int. Spec. Conf. on Cold-Formed Steel Structures. Univ. of Missouri-Rolla, Rolla, MO, pp. 437−466.

Chen, J., Young, B., Uy, B., 2006. Behavior of high strength structural steel at elevated temperatures. J. Struct. Eng. 132 (12), 1948−1954.

Chiew, S.P., Zhao, M.S., Lee, C.K., 2014. Mechanical properties of heat-treated high strength steel under fire/post-fire conditions. J. Constr. Steel Res. 98, 12−19.

E21-09, 2009. ASTM. Standard Test Methods for Elevated Temperature Tension Tests of Metallic Materials. West Conshohocken, United States.

EN 10002-5, 1992. CEN. Metallic Materials-Tensile Testing -Part 5: Method of Testing at Elevated Temperature.

EN 10025-3, 2004. CEN. Hot Rolled Products of Structural Steels - Part 3: Technical Delivery Conditions for Normalized/normalized Rolled Weldable Fine Grain Structural Steels.
EN 10025-6, 2009. CEN. Hot Rolled Products of Structural Steels – Part 6: Technical Delivery Conditions for Flat Products of High Yield Strength Structural Steels in the Quenched and Tempered Condition.
EN 1993-1-2, 2005. EC3. Design of Steel Structures -Part 1–2: General Rules-Structural Fire Design. European Committee for Standardization, Brussels, Belgium.
GB/T 2105, 1991. Standardization Administration of the People's Republic of China. Metallic Materials-Standard Test Method for the Young's Modulus, Shear Modulus and Poisson's Ratio (Dynamic Method). China Standard Press, Beijing.
GB/T 228, 2002. Standardization Administration of the People's Republic of China. Metallic Materials-Tensile Testing at Ambient Temperature. China Standard Press, Beijing.
GB/T 4338, 2006. Standardization Administration of the People's Republic of China. Metallic Materials-Tensile Testing at Elevated Temperature. China Standard Press, Beijing.
GB/T 700, 2006. Standardization Administration of the People's Republic of China. Carbon Structural Steel. China Standard Press, Beijing.
GB/T228.1, 2010. Standardization Administration of the People's Republic of China. Metallic Materials Ensile Testing. 1: Method of Test at Room Temperature. China Standard Press, Beijing.
GB/T4338, 2006. Standardization Administration of the People's Republic of China. Metallic Materials-Tensile Testing at Elevated Temperatures. China Standard Press, Beijing.
JIS G3106-SM41, 1999. Japanese Industrial Standard. Japanese Industrial Standard—Rolled Steels for Welded Structure. Japanese Standards Association, Tokyo.
Kodur, V.K.R., Dwaikat, M.M.S., Fike, R.S., 2010. High-temperature properties of steel for fire resistance modeling of structures. ASCE J. Mater. Civil Eng. 5, 423–434.
Kodur, V.K.R., Harmathy, T.Z., 2008. In: DiNenno, P.J. (Ed.), Properties of Building Materials: SFPE Handbook of Fire Protection Engineering, fourth ed. National Fire Protection Association, Quincy, MA. 1.167-1.195.
Li, G.Q., Zhang, X.J., 2001. Experimental studies of the material properties of SM41 steel at elevated temperatures. Ind. Constr. 31 (16), 57–59.
Ma, J.L., Chan, T.M., Young, B., 2015. Material properties and residual stress of cold-formed high strength steel hollow sections. J. Constr. Steel Res. 109, 152–165.
Outinen, J., Kaitila, O., Mäkeläinen, P., 2001. High-Temperature Testing of Structural Steel and Modelling of Structures at Fire Temperatures. Research Rep. No. TKK-TER-23. Helsinki Univ. of Technology Laboratory of Steel Structures, Helsinki, Finland.
Outinen, J., 2007. Mechanical Properties of Structural Steels at High Temperatures and after Cooling Down [Ph.D. Thesis]. Helsinki University of Technology, Helsinki, Finland.
Poh, K.W., 2001. Stress-strain-temperature relationship for structural steel. J. Mater. Civ. Eng. 13, 371–379.
Qiang, X.H., Bijlaard, F.S.K., Kolstein, H., 2012. Dependence of mechanical properties of high strength steel S690 on elevated temperatures. Steel Constr. 30, 73–79.
Qiang, X.H., Bijlaard, F.S.K., Kolstein, H., 2012a. Deterioration of mechanical properties of high strength structural steel S460N under steady state fire condition. Mater. Des. 36, 438–442.
Qiang, X.H., Bijlaard, F.S.K., Kolstein, H., 2012b. Post-fire mechanical properties of high strength structural steels S460 and S690. Eng. Struct. 35, 1–10.

Qiang, X.H., Jiang, X., Bijlaard, F.S.K., Kolstein, H., 2016. Mechanical properties and design recommendations of very high strength steel S960 in fire. Eng. Struct. 112, 60–70.
Tan, W., 2000. Experiments and research of steel material properties at elevated temperature. Ind. Constr. 30 (10), 61–63, 67.
Wang, W.Y., Liu, B., Kodur, V.K.R., 2013. Effect of temperature on strength and elastic modulus of high strength steel. J. Mater. Civ. Eng. 25 (2), 174–182.

CHAPTER 4

Residual mechanical properties of steels after fire exposure

Contents

Material Properties of Steel in Fire Conditions
ISBN 978-0-12-813302-6
https://doi.org/10.1016/B978-0-12-813302-6.00004-7

4.1 Introduction

After a fire event, the exposure to extreme temperature variations could have reduced the load-bearing capacities of steel members. The main reason for this is the reduction in postfire mechanical properties (yield strength, elastic modulus, ultimate strength, and elongation) of steels. After a fire attack, some steel buildings may collapse (Fig. 4.1A) and some may not have serious damage (Fig. 4.1B). On cooling from elevated temperatures, the structural engineer then has to decide if the residual strength of the load carrying members were still adequate for future use. In this chapter, tension test and results on residual strength and elastic modulus of steels after fire exposure are presented.

4.2 Residual properties of hot-rolled Q420 steel

An experimental investigation was conducted by Lu et al. (2016) to reveal the post-fire mechanical properties of hot-rolled Q420. Specimens were heated to various preselected temperatures up to 1000 °C and subsequently cooled downto ambient temperature via two different methods, namely, air and water cooling. Tensile coupon tests were performed to obtain the post-fire stress—strain curves, elastic moduli, yield strengths, ultimate strengths, and ductility.

4.2.1 Specimen details

The specimens considered in this test were cut in the longitudinal direction from hot-rolled steel sheet of grades Q420. The steel sheets utilized comply with GB/T 700 (2006) and GB/T 1591 (2008). Q420 is the abbreviations of the grade designations of this steel; Q refers to yield strength (in Chinese Pinyin) and 420 denotes the corresponding minimum nominal yield strength of 420 N/mm^2. The shapes and dimensions of the test specimens

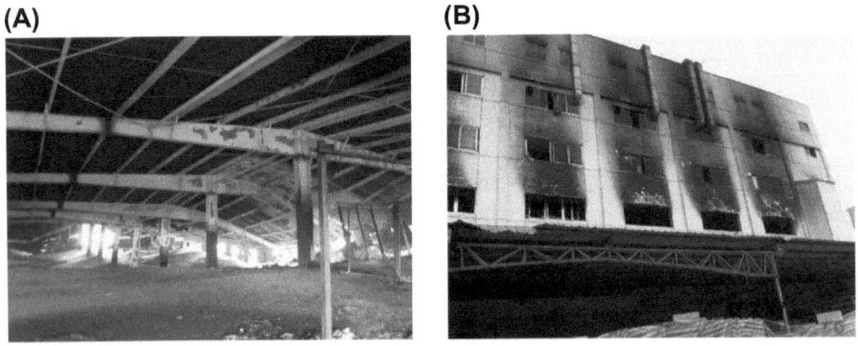

Figure 4.1 Steel buildings after fire attack. (A) Building that collapsed and (B) building that did not collapse.

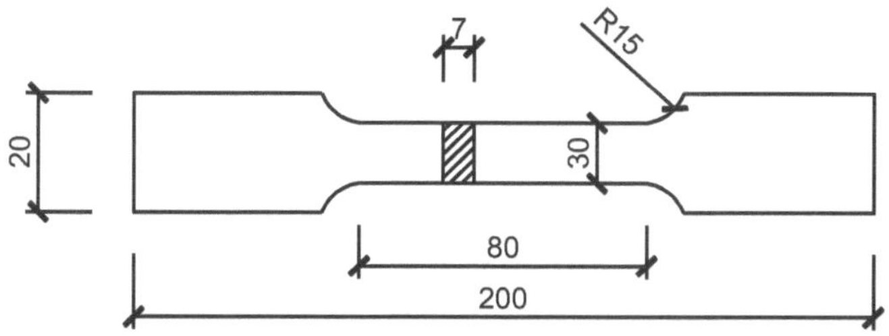

Figure 4.2 Shape and dimensions of the specimen.

in accordance with GB/T 228.1-2010 (2010) and GB/T 4338-2010 (2010) are shown in Fig. 4.2. The dimensions of each specimen were measured with a Vernier caliper at three points within the gauge length. The average values of the measured dimensions were used to calculate the mechanical properties of the steels.

4.2.2 Test equipment and procedure

The entire procedure of the experiment mainly comprises two steps. In the first step, the specimens were initially heated to the preselected elevated temperatures and subsequently cooled down to ambient temperature. In the second step, a tensile coupon test was conducted on the specimens at ambient temperature. The heating process was accomplished by a temperature-controlled electric furnace. The thermocouple located inside the furnace measured the air temperature in the furnace and relayed this information to the control system to facilitate the adjustment of the heating rate; thus, a closed control loop was formed.

In this test, 10 elevated temperatures were set for the Q420 specimens, i.e., 100, 200, 300, 400, 500, 600, 700, 800, 900, and 1000°C. In the heating process, the furnace temperature was initially increased at a rate of 15°C/min to a temperature of 50°C less than the target temperature; this temperature was maintained for 10 min. Subsequently, the furnace temperature was raised to the target temperature at a rate of 5°C/min and held for another 20 min. This heating process ensures the uniform temperature distribution in the specimens and avoids the exceeding of actual temperature from the target temperature. The influence of heating duration can be ignored according to Tao et al. (2012); Smith et al. (1981). Subsequently, the specimens were removed from the furnace and cooled down to ambient temperature. Both air and water cooling methods were considered; the specimens for the air cooling method were exposed to air and allowed to cool down at their own rates to simulate the situation in which a fire dies out naturally. The specimens for the water cooling method were cooled down by water spraying using a water jet to simulate the scenario in which fire is extinguished by fire guns. The entire heating–cooling procedure is plotted in Fig. 4.3.

The water volume adopted in the experiment was determined based on the principle that the volume sprayed on each unit area of the specimen surface is proportional to the volume sprayed on the members in actual fire extinguishment.

The tensile test was conducted via a universal material testing machine, which can exert a maximum force of 600 kN in tension. An electronic extensometer was attached to the specimens to measure strain precisely. The electronic extensometer was removed from the Q420 specimens when the stress–strain curves reached the maximum stress value owing to the limited measurement range; nonetheless, the strain range recorded is sufficient for design purposes. The tensile load was applied at a constant stress rate of 5 MPa/s during the elastic stage and, subsequently, at a constant strain rate of 0.001/s; this application satisfied the requirement of GB/T 228.1-2010 (2010). All the data obtained from the experiment, such as force, stroke, stress, strain, and stress–strain curves, were recorded by a computer that was compatible with the testing machine. For comparison, tensile tests were also conducted at ambient temperature on specimens that had not been exposed to elevated temperatures. Three specimens that were cooled down from the same elevated temperature with the same cooling method were tested and regarded as a group to minimize accidental errors. The average value of the three specimens in each group was adopted to analyze the postfire mechanical properties of the steels.

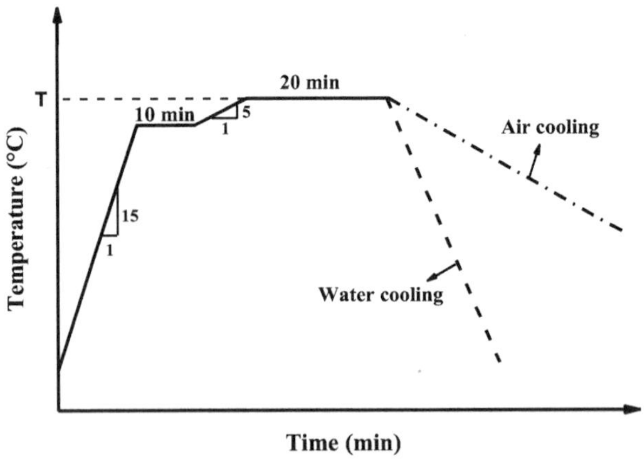

Figure 4.3 Heating and cooling procedure.

4.2.3 Failure modes

Tensile coupon tests were conducted on the steel specimens cooled down from elevated temperatures until fracture. The failure modes of the specimens were then obtained and shown in Fig. 4.4. The failure modes of the specimens that were not exposed to elevated temperatures were also included for comparison purposes. The hot-rolled steel Q420 exhibited obvious ductile failure with necking, and no brittle failure was observed regardless of the exposure temperature and the cooling method used. These observations indicate that the hot-rolled Q420 steel is able to avoid brittle failure and keep their original ductility after fire exposure, which are positive for their reuse after fire events.

4.2.4 Stress–strain relationships

The postfire stress–strain relationships obtained from the experiment are plotted in Fig. 4.5 (each curve plotted is one of three curves derived from the specimens in a group). The postfire mechanical properties of the steel were studied based on these curves. In general, the stress–strain curves of most of the specimens showed an obvious plateau in yield at ambient temperature and after cooling down from various elevated temperatures. The stress–strain curves of hot-rolled steel Q420 after exposure to temperatures lower than 700°C were similar to those obtained at ambient temperature. When the exposure temperatures exceeded 700°C, the stress–strain curves changed significantly. The influences of different cooling methods also contributed significantly to the changes in the postfire mechanical properties of the hot-rolled Q420 steel, particularly when the exposure temperature exceeded 800°C. In general, after cooling down

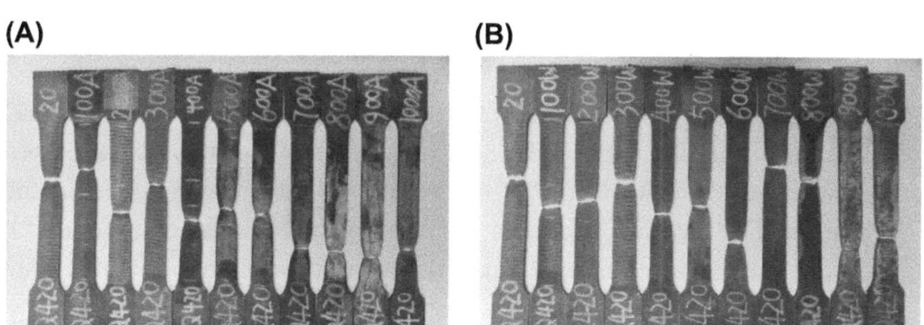

Figure 4.4 Failure modes of the test specimens. (A) Q420 specimens after air cooling and (B) Q420 specimen after water cooling.

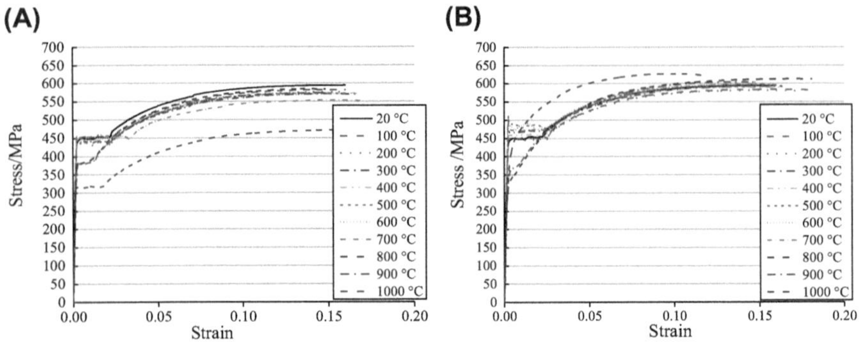

Figure 4.5 Stress–strain curves after exposure to various fire temperatures. (A) Q420 steel after air cooling and (B) Q420 steel after water cooling.

from high fire temperatures, the steels cooled by water spraying showed higher yield and ultimate strengths than those cooled in air.

4.2.5 Elastic modulus

The postfire elastic modulus of the steel tested in this experiment is defined as the initial slopes of their stress–strain curves. The method adopted to calculate the initial slope accorded with GB/T22315-2008 (2008). To describe the deterioration in the mechanical properties of steel after fire, the elastic modulus residual factor is defined as the ratio of the elastic modulus after cooling down from elevated temperatures (E_{PT}) to that at ambient temperature (E) without fire exposure. Table 4.1 lists the postfire elastic modulus and the corresponding residual factors obtained from the experiment for Q420 steel. The residual factors are also plotted as the function of exposure temperature in Fig. 4.6 (each listed and plotted value is the average value of the three specimens in a group). The elastic modulus of high-grade Q420 was not affected after exposure to temperatures below 700°C; subsequently, however, this value decreased suddenly. Nonetheless, at least 85% of the elastic modulus was regained after exposure to elevated temperatures up to 1000°C for all of the hot-rolled steels included in the experiment. The postfire modulus of the Q420 steel indicated that the stiffness did not decrease significantly after exposure to fire.

4.2.6 Yield stress

The yield strength of hot-rolled Q420 steel was determined with the 0.2% proof stress method although the stress–curves of most of the test

Table 4.1 Postfire elastic moduli and residual factors after air cooling and water cooling.

	Air cooling		Water cooling	
Temperature	Elastic moduli (GPa)	Residual factors	Elastic moduli (GPa)	Residual factors
20	204.2	1	204.2	1
100	202.2	0.99	209.5	1.026
200	206.1	1.009	201.9	0.989
300	208.9	1.023	206.4	1.011
400	205.5	1.006	205.9	1.008
500	207.5	1.016	204.3	1
600	206.8	1.013	202.6	0.992
700	196	0.96	201.9	0.989
800	178.3	0.873	186.3	0.912
900	176	0.862	171.4	0.839
1000	176.4	0.864	174	0.852

specimens exhibited clearly defined yield plateaus. The 0.2% proof stress $f_{0.2}$ was determined from the intersection of the stress–strain curve and proportional line offset by 0.2% strain level (Fig. 4.5). The yield strength residual factor is defined as the ratio of the yield strength after cooling down from elevated temperatures (f_{yPT}) to that at ambient temperature without the heating–cooling (f_y) processes. Table 4.2 lists the yield strengths and residual factors derived from the experiment. The residual factors are also plotted as the function of exposure temperature in Fig. 4.7 (each listed and plotted value is the average value of the three specimens in a group). The

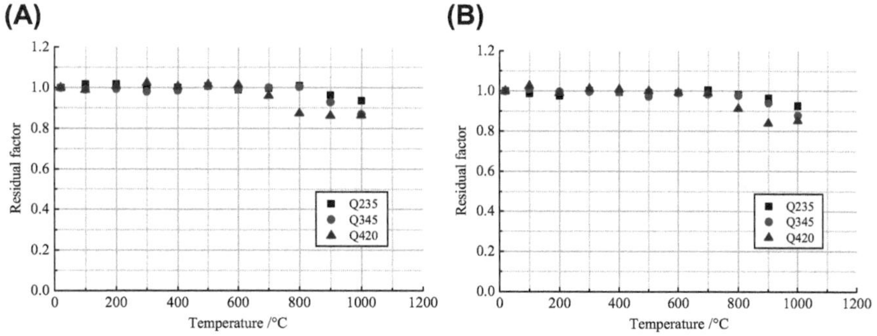

Figure 4.6 Postfire elastic modulus residual factors. (A) Hot-rolled steels after air cooling and (B) hot-rolled steels after water cooling.

Table 4.2 Postfire yield strengths and residual factor after air cooling and water cooling.

Temperature	Air cooling		Water cooling	
	Yield strengths (MPa)	Residual factors	Yield strengths (MPa)	Residual factors
20	447	1	447	1
100	446	0.998	448	1.002
200	441	0.987	449	1.004
300	453	1.013	468	1.047
400	451	1.009	459	1.027
500	437	0.978	489	1.094
600	446	0.998	482	1.078
700	450	1.007	476	1.065
800	382	0.855	337	0.754
900	386	0.864	365	0.817
1000	314	0.702	409	0.915

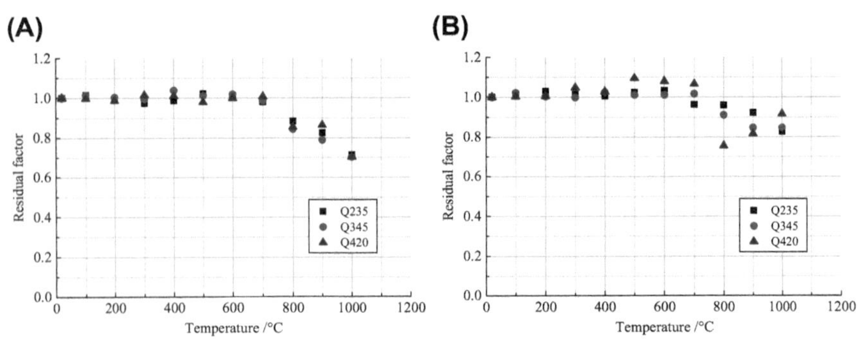

Figure 4.7 Postfire yield strength residual factors. (A) Air cooling and (B) water cooling.

yield strengths of hot-rolled steel obviously declined after exposure to certain elevated temperatures. The yield strengths of hot-rolled Q420 steel were almost unaffected until the temperature reached 700°C and decreased at a rapid rate thereafter. A reduction of 30% was observed after exposure to temperatures reaching 1000°C. It is worth noting that when adopting water cooling method, Q420 showed an obvious rebound in yield strength after cooling from temperatures ranging from 800 to 1000°C, which can be attributed to the change in the microstructures of the steels as a result of rapid cooling.

The influences of different cooling methods on residual yield strength were significant, particularly when the steel was cooled down from high exposure temperatures. The water-cooled specimens of hot-rolled Q420 steel regained more of their original yield strengths than the air-cooled specimens did. This difference is attributed to the strengthening effect induced by water cooling, which is equivalent to quenching. This observation indicates that Q420 steel can preserve more of their yield strengths after exposure to fires extinguished by fire guns.

4.2.7 Tensile strength

The ultimate strength residual factor is defined as the ratio of the ultimate strength after cooling down from elevated temperatures (f_{uPT}) to that at ambient temperature without exposure to fire (f_u). Table 4.3 presents the ultimate strengths and their corresponding residual factors after air and water cooling, respectively. The residual factors are also plotted in Fig. 4.8 as the function of exposure temperatures (each value listed and plotted is the average value of the three specimens in a group). Ultimate strength showed a significantly gentler decreasing trend with respect to exposure temperature than yield strength did. At least 80% of the original ultimate strengths of the hot-rolled Q420 steel could be regained after air cooling from elevated temperatures up to 1000°C. In underwater cooling conditions, however, hot-rolled Q420 steel displayed almost no reduction in ultimate

Table 4.3 Postfire ultimate strengths and residual factors after air cooling and water cooling.

	Air cooling		Water cooling	
Temperature	Ultimate strength (MPa)	Residual factors	Ultimate strength (MPa)	Residual factors
20	595	1	595	1
100	583	0.98	583	0.98
200	577	0.97	581	0.976
300	571	0.96	593	0.997
400	573	0.963	589	0.99
500	572	0.961	596	1.002
600	571	0.96	597	1.003
700	554	0.931	598	1.005
800	585	0.983	613	1.03
900	575	0.966	604	1.015
1000	474	0.797	626	1.052

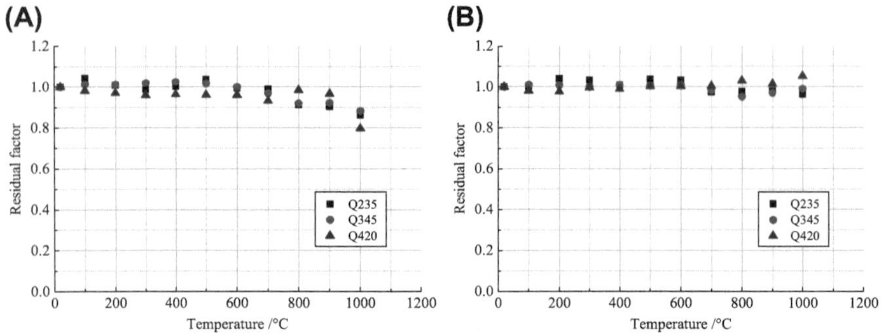

Figure 4.8 Postfire ultimate strength residual factors. (A) Air cooling and (B) water cooling.

strength regardless of the temperature it had been exposed to. Furthermore, the ultimate strength of high-grade Q420 even increased slightly after exposure to fire temperatures above 700°C. The influences of different cooling methods on postfire ultimate strength were similar to those on yield strength. The hot-rolled Q420 specimens exhibited higher ultimate strengths than air-cooled specimens did as a result of the aforementioned strengthening effect of quenching.

4.2.8 Ductility

The ductility of steel is an indicator reflecting its plastic deformation capacity, which is defined in this study according to the percentage elongation of fracture. The percentage elongation of fracture is calculated as the ratio of the gauge length after fracture to the original gauge length. The postfire ductility residual factor is defined as the ratio of the percentage elongation of fracture after cooling down from elevated temperatures ($\delta_{u,PT}$) to that at ambient temperature without exposure to fire (δ_u). Table 4.4 presents the postfire percentage elongation of fracture obtained through the experiment and the corresponding residual factors for air and water cooling, respectively. The residual factors are also plotted in Fig. 4.9 as the function of exposure temperature (each listed and plotted value is the average value of the three specimens in a group).

The postfire ductility of the hot-rolled Q420 steel was significantly influenced by cooling methods. Under air cooling, the ductility of Q420 remained almost unchanged after fire exposure. Underwater cooling, the ductility obviously decreased after exposure to temperatures exceeding

Table 4.4 Postfire ultimate strengths and residual factors after air cooling and water cooling.

	Air cooling		Water cooling	
Temperature	Percentage elongation of fracture (%)	Residual factors	Percentage elongation of fracture (%)	Residual factors
20	31.3	1	31.3	1
100	30.3	0.97	29.8	0.95
200	30.3	0.97	30.3	0.97
300	30.5	0.98	29	0.93
400	30.3	0.97	30.5	0.97
500	31.3	1	31.1	1
600	31	0.99	29.2	0.93
700	30.3	0.97	29.5	0.94
800	28.2	0.9	26.1	0.84
900	28.7	0.92	26	0.83
1000	31	0.99	22.9	0.73

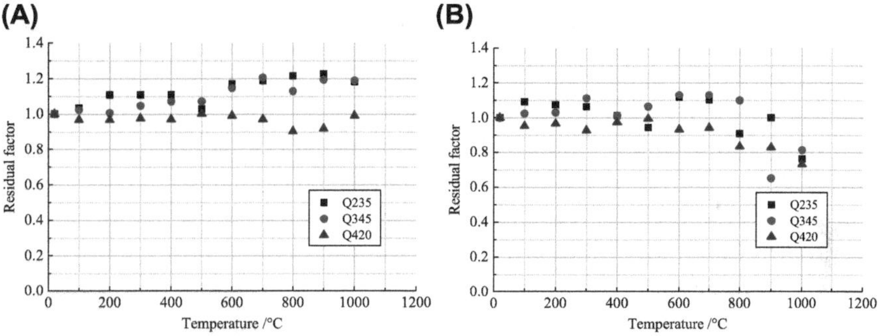

Figure 4.9 Postfire ductility residual factors. (A) Air cooling and (B) water cooling.

800°C for Q420 steel. Therefore, great caution should be given in reusing hot-rolled Q420 steel cooled by fire guns.

4.3 Residual properties of high strength Q460 steel

4.3.1 Experimental program

Tensile coupon tests were conducted to obtain the residual stress–strain curves and mechanical properties of high strength Q460 steel after experiencing preselected temperatures up to 900°C.

4.3.1.1 Test specimen

Test specimens were cut in the longitudinal direction of high strength Q460 steel sheet of 8 mm thickness. Their shapes and sizes were in accordance with GB/T 228.1-2010 (2010) (Fig. 4.10). The base metal thickness and width of each specimen were measured at three points within the gauge length using a micrometer. The averages of these measured dimensions were used in the calculations of mechanical properties. Totally 45 specimens were fabricated. At every temperature, including ambient temperature, three specimens were tested to obtain reliable results by averaging the test data.

4.3.1.2 Test setup

The electric furnace shown in Fig. 4.11A was used in this experimental study to achieve the desired elevated temperatures. The thermocouple located inside the furnace gave the air temperature of the furnace on the screen of temperature control system (Fig. 4.11B). Three additional thermocouples were placed inside the furnace to measure the temperature independently. One thermocouple is used to measure the temperature of furnace and the other two are used to record the temperature of specimens. Eight different temperatures experienced in steel were selected in this study, namely 20, 300, 400, 500, 600, 700, 800, and 900°C. The steel becomes very soft at temperatures above 900°C, and hence, temperatures higher than 900°C were not considered in this study.

Fig. 4.11C shows the SANS tension testing machine used to obtain the postfire mechanical properties. The specimen was connected to the top and bottom grips, which were accurately aligned with each other. The bottom end was fixed while the top end was free to move upwards. A tensile load was then applied at a constant stress rate of 10 MPa/s during elastic stage. During yield stage, the load was applied at a constant strain rate of 0.001/s

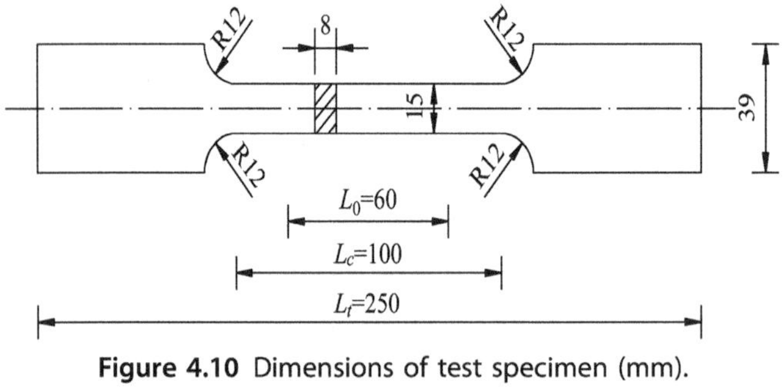

Figure 4.10 Dimensions of test specimen (mm).

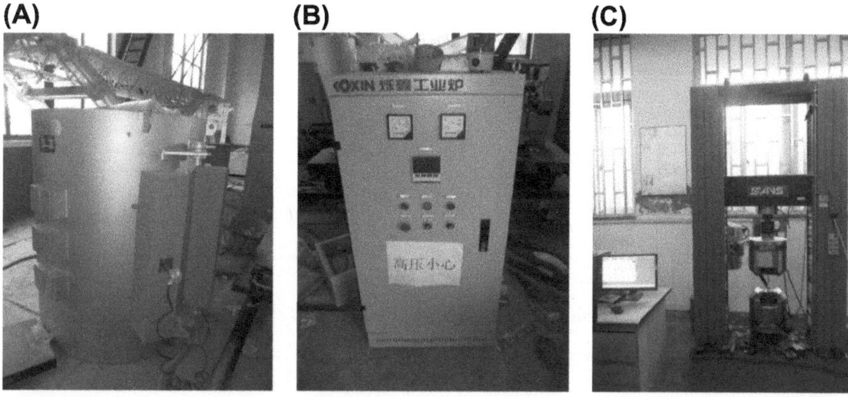

Figure 4.11 Test setup. (A) Furnace; (B) temperature control system; and (C) tension test machine.

and then the load was applied at a constant displacement rate of 10 mm/min until failure. An extensometer was connected on the middle part of specimen to obtain the extension of gauge length. The data acquisition system was used to record the load and displacement. After the specimen had failed, a micrometer was used to measure the ultimate elongation of specimen.

4.3.1.3 Heating and cooling procedures

Two cooling methods were considered in this experiment, namely natural air cooling and cooling by water. Natural air cooling is defined as the specimen cools down in ambient condition, and cooling by water is defined as the specimen cools down in water. These two cooling methods simulate the two possible environments in a fire event, this is, the fire burns up naturally and is then extinguished by water.

In the test, initially, the specimens were placed inside the furnace using steel wire. Every time, two groups (six specimens) were placed into the furnace. The furnace temperature was then increased to target value using a heating rate of 15°C/min. After reaching this temperature, it was left for 20 min to ensure a uniform temperature distribution in steel. The specimens were then removed from the furnace. One group was placed in a tray to air cool at its own rate. Another group was placed into water and was taken out after the temperature of specimen decreased to ambient temperature. The temperature–time curves for specimens with target temperatures of 600–900°C are shown in Fig. 4.12. As can be seen from the

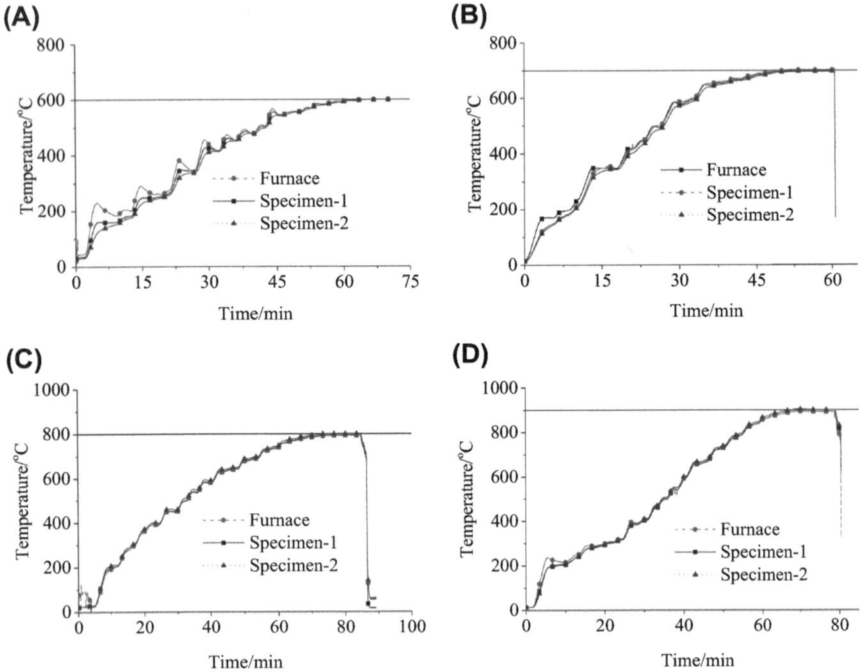

Figure 4.12 Temperature—time curve of specimens. (A) Target temperature of 600°C; (B) target temperature of 700; (C) target temperature of 800°C; and (D) target temperature of 900°C.

plots, the temperatures in steel are almost similar to that in furnace because the specimens are very thin in thickness. Three days later, all the specimens were installed into the tension testing machine to obtain the mechanical properties.

4.3.2 Test results

The test results include visual observation of specimen before and after failure, stress—strain relationship, yield strength, ultimate strength, elastic modulus, and ultimate elongation.

4.3.2.1 Visual observation

Fig. 4.13 shows the test specimens after experiencing different temperatures and cooling down. The target temperature for the specimens experienced were 20, 300, 400, 500, 600, 700, 800, and 900°C from left to right. As can be seen from the photo, at temperature of 300°C, the color is blue, and this

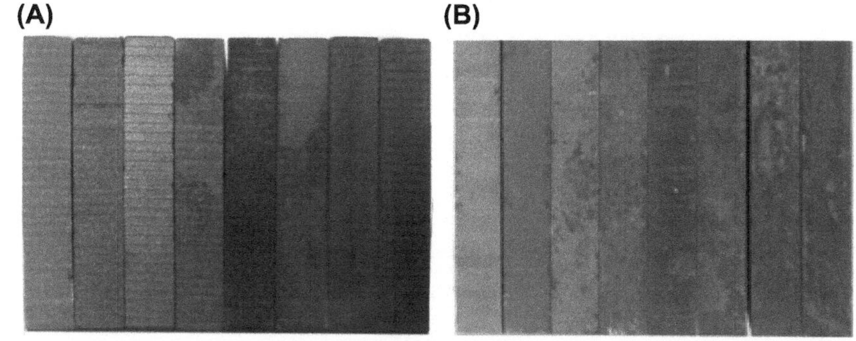

Figure 4.13 Surface characteristic of specimens after experiencing elevated temperatures. (A) Natural cooling and (B) water cooling.

phenomenon is called blue brittleness. In addition, the specimens deteriorated quite steadily up to 500°C. Beyond this temperature, the effect of heat caused visible damage to steel specimens. For the experienced temperatures in steel above 600°C, the surface of specimen becomes quite dark. This color would be a useful indicator for approximate fire intensity in a practical scenario. For the groups of natural air cooling, the surface of specimen is clean. However, for the groups of cooling by water, there is obvious rust on the surface of specimens.

Fig. 4.14 shows the failed steel specimens after the tensile load tests. In the Figures, the star "*" means numbers from 1 to 7. Almost all the specimens were damaged within the gauge length, and obvious necking phenomenon was observed. The fracture section in the specimens is smooth when the experienced temperature is up to 400°C. Beyond this temperature, the fracture section of the specimen becomes rough, and at the boundary of fracture section, some surfaces are at an angel of 45 degree to the cross section. At temperature of 900°C, the center of fracture section is very flat and smooth, both for natural cooling and water cooling.

4.3.2.2 Stress–strain curves

Fig. 4.15A–F show the comparison of stress–strain curves of high strength Q460 steel for natural air cooling and cooling by water after experiencing temperatures in the range of 20–900°C. The stress–strain curves of both natural air cooling and cooling by water show a linear elastic region followed by a well-defined yield plateau at ambient and different experienced temperatures except for the specimen cooling by water when the

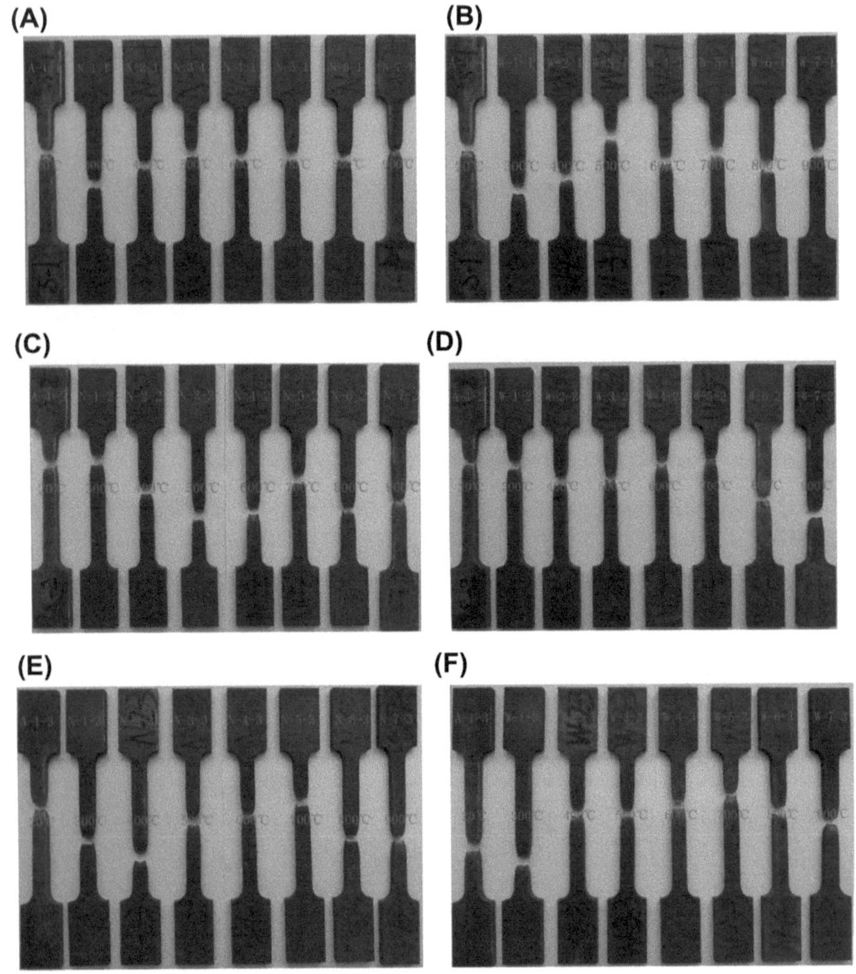

Figure 4.14 Failure modes of test specimens. (A) Group N-*-1; (B) group W-*-1; (C) group N-*-2; (D) group W-*-2; (E) group N-*-3; and (F) group W-*-3.

experienced temperature exceeds 800°C. For the specimens that have yield plateau, the yield strength is defined as lower boundary of yield plateau. For the specimens that do not have yield plateau, the yield strength is evaluated at strain level of 0.2% offset. The 0.2% offset yield strength is the intersection point of the stress–strain curve and the proportional line offset by 0.2% strain. For the natural air cooling specimens at the temperatures above 800°C, the ultimate strain is much bigger than the other experienced temperatures, i.e., more ductile after experiencing higher temperatures.

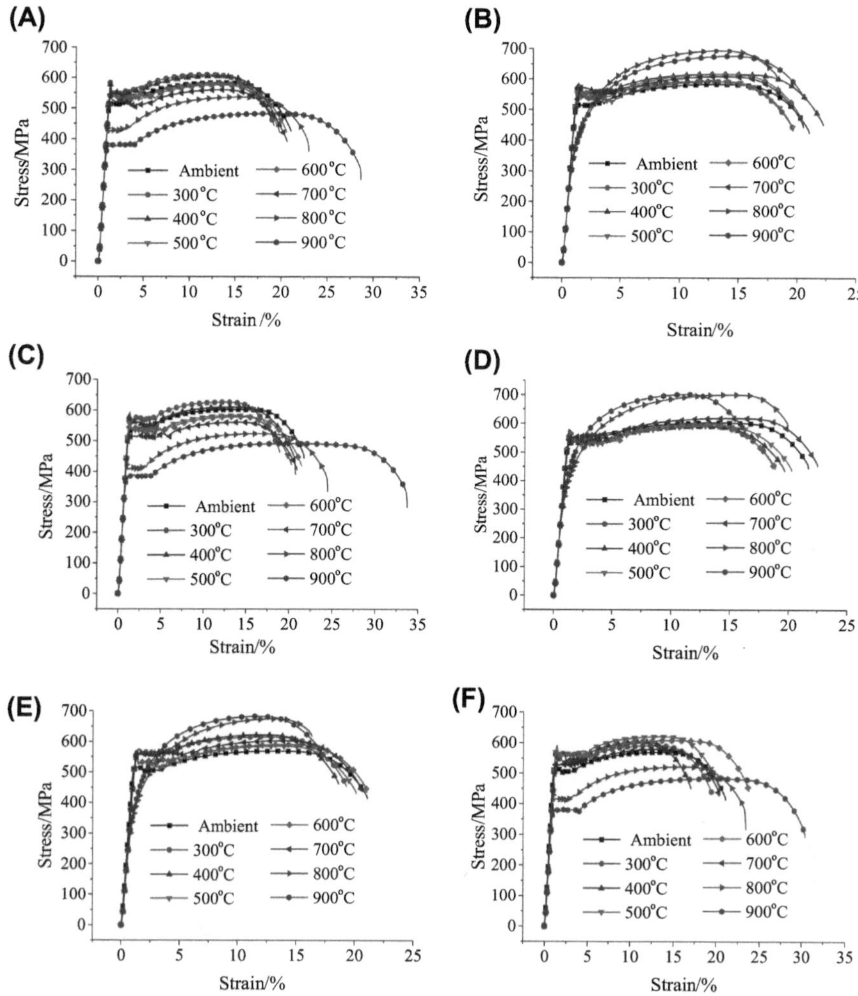

Figure 4.15 Stress—strain curves for different experienced temperatures and cooling methods. (A) Group N-*-1; (B) group W-*-1; (C) group N-*-2; (D) group W-*-2; (E) group N-*-3; and (F) group W-*-3.

Different from the specimens with air cooling, the ultimate strength of specimens with cooling by water at the temperatures exceeding 800°C is much higher than that at other experienced temperatures.

4.3.2.3 Residual yield strength

As defined above, the yield strength after experiencing elevated temperatures are tabulated in Table 4.5 by averaging the three specimens at the

Table 4.5 Postfire mechanical properties of high strength Q460 steel.

Temperature/°C	f_{yPT}/MPa		f_{uPT}/MPa		E_{PT}/GPa		ε_{PT}/%	
	Natural	Water	Natural	Water	Natural	Water	Natural	Water
Ambient	513.3	513.3	586.7	586.7	214.3	214.3	25.77	25.77
300	545.0	540.0	608.3	600.0	218.4	211.7	22.96	23.42
400	538.3	543.3	606.7	609.0	214.2	220.7	22.57	23.15
500	538.3	521.7	595.0	591.7	216.2	217.3	23.64	23.11
600	531.7	535.0	588.3	601.7	215.7	217.9	24.72	23.13
700	513.3	553.3	566.7	611.7	215.0	223.5	24.73	24.94
800	411.7	361.7	526.7	690.0	215.6	213.9	28.01	19.07
900	376.7	370.0	486.7	686.7	203.9	203.5	34.07	20.75

same temperature and with same cooling method. These residual factors are given in Table 4.6 and are also plotted in Fig. 4.16 as a function of experienced temperature. Fig. 4.16 shows that the yield strength residual characteristics of natural air cooling and cooling by water are different. The yield strengths of all the specimens did not decrease much up to 700°C. Specimens with cooling by water lost their yield strength more rapidly than natural air cooling specimens in the experienced temperature range of 700–800°C, with a strength reduction of approximately 35%. However, the yield strength increased a little at the temperature of 900°C. In contrast, the yield strength of natural air cooling specimen reduced gradually above the temperature of 600°C.

4.3.2.4 Residual ultimate strength

The ultimate strength f_{uPT} after fire exposure is measured as the maximum value of stress in the stress–strain curves and tabulated in Table 4.5. These

Table 4.6 Residual factor of postfire mechanical properties of high strength Q460 steel.

Temperature/°C	f_{yPT}/f_y		f_{uPT}/f_u		E_{PT}/E		$\varepsilon_{PT}/\varepsilon$	
	Natural	Water	Natural	Water	Natural	Water	Natural	Water
Ambient	1.00	1.00	1.00	1.00	1.00	1.00	1.00	1.00
300	1.06	1.05	1.04	1.02	1.02	0.99	0.89	0.91
400	1.05	1.06	1.03	1.04	1.00	1.08	0.88	0.90
500	1.05	1.02	1.01	1.01	1.01	1.01	0.92	0.90
600	1.04	1.04	1.00	1.03	1.01	1.02	0.96	0.90
700	1.00	1.08	0.97	1.04	1.00	1.04	0.96	0.97
800	0.80	0.70	0.90	1.18	1.01	1.00	1.09	0.74
900	0.73	0.72	0.83	1.17	0.95	0.95	1.32	0.81

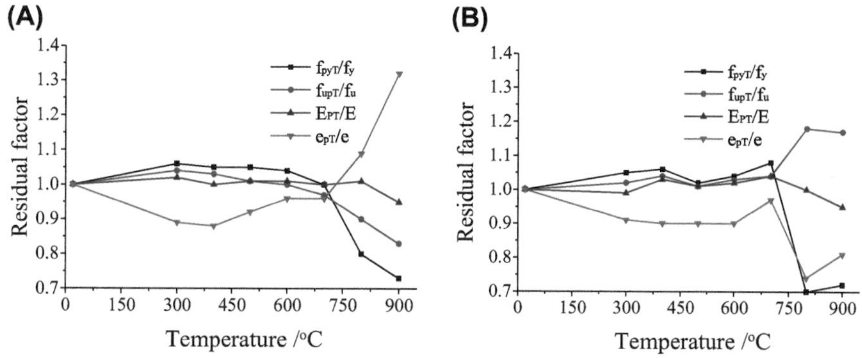

Figure 4.16 Residual factor of postfire mechanical properties of high strength Q460 steel. (A) Natural air cooling and (B) cooling by water.

residual factors are given in Table 4.6 and are plotted in Fig. 4.16 as a function of experienced temperature. As can be seen from Fig. 4.16, the reduction in ultimate strength follows a similar trend to that observed for yield strength at temperatures up to 700°C. However, the reduction in ultimate strength begins to increase for specimens cooling by water when the temperature exceeds 700°C as shown in Fig. 4.16B.

4.3.2.5 Elastic modulus

Elastic modulus was calculated from the initial slope of the stress–strain curve as shown in Fig. 4.15. The average values of the elastic modulus for three specimens at the same experienced temperature are tabulated in Table 4.5. The elastic modulus residual factor for experiencing temperatures was then calculated as the ratio of the residual elastic modulus of steel E_{PT} after experiencing an elevated temperature T to the original elastic modulus at ambient temperature E. These factors are given in Table 4.6 and are also plotted in Fig. 4.16 as a function of experienced temperature. Fig. 4.16 shows that the elastic modulus remained relatively unchanged up to 800°C for both kinds of cooling methods. There was a 5% reduction in elastic modulus for experiencing temperature of 900°C. This suggests that high strength Q460 steel does not lose its stiffness even after being exposed to a very high temperature in a fire.

4.3.2.6 Ultimate elongation

Ultimate elongation of steel is defined based on the level of deformation that steel can undergo plastically before fracture. In this study, extension of gauge length was measured until fracture and tabulated in Table 4.5 The ultimate elongation residual factor for experiencing temperatures was then calculated as the ratio of the ultimate elongation of steel ε_{PT} after experiencing an elevated temperature T to the original ultimate elongation at ambient temperature ε. Effects of experienced temperature and cooling method on the residual factor of ultimate elongation of high strength Q460 steel were studied by comparing the extension of gauge length at fracture, as show in Fig. 4.16. It can be observed that the ultimate elongation of high strength Q460 steel slightly decreases for experienced temperatures up to 700°C for both kinds of cooling methods but increased considerably as for natural air cooling specimens for temperatures beyond 800°C. However, for the specimens' experience cooling by water, the ultimate elongation decreases sharply for experienced temperatures exceeding 700°C.

4.3.3 Proposed equations

Comparison of residual factors for high strength Q460 steel with mild Q235 steel and high strength S460 steel showed that the predictions for Q235 and S460 are not applicable to predict the residual factors for high strength Q460 steel. Therefore, new predictive equations were proposed for the mechanical properties residual factors obtained in this study. Test results from this study also showed that the cooling method has some influence on the mechanical properties residual factors. Hence, two separate sets of predictive equations were developed for natural air cooling and cooling by water.

4.3.3.1 *Yield strength*

Experimental results have shown that high strength Q460 steel can keep the original yield strength after they experienced to temperatures up to 700°C. However, predictive equations are useful to calculate the residual yield strength accurately for different experienced temperatures, which are expressed in Eqs. (4.1) and (4.2) by curve fitting to the test data. The predictions from these equations are compared with the test results from this study in Fig. 4.17A. This figure shows that there is a good agreement between the proposed equations and the test results.

Natural air cooling:

$$f_{y,T}/f_y = -1.17\times 10^{-9}T^3 + 5.54\times 10^{-7}T^2 + 1.33\times 10^{-4}T + 1.0 \tag{4.1}$$

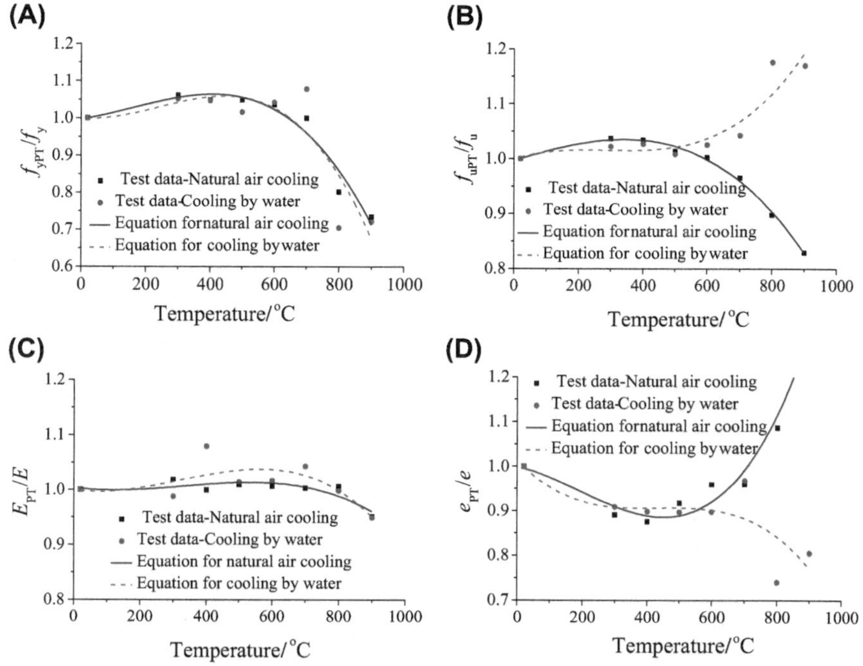

Figure 4.17 Comparison of proposed equations with test data. (A) Yield strength; (B) ultimate strength; (C) elastic modulus; and (D) elongation.

Cooling by water:

$$f_{y,PT}/f_y = -1.73 \times 10^{-9}T^3 + 1.25 \times 10^{-6}T^2 - 8.05 \times 10^{-5}T + 1.0 \tag{4.2}$$

where f_{yPT} is the residual yield strength after experiencing an elevated temperature T, f_y is the yield strength at ambient temperatures, and T is an experienced temperature in steel.

4.3.3.2 Ultimate strength

Eqs. (4.3) and (4.4) present the proposed equations for the ultimate strength residual factors of high strength Q460 steel with different cooling methods. The predictions from these equations are compared with the test results from this study in Fig. 4.17B. This figure shows that there is a good agreement between the proposed equations and the test results.

Natural air cooling:

$$f_{u,T}/f_u = -3.81\times10^{-10}T^3 - 6.36\times10^{-8}T^2 + 1.79\times10^{-4}T + 1.0 \tag{4.3}$$

Cooling by water:

$$f_{u,PT}/f_u = 8.11\times10^{-10}T^3 - 7.03\times10^{-7}T^2 + 1.93\times10^{-4}T + 1.0 \tag{4.4}$$

where f_{uPT} is the residual ultimate strength after experiencing an elevated temperature T and f_u is the ultimate strength at ambient temperatures.

4.3.3.3 Elastic modulus

Experimental results have shown that the elastic modulus varies when high strength Q460 steel is experiencing elevated temperatures. As a simple guide, it can be assumed that high strength Q460 steels can keep the same as of the original elastic modulus after they experience temperatures up to 800°C. However, with the availability of accurate elastic modulus residual factors of different cooling method for high strength Q460 steel, it was considered important to develop predictive equations that are suitable for high strength Q460 steel. Two separate sets of predictive equations (Eq. 4.5 and Eq. 4.6) were developed for natural air cooling and cooling by water. Fig. 4.17C shows that the test results from this study agree well with the proposed equations.

Natural air cooling:

$$E_T/E = -4.0\times10^{-10}T^3 + 3.93\times10^{-7}T^2 - 7.79\times10^{-5}T + 1.0 \tag{4.5}$$

Cooling by water:

$$E_{PT}/E = -7.15\times10^{-10}T^3 + 6.86\times10^{-7}T^2 - 9.27\times10^{-5}T + 1.0 \tag{4.6}$$

where E_{PT} is the residual elastic modulus after experiencing an elevated temperature T, and E is the elastic modulus at ambient temperatures.

4.3.3.4 Ultimate elongation

Eqs. (4.7) and (4.8) present the proposed equations for the ultimate elongation reduction factors of high strength Q460 steel with different cooling methods. The predictions from these equations are compared with the test results from this study in Fig. 4.17D. This figure shows that there is a good agreement between the proposed equations and the test results.

Natural air cooling:

$$\varepsilon_{PT}/\varepsilon = 1.68 \times 10^{-9} T^3 - 9.55 \times 10^{-7} T^2 - 1.62 \times 10^{-4} T + 1.0 \quad (4.7)$$

Cooling by water:

$$\varepsilon_{PT}/\varepsilon = -1.37 \times 10^{-9} T^3 + 1.78 \times 10^{-6} T^2 - 7.62 \times 10^{-4} T + 1.01 \quad (4.8)$$

where ε_{PT} is the residual elastic modulus after experiencing an elevated temperature T, and ε is the elastic modulus at ambient temperatures.

4.4 Residual properties of high strength Q690 steel

The tension tests being carried out in this study to obtain the mechanical properties of Q690 steel after being exposed to the elevated temperature up to 1000°C are described as follow.

4.4.1 Specimen details

Q690 steel is a typical high strength structural steel with nominal yield strength of 690 MPa. Its high strength is developed by thermal mechanical rolling and inclusion of alloys, such as niobium and chromium. The chemical composition of Q690 steel compared with other high strength steels (HSSs) is tabulated in Table 4.7 It is indicated that the percentage of niobium and chromium in HSSs is much higher than those in lower strength steels. Niobium and chromium are the two alloys to enhance the strength of alloy steels.

The tension specimens were cut from 20 mm-thick Q690 steel plates. The dimensions and shapes of the specimens are shown in Fig. 4.18. The specimens were fabricated in accordance with GB/T 228.1-2010 (2010). The gauge length is 50 mm. To obtain more reliable results, at each elevated temperature and the adopted cooling method, three identical specimens were tested as a group, and if the difference of the results exceeds 10% in a group, additional specimen was tested and the result associated with the maximum difference was excluded. Otherwise, the average value of a group was used to represent the test result.

Table 4.7 Comparison of the chemical composition between various steels based on standard recommendations.

Chemical composition	Q690 steel	BISPLATE80 steel	S690 steel	Q460 steel	Q345 steel
Carbon	≤0.18	≤0.16	≤0.2	≤0.2	≤0.2
Silicon	≤0.6	≤0.25	≤0.8	≤0.55	≤0.5
Manganese	≤2.0	≤1.5	≤1.7	1.0 −1.7	≤1.7
Sulfur	≤0.03	≤0.008	≤0.025	≤0.035	≤0.03
Phosphor	≤0.03	≤0.01	≤0.015	≤0.035	≤0.03
Niobium	≤0.11	–	≤0.06	0.015 −0.06	≤0.07
Vanadium	≤0.12	–	≤0.12	0.02 −0.2	≤0.15
Titanium	≤0.2	–	≤0.05	0.02 −0.2	≤0.20
Chromium	≤1.0	0.25	≤1.5	≤0.7	≤0.3
Nickel	≤0.8	–	≤2.0	≤0.7	≤0.5
Molybdenum	≤0.3	≤0.25	≤0.7	–	≤0.10
Boron	≤0.004	≤0.002	–	–	–

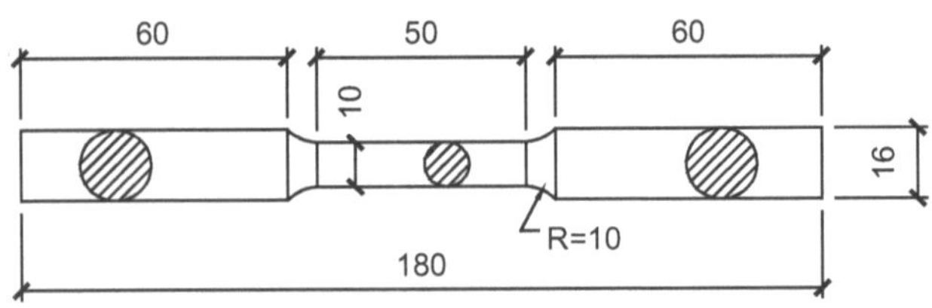

Figure 4.18 Details of test specimen (mm).

4.4.2 Heating and cooling

The heating process of the specimens was performed in a furnace with a support system as shown in Fig. 4.19A. The maximum temperature and heating rate of the furnace are 1200 and 20°C/min, respectively. The furnace temperature is controlled by a smart system, and the heating rate is adjusted automatically when the temperature approaches preset target temperature. Eight target temperatures were predetermined ranging from 300 to 1000°C with an interval of 100°C. For each target temperature,six identical specimens were placed on the support system and heated from room temperature (*RT*) to the target temperature as shown in Fig. 4.19B. Three thermocouples were attached on three specimens of a group to measure the actual temperature of the specimens. Once a target

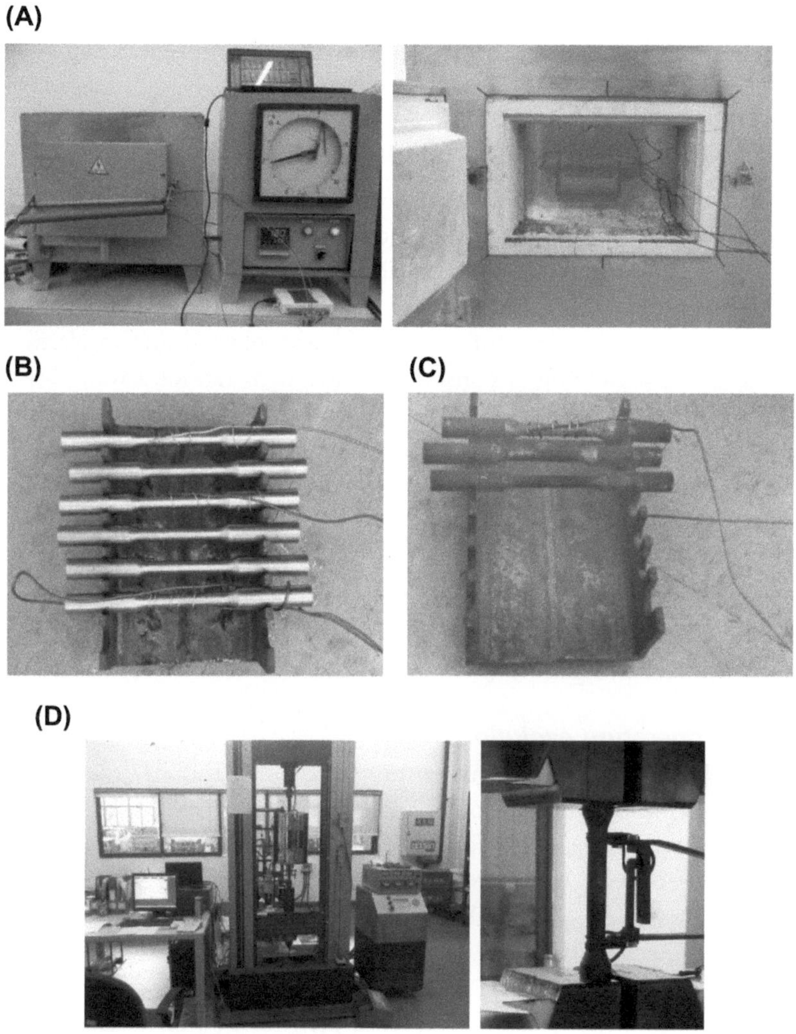

Figure 4.19 Postfire material property tests. (A) Heating furnace; (B) specimens and support system; (C) specimens cooled by air; and (D) material tensile testing system and test setup.

temperature was reached, the temperature was maintained for 30 min to ensure the temperature in the specimen was uniformly distributed.

Two cooling methods, namely air cooling and water cooling, were adopted in the test. The thermocouples were connected to the specimens during both of the heating and cooling processes. Once a target temperature is being reached, the temperature was maintained for 30 min, and then, three specimens were cooled down by air (shown in Fig. 4.19C) and the other three were cooled down by water.

4.4.3 Tension setup and procedures

After being exposed to the elevated temperature and cooled down to *RT*, the specimens were tested by an electric universal testing machine as shown in Fig. 4.19D, to obtain the postfire stress—strain relationship for a specific target temperature. According to the requirement in GB/T 228.1-2010 (2010), a constant strain rate of $0.00025s^{-1}$ was selected during elastic and yield stage. After yielding, a constant displacement rate of 3 mm/min was used until the fracture of the specimen. Both tension load and displacement were recorded by a data acquisition system. After the specimen fails, the ultimate elongation of the specimen was gauged by a micrometer.

4.4.4 Test results

The test results concerning to the temperature evolution, failure mode, stress—stain relationship, yield, and tensile strength, elastic modulus and ratio of ultimate elongation are reported in below.

4.4.4.1 Temperature evolution

As previously discussed, the temperature variation of the specimen includes heating and cooling phases. Four typical temperature evolutions of specimens are shown in Fig. 4.20, for target temperature of 600, 700, 900, and 1000°C, respectively. As can be seen from the figure, the specimen temperature increased quickly at the beginning with the heating rate around 20°C/min. The heating ratio decreased gradually when the temperature approached target temperature. Once the temperature reached the target temperature, the heating rate was kept at zero for 30 min to maintain the target temperature. At the cooling phase, the temperature declined sharply both for air cooling and water cooling. For air cooling, the cooling rate is greater at higher temperature. For the water cooling, the temperature went down to *RT* just in few seconds.

4.4.4.2 Color spectrum and failure mode

For different temperature exposures and cooling methods, the surface colors of the specimens are quite different. The color spectrum of the specimens after cooling is shown in Fig. 4.21. The exposed temperature from left to right is 300—1000°C with an interval of 100°C. The specimen

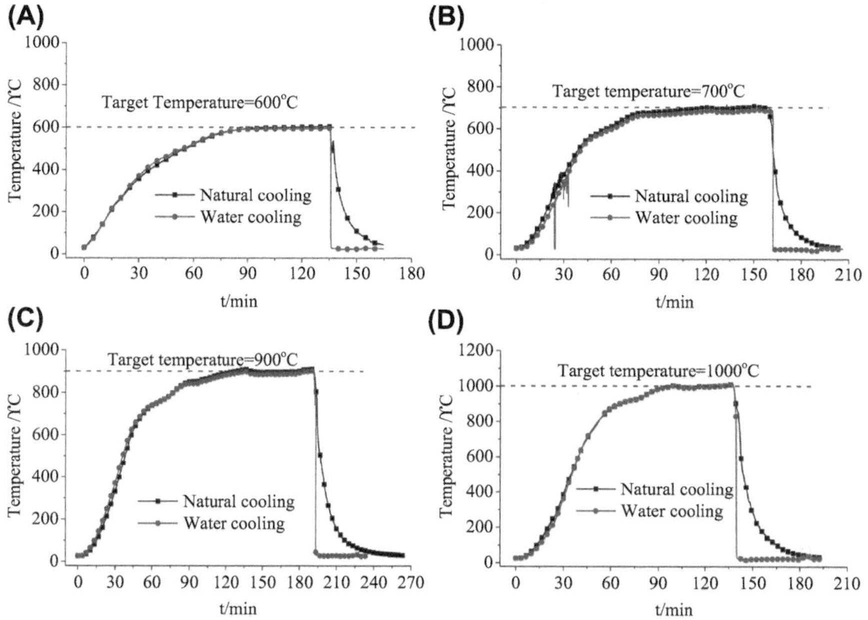

Figure 4.20 Temperature variation during heating and cooling. (A) 600°C; (B) 700°C; (C) 900°C; and (D) 1000°C.

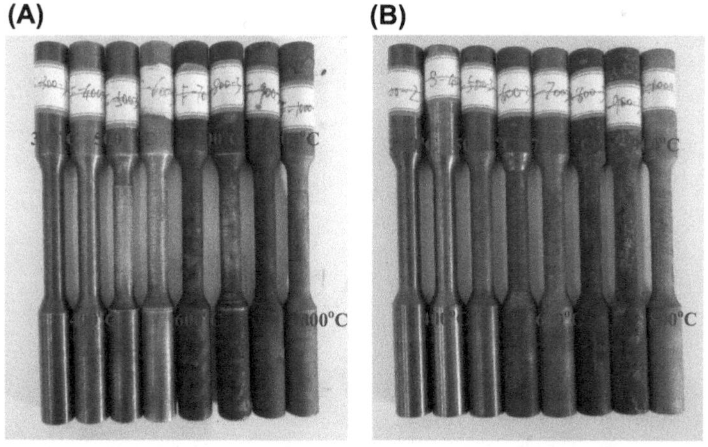

Figure 4.21 Color spectrum of specimens after exposure to high temperatures. (A) Air cooling and (B) water cooling.

color at *RT* is silvery white, as shown in Fig. 4.19B. The surface color of the specimens by both air and water cooling is brown for the exposed temperature 300°C. While the specimens exposed to target temperature from 400 to 900°C, the surface color of the specimens varies with the increase of the exposed temperature when cooled by air, whereas the color changes from light blue to darker colors with the increase of exposed temperature for specimens cooled by water. The visual observation on surface color of specimens exposed to temperatures from 400 to 900°C is similar to those reported (Li et al., 2017). For the specimen experiencing temperature of 1000°C, the surface color is gray and the surface becomes smooth for the both specimens cooled by air and water. Surface flaking was observed for the specimens experiencing temperature of 900°C due to the reason of the oxide.

The failure mode of the specimens is shown in Figs. 4.22 and 4.23. Necking phenomenon are clearly observed for all specimens. The fracture occurs within the range of gauge length, and the failure location is very close to one end of gauge length. This can be attributed to the fact that the temperature along the length of the specimen is not uniformly distributed in cooling phase as the diameter at ends of the specimen are larger than that of middle portion. The slower cooling yields higher temperature that consequently results in the lower strength at the ends of the specimen. For the specimens exposed to temperature of 800°C and above with water cooling, a zigzag shape of fractured cross section is observed, where the

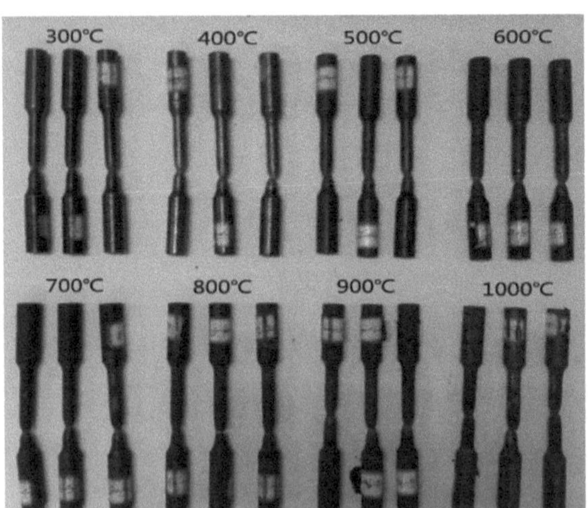

Figure 4.22 Failure modes of test specimens (air cooling).

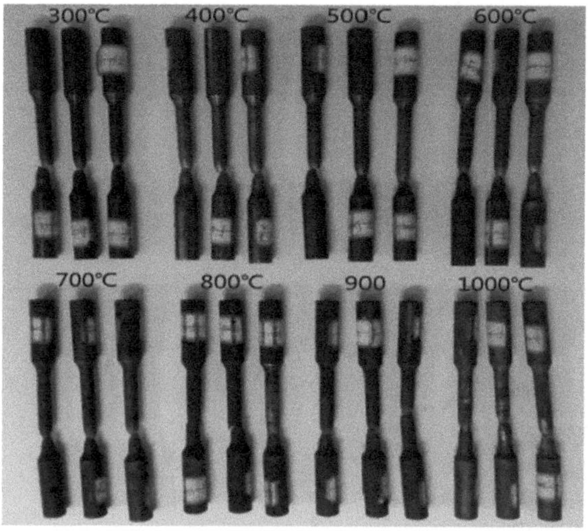

Figure 4.23 Failure modes of test specimens water cooling.

shape of the fractured cross section for other specimens is flat. This can be attributed to that the steel becomes brittle and the ductility deteriorates seriously associated with water cooling.

4.4.4.3 Stress–strain relationships

The stress–strain relationships of the specimens are presented in Fig. 4.24. As previously discussed for each target temperature, three identical specimens were tested to produce reliable results. In the case that the difference of the three test results exceeded 10%, the result with the largest discrepancy was eliminated and an additional test was carried out. Therefore, for each target temperature, there are three series of data for air cooling (noted as series A-1, series A-2, and series A-3) and three series of data for water cooling (noted as series W-1, series W-2, and series W-3). It can be observed from the figure that both the shape and scale of the three curves corresponding to the same target temperature and cooling method are very similar.

As can be seen from Fig. 4.24, at ambient condition and exposed temperature range of 300–700°C, there is obvious yield plateau in the stress–strain curves and the strain hardening is not significant for both air and water cooling. The stress–strain curves for the specimens experiencing target temperature of 300–600°C are similar to that at ambient temperature, and no significant variation was observed. For the specimens exposed

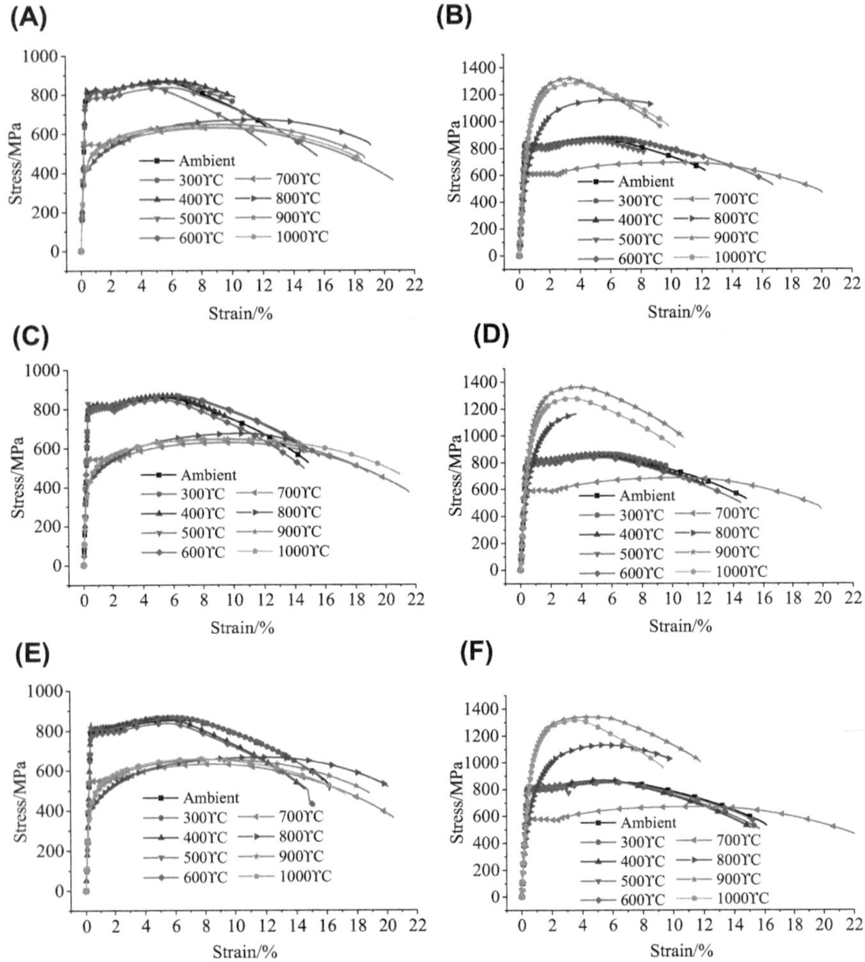

Figure 4.24 Stress-—strain curves for different exposed temperatures and cooling methods. (A) Series A-1; (B) series W-1; (C) series A-2; (D) series W-2; (E) series A-3; and (F) series W-3.

to temperature of 700°C, the yield strength deteriorated notably and the ultimate strain enlarged.

In addition, for the specimens experiencing target temperature of 800—1000°C, the yield plateau vanished, and the scale of the strength was quite different between air and water cooling. For air cooling, the magnitudes of the strength further reduced, and the ultimate strain was greater than that at ambient temperature. However, in the case of water cooling, the strength increased significantly, and the ultimate strain decreased. This phenomenon corresponds to effect of quench hardening process, a mechanical process in which steels and alloys are strengthened and hardened. For metals consist of ferrous metals and alloys, quench hardening can be achieved by heating the metals to a certain temperature, dependent upon the type of metal, and then rapidly cooling the metal, such as using water. This process results in hardening material by either surface hardening or through hardening depending on the cooling rate. The quenched metal is then often tempered to reduce the brittleness induced in the quench hardening process.

4.4.4.4 Yield and tension strength

According to the recorded stress—strain curves of high strength Q690 steel after high temperature exposure, the corresponding yield strength (f_{yPT}) and tension strength (f_{uPT}) can be generated. For the curves having the obvious yield plateau, the yield strength was determined as the stress level at the low yield plateau. For the curves without the yield plateau, the yield strength is determined by commonly used 0.2% offset method. The tension strength is obtained by selecting the maximum value of stress level in the stress—strain curves. The yield and tension strength of Q690 steel after being exposed to elevated temperature are tabulated in Table 4.8 To clearly observe the evolution of strength with the exposed temperature presented in Table 4.9, the residual factors are calculated as the ratio of the strength after high temperature exposure to the origin value of specimens at ambient temperature.

Table 4.8 Mechanical properties of high strength Q690 steel after high temperature exposure.

Temperature/°C	f_{yPT}/MPa		f_{uPT}/MPa		E_{PT}/GPa		ε_{PT}/%	
	Air	Water	Air	Water	Air	Water	Air	Water
25	797.5	797.5	859.2	859.2	210.7	210.7	16	16
300	808.7	809.1	86.58	865.8	215.6	215.3	17.3	16.9
400	810.0	808.1	867.3	866.3	218.4	218.8	15.6	15.6
500	800.3	804.5	858.9	850.8	218.0	215.8	14.0	14.1
600	780.0	786.8	839.9	847.7	221.1	219.6	15.4	16.5
700	543.8	584.2	630.3	681.5	215.6	218.0	20.9	20.8
800	416.2	609.2	671.0	1151.4	197.2	203.5	17.8	9.21
900	426.0	948.1	649.5	1340.5	191.3	203.1	19.9	10.6
1000	434.2	909.7	644.9	1292.6	188.2	202.7	19.47	10.4

Table 4.9 Residual factor of mechanical properties of Q690 steel after high temperature exposure.

Temperature/°C	f_{yPT}/f_y		f_{uPT}/f_u		E_{PT}/E		$\varepsilon_{PT}/\varepsilon$	
	Air	Water	Air	Water	Air	Water	Air	Water
25	1.00	1.00	1.00	1.00	1.00	1.00	1.00	1.00
300	1.01	1.01	1.01	1.01	1.02	1.02	1.08	1.05
400	1.02	1.01	1.01	1.01	1.04	1.04	0.97	0.97
500	1.00	1.01	1.00	0.99	1.03	1.02	0.87	0.88
600	0.98	0.99	0.98	0.99	1.05	1.04	0.96	1.03
700	0.68	0.73	0.73	0.79	1.02	1.03	1.30	1.30
800	0.52	0.76	0.78	1.34	0.94	0.97	1.11	0.58
900	0.53	1.19	0.76	1.56	0.91	0.96	1.24	0.66
1000	0.54	1.14	0.75	1.50	0.89	0.96	1.21	0.65

The variations of the residual factors associated with the yield and tension strength corresponding to the different elevated temperature are plotted in Fig. 4.25. It is clear from the figure that the deterioration of the yield and tension strength of Q690 steel after high temperature exposure are not significant being within 3% for the exposed elevated temperature up to 600°C, no matter which cooling method is applied. After being exposed to a temperature of 700°C, the strength reduction for both the yield and tension strength of Q690 steel is approximately 30% and 25% associated with air and water cooling, respectively. With the air cooling being applied, although the tension strength of the steel being exposed to temperature 800−1000°C is comparable to that of being exposed to temperature 700°C, the reduction of the yield strength is abort 50% after being exposed to temperature 800−1000°C. In the case of water cooling, both the yield

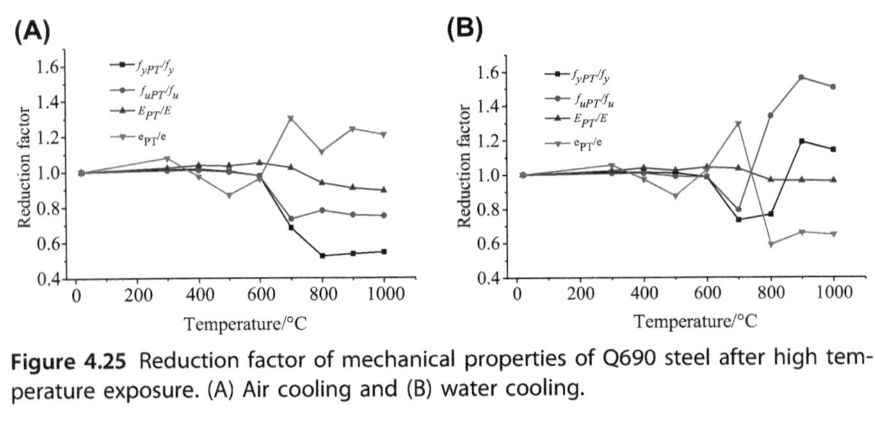

Figure 4.25 Reduction factor of mechanical properties of Q690 steel after high temperature exposure. (A) Air cooling and (B) water cooling.

and tension strength increase approximately 30%–60% after being exposed to temperature 800–1000°C due to the effect of quench hardening.

4.4.4.5 Elastic modulus

The values of elastic modulus of Q690 steel after high temperature exposure (E_{PT}) are listed in Table 4.8 The corresponding residual factors are tabulated in Table 4.9 and illustrated in Fig. 4.25. Generally, the elastic modulus associated with the temperatures below 600°C is almost the same as that at ambient temperature with little fluctuation, no matter which cooling method is applied. After being exposed to temperature above 600°C, the values of elastic modulus oscillate considerably with the maximum fluctuation being less than 10%. This conclusion agrees well with that reported by Li et al. (2017), even though the different thickness of steel plate was tested.

4.4.4.6 Ratio of ultimate elongation

As mentioned above, the ratio of ultimate elongation (ε_{PT}) is obtained by measuring the elongated gauge length of the fractured specimen, and it is calculated as the ratio of the elongated gauge to the original gauge. Therefore, the ultimate elongation cannot be directly obtained from the ultimate strain in the stress–strain curve due to the shortening of the specimen associated with diminishing of tensile stress after the fracture occurs. The measured ultimate elongation of Q690 steel specimens after high temperature exposure are tabulated in Table 4.8. The residual factors corresponding to the ratios of ultimate elongation are listed in Table 4.9 and presented in Fig. 4.25. It is noticed that the variation of the ratios of ultimate elongation associated with air and water cooling is similar to that for the exposed temperature up to 700°C. But once the exposed temperature reaches 800°C, the effect of cooling method on the ratio becomes quite differently signified by the abrupt decrease in the ratio of ultimate elongation when water cooling is applied due to the influence of quench hardening.

4.4.5 Proposed prediction equations

As the exposed elevated temperature and cooling methods are the two key factors influencing mechanical properties of high strength Q690 steel after high temperature exposure, equations to predict strength, elastic modulus, and elongation were developed as a function of exposed temperature, for air and water cooling, respectively. The comparison of predicted and tested results shown in Fig. 4.26 demonstrated a good agreement with the two results, and therefore, the proposed equations can accurately predict the residual mechanical properties of high strength Q690 steel after high temperature exposure.

4.4.5.1 Yield strength

For air and water cooling,

$$f_{Y,T}/f_Y = 1, \quad T \leq 600^\circ\text{C} \tag{4.9a}$$

For air cooling,

$$f_{YPT}/f_Y = 1.23 \times 10^{-6}T^2 - 2.44 \times 10^{-3}T + 2.1 \quad 600^\circ\text{C} < T \leq 800^\circ\text{C} \tag{4.9b}$$

$$f_{YPT}/f_Y = 0.54 \quad T \geq 800^\circ\text{C} \tag{4.9c}$$

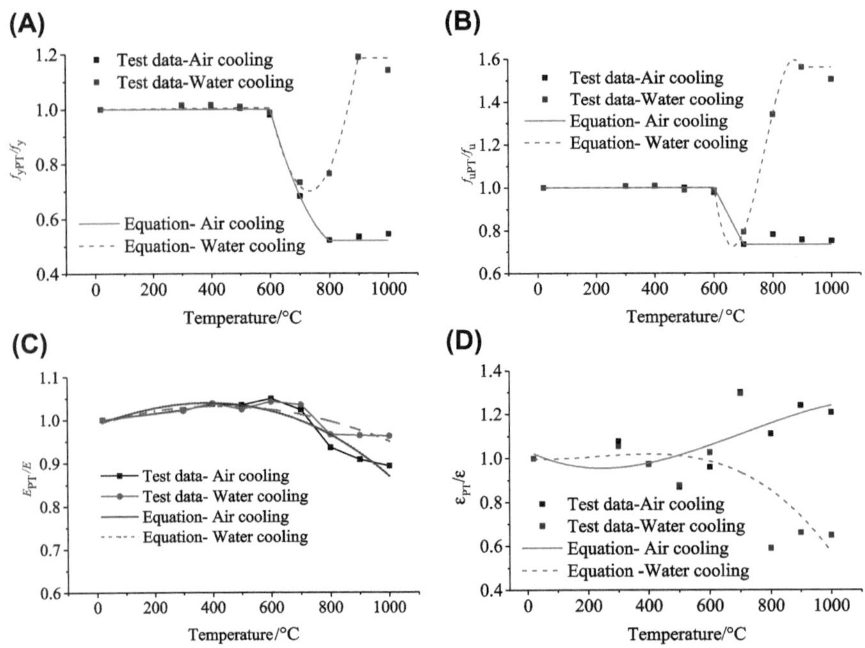

Figure 4.26 Comparison of results obtained from proposed equations with tests. (A) Yield strength; (B) tension strength; (C) elastic modulus; and (D) ultimate elongation.

For water cooling,

$$f_{yPT}/f_y = 1.634 \times 10^{-5}T^2 - 2.4 \times 10^{-2}T + 9.54 \quad 600^\circ C < T \leq 900^\circ C \tag{4.9d}$$

$$f_{yPT}/f_y = 1.19 \quad T \geq 900^\circ C \tag{4.9e}$$

where f_{yPT} is the residual yield strength after high temperature exposure, f_y is the yield strength at ambient temperature, and T is the elevated temperature of steel being exposed to.

4.4.5.2 Tension strength

For air and water cooling,

$$f_{u,T}/f_u = 1.0 \quad 20^\circ C \leq T \leq 600^\circ C \tag{4.10a}$$

For air cooling,

$$f_{uPT}/f_u = -2.27 \times 10^{-2}T + 2.6 \quad 600^\circ C \leq T < 700^\circ C \tag{4.10b}$$

$$f_{uPT}/f_u = 0.75 \quad T > 700^\circ C \tag{4.10c}$$

For water cooling,

$$f_{uPT}/f_u = -1.8 \times 10^{-7}T^3 - 4.16 \times 10^{-4}T^2 - 0.314T + 78.57 \quad 600^\circ C < T \leq 900^\circ C \tag{4.10d}$$

$$f_{uPT}/f_u = 1.56 \quad T > 900^\circ C \tag{4.10e}$$

where f_{uPT} is the residual tensile strength after high temperature exposure and f_u is the tensile strength at ambient temperature.

4.4.5.3 Elastic modulus

For air cooling,

$$E_T/E = 0.99 + 2.4 \times 10^{-4}T - 2.5 \times 10^{-7}T^2 \quad 20^\circ C \leq T \leq 1000^\circ C \tag{4.11a}$$

For water cooling,

$$E_{PT}/E = 0.99 + 2.0 \times 10^{-4}T - 2.4 \times 10^{-7}T^2 \quad 20^\circ C \leq T \leq 1000^\circ C \tag{4.11b}$$

where E_{PT} is the residual elastic modulus after high temperature exposure and E is the elastic modulus at ambient temperature.

4.4.5.4 Ultimate elongation

For air cooling,

$$\varepsilon_T/\varepsilon = -8.2\times10^{-10}T^3 + 1.72\times10^{-6}T^2 - 7\times10^{-4}T + 1.04 \quad 20^\circ\mathrm{C} \leq T \leq 1000^\circ\mathrm{C} \tag{4.12a}$$

For water cooling,

$$\varepsilon_{PT}/\varepsilon = -1.1\times10^{-9}T^3 + 7.46\times10^{-7}T^2 - 7.07\times10^{-5}T + 1.0 \quad 20^\circ\mathrm{C} \leq T \leq 1000^\circ\mathrm{C} \tag{4.12b}$$

where ε_{PT} is the residual ultimate elongation after high temperature exposure and ε is the ultimate elongation at ambient temperature.

4.5 Residual properties of high strength S460 steel

A research project based on experimental study was carried out by Qiang et al. (2012) to investigate the deterioration of mechanical properties of high strength S460 steel after fire. Tensile coupon tests were undertaken after cooling down from elevated temperatures up to 1000 °C. The post-fire elastic modulus, yield and ultimate strengths, ductility and stress—strain curves were obtained from this study. The results of the test are introduced below.

4.5.1 Experimental investigation

4.5.1.1 Test material and specimen

The test specimens were cut from S460NL steel sheet with nominal thicknesses of 5 mm S460NL is weldable normalized steel for steel constructions, with high yield strength and high toughness, produced in compliance with EN 10025-3 CEN (2004). S460NL is the grade designation abbreviation of this steel, where *S* means structural steel; 460 is its minimum yield strength; *N* means normalized rolled delivery condition; and *L* means low notch toughness testing temperature. The chemical composition of the tested HSS S460 is shown in Table 4.10 The dimensions of the specimens were prepared in accordance with EN 10002-5 (CEN, 2009) and ASTM standard E21-09 (2009). As shown in Fig. 4.27, one hole was provided at each end of the specimen to fix it to the grips of Gleeble 3800 using two steel bars.

Table 4.10 Chemical composition of high strength steel S460NL and S690QL material (%).

Chemical element	C	Si	Mn	P	S	Cr	Cu
	0.172	0.483	1.5	0.012	0.005	0.02	0.025
Chemical element	N	Nb	Ni	Ti	V	Al	Mo
	0.0051	0.046	0.018	0.002	0.087	0.037	0.002

(A) (B)

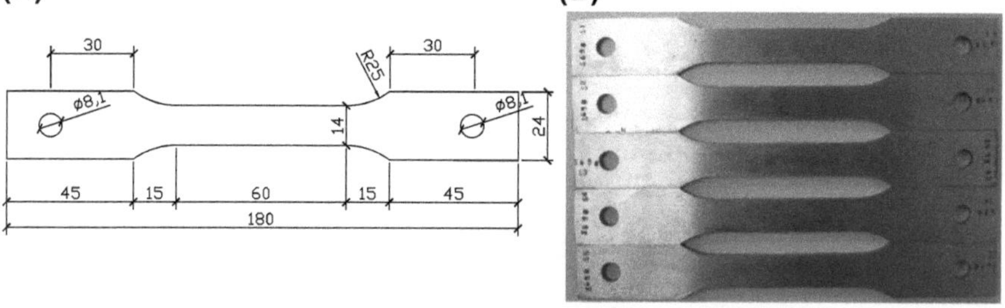

Figure 4.27 Tensile test coupon and dimension. (A) Dimension and (B) test coupons.

4.5.1.2 Test equipment and procedure

The heating of specimens were conducted in a temperature controlled electrical furnace, as shown in Fig. 4.28. In the furnace, the specimens were heated from ambient temperature to a preselected temperature under heating rate of 10°C/min, which corresponded to a natural fire. In this study, 11 elevated temperatures were selected for investigating S460, i.e., 300, 400, 500, 600, 650, 700, 750, 800, 850, 900, and 1000°C. After reaching every preselected elevated temperature, approximately 10 min holding time was required for the uniform temperature achieved in the whole specimen. And then the specimens were cooled to ambient temperature, after which the tensile tests were applied to them at ambient temperature.

The tensile tests were conducted using Gleeble 3800 System, which is a fully integrated digital closed loop control thermal and mechanical testing system, as shown in Figs. 4.29 and 4.30. Gleeble 3800 has a digital control system, which provides all the signals necessary to control thermal and mechanical test variables simultaneously through the digital closed-loop thermal and mechanical servo systems. The Gleeble 3800 mechanical system is an integrated hydraulic servo system capable of exerting as much as

Figure 4.28 Temperature-controllable furnace.

Figure 4.29 Test rig—Gleeble 3800 System.

Figure 4.30 Tensile test setup in Gleeble 3800 System.

20 tons of static force in compression or 10 tons in tension. In the tensile rig of Gleeble 3800, the tension is applied by using two loading rods as shown in Fig. 4.30. The right loading rod is fixed while the left loading rod moves left to apply tension to the specimens.

At the end of both loading rods, one pair of clamps and a steel bar are used to fix the specimens. All the tensile tests were conducted at ambient temperature until failure, and the air temperature inside the rig was obtained by Gleeble 3800 System automatically. In the tensile tests, strain control was used, and the strain rate used herein was 0.005/min, which satisfies the requirement of ASTM Standard E21-09 (2009). The experimental process was controlled by some predefined programming options using QuikSim Software. All the data obtained from the tests (i.e., force, stroke, strain, temperature, and time) were recorded by a computer and could be monitored during testing. For comparison, tensile tests were also conducted at ambient temperature on specimens that were not exposed to elevated temperatures.

4.5.2 Experimental results

The postfire stress–strain relationships obtained from this experimental study are plotted in Fig. 4.31. Their postfire mechanical properties were investigated based on the stress–strain curves, as shown in this section.

4.5.2.1 Elastic modulus

At elevated temperatures, the elastic modulus of steel is determined from the stress–strain curve, based on the initial slope. Similarly, the elastic modulus of postfire steel is also obtained from the stress–strain curve of steel after exposed to certain elevated temperature. The deterioration of steel properties at elevated temperatures is represented by reduction factors

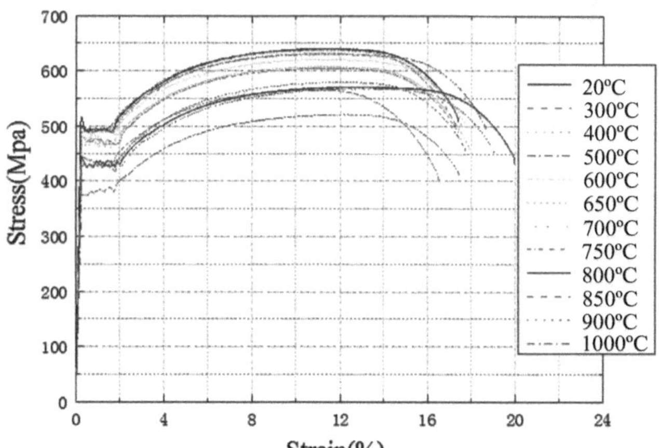

Figure 4.31 Stress–strain curves of S460 after cooling down from various fire temperatures.

at corresponding temperature, while the postfire material properties of steel are representative by residual factors. The residual factor is calculated as the ratio of the elastic modulus after cooling down from some elevated temperature (E_{PT}) to that at ambient temperature without being exposed to fire (E_{20}). The elastic modulus and residual factors of HSS after exposed to fire obtained from this investigation are given in Table 4.11. Fig. 4.32 presents the variation of elastic modulus residual factors after cooling down from various elevated temperatures up to 1000°C for S460. As seen in Fig. 4.32, when S460 is cooled down after being exposed to elevated temperatures below 600°C, it can regain its elastic modulus. But beyond 600°C, there is a considerable degradation of its elastic modulus for S460. It is interesting to note that at least 75% of elastic modulus is regained after cooling from temperatures up to 800°C for S460 and at least 60% for temperatures up to 1000°C.

4.5.2.2 Yield strength

To determine the yield strength of steel at ambient temperature, the 0.2% proof stress method is normally used, while for elevated temperatures, the 0.5%, 1.5%, and 2.0% total strain methods are usually used as well (Outinen

Table 4.11 Postfire elastic modulus, ultimate strengths, and the residual factors.

	Elastic modulus		Ultimate strength	
Temperature (°C)	E-modulus (MPa)	Residual factor	Ultimate strength (MPa)	Residual factor
20	202,619	1	640	1
100	—	—	—	—
200	—	—	—	—
300	202,300	0.998	638	0.996
400	198,870	0.981	637	0.995
500	194,910	0.962	631	0.985
600	191,320	0.944	621	0.97
650	190,000	0.938	608	0.95
700	186,140	0.919	605	0.945
750	176,184	0.87	566	0.884
800	173,660	0.857	571	0.892
850	170,780	0.843	580	0.906
900	163,980	0.809	568	0.887
1000	144,470	0.713	521	0.814

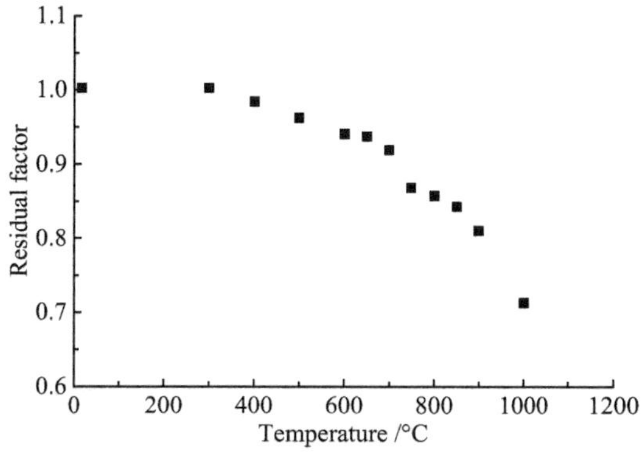

Figure 4.32 Elastic modulus residual factors.

and Makelainen, 2004; Outinen, 2007; Chen et al., 2006). In this study, the yield strengths were determined using the 0.2% proof stress method as well as the 0.5%, 1.5%, and 2% total strain methods. The yield strength based on 0.2% proof stress method is measured from the intersection of the stress–strain curve and the proportional line offset by 0.2% strain level. The strengths at 0.5%, 1.5%, and 2.0% strain levels are measured from the intersection of stress–strain curve and the nonproportional vertical lines at given strain values. After being exposed to various elevated temperatures, the postfire residual yield strengths of S460 are presented in Table 4.12 The yield strength residual factors after cooling down from elevated temperatures were calculated as the ratio of yield strength at elevated temperatures (f_{yPT}) to that at ambient temperature (f_{y20}), and the results based on different strain methods are given in Table 4.13. Fig. 4.33 shows the deterioration of residual yield strength with temperatures for S460. As seen in Fig. 4.33, the postfire yield strengths of S460 are almost not affected until they have achieved temperatures above 600°C. For S460, at least 75% of its yield strength can be preserved after cooling from temperatures up to 1000°C. Based on this study, S460 can regain its nominal yield strength if it is exposed to temperatures below 600°C, which is positive for the reuse of HSS structures after fire. For safe consideration, a conservative suggestion is recommended to use 90% of the nominal yield strength for postfire S460 if it is exposed to fire temperature below 600°C. Therefore, if members made of S460 are exposed to fire temperature below 600°C and their distortions remain within the tolerances for straightness and shape, they are reusable after fire.

Table 4.12 Postfire residual yield strengths (MPa).

Temperature (°C)	0.2%	0.5%	1.5%	2.0%
20	490	495	490	524
100	—	—	—	—
200	—	—	—	—
300	490	496	495	521
400	489	489	495	498
500	494	496	490	509
600	480	478	492	506
650	466	472	474	473
700	474	474	471	472
750	442	439	430	455
800	428	425	437	432
850	427	427	428	441
900	427	424	425	426
1000	374	375	387	400

Table 4.13 Postfire yield strength residual factors.

Temperature (°C)	0.2%	0.5%	1.5%	2.0%
20	1	1	1	1
100	—	—	—	—
200	—	—	—	—
300	1	1.002	1.01	0.995
400	0.997	0.988	1.011	0.951
500	1.007	1.002	1	0.972
600	0.98	0.966	1.005	0.965
650	0.95	0.954	0.968	0.903
700	0.968	0.959	0.96	0.901
750	0.901	0.887	0.876	0.868
800	0.874	0.858	0.891	0.824
850	0.871	0.863	0.873	0.843
900	0.871	0.858	0.867	0.813
1000	0.763	0.758	0.79	0.764

4.5.2.3 Ultimate strength

The tensile tests were carried out until specimen failed; therefore, the residual factors for ultimate strengths were calculated as the ratio of ultimate strength after cooling down from elevated temperature (f_{uPT}) to that at ambient temperature (f_{u20}). Table 4.11 and Fig. 4.34 present the residual ultimate strengths after being exposed to elevated temperatures up to 1000°C. The results show when S460 is exposed to elevated temperatures

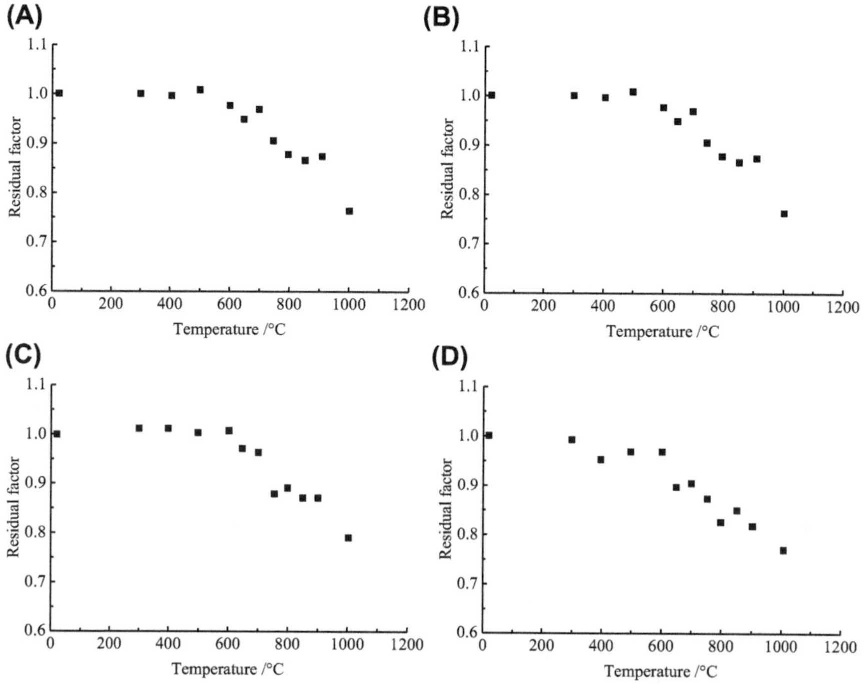

Figure 4.33 Yield strength residual factors based on various strain methods. (A) 0.2% proof stress method; (B) 0.5% total strain method; (C) 1.5% proof stress method and (D) 2.0% total strain method.

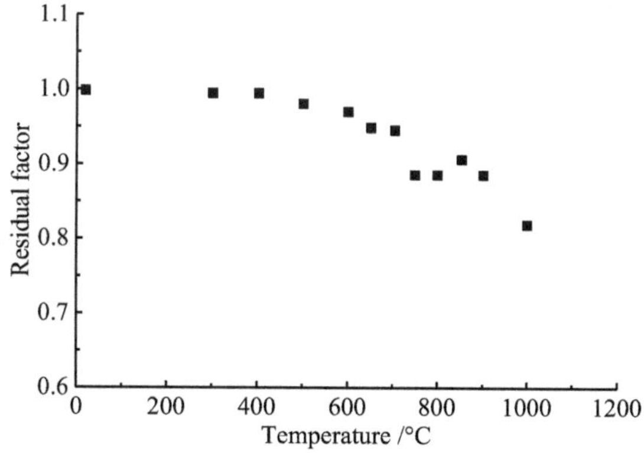

Figure 4.34 Postfire ultimate strength residual factors.

below 600°C, it can regain its nominal ultimate strength after cooling down. For S460 even when it is exposed to temperatures as high as 1000°C, it can regain at least 80% of its ultimate strength.

4.5.2.4 Ductility

Ductility of steel is defined based on the deformation that steel can undergo before fracture. The postfire ductility of HSS is studied by comparing the strain values at fracture, as shown in Figs. 4.35 and 4.36. S460 shows higher ductility at ambient temperature and after cooling down from lower fire temperatures. Fig. 4.35 shows the failure modes of HSS tensile test specimens conducted at ambient temperature after cooling down from elevated temperatures up to 1000°C; for comparison, the failure modes of HSS without being exposed to fire are also included. It is seen that every specimen shows ductile failure with necking, and no brittle failure is observed for S460 no matter how high temperature the specimen is exposed to. This observation also indicates that lack of ductility is not a concern for HSS S460 after being exposed to fire, which is very promising for the safety of steel structures with S460 going through fire disaster.

4.5.3 Predictive equations for residual properties

As temperature was the main reason causing the deterioration of material properties of steel, the equations were developed as a function of the highest temperature T, which specimens have been exposed to. For S460, two options of predictive equations were developed. One is recommended for simple use; another one is recommended for very precise evaluation.

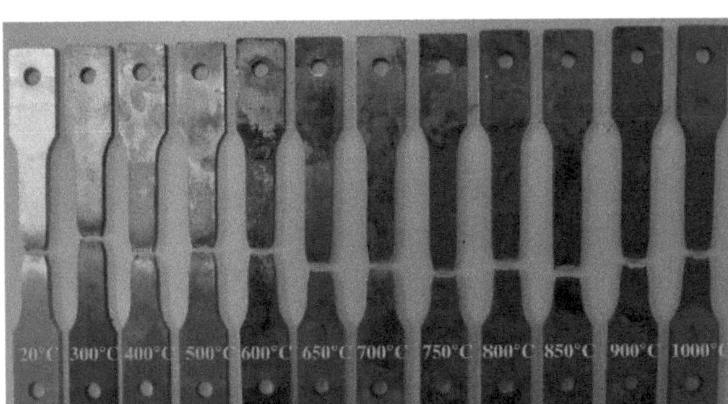

Figure 4.35 Failure modes of S460 tensile specimens after being exposed to various elevated temperatures.

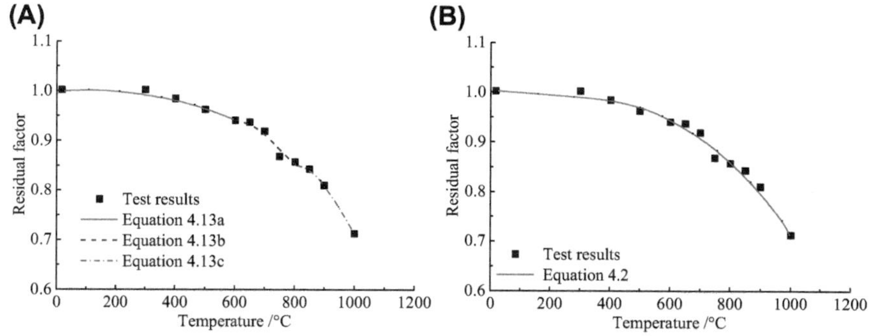

Figure 4.36 Comparison of predicted elastic modulus residual factor from (A) Eq. (4.13) and (B) Eq. (4.14) with test results.

Figs. 4.36 and 4.37 show that the proposed predictive equations agree well with the experimental results and therefore is capable to accurately predicting the postfire mechanical properties of S460.

4.5.3.1 Elastic modulus

New predictive equations were proposed as follows to present the deterioration of residual elastic modulus after being exposed to elevated temperatures. There are two options for S460.

Option 1:

$$20 \leq T \leq 600, \quad \frac{E_{PT}}{E_{20}} = -2.69 \times 10^{-7} T^2 + 6.55 \times 10^{-5} T + 0.999 \tag{4.13a}$$

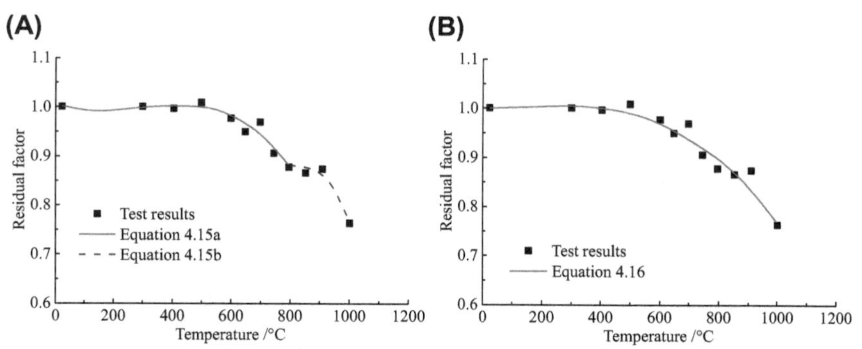

Figure 4.37 Comparison of predicted yield strength residual factor from equations with test results. (A) Eq. (4.15) and (B) Eq. (4.16).

$$600 < T \leq 800 \quad \frac{E_{PT}}{E_{20}} = 0.947 - \frac{(T-600)^{1.618}}{68.84\theta} \tag{4.13b}$$

$$600 < T \leq 800 \quad \frac{E_{PT}}{E_{20}} = -2.545 \times 10^{-6} T^2 + 3.856 \times 10^{-3} T - 0.598 \tag{4.13c}$$

The predictions from Eq. (4.13) are compared with test results obtained from this study in Fig. 4.36A. This figure shows that the agreement between the proposed equations and the test results is good.

Option 2:

$$20 \leq \theta \leq 1000 \quad \frac{E_{PT}}{E_{20}} = -3.84 \times 10^{-10} T^3 + 1.43 \times 10^{-7} T^2 - 4.18 \times 10^{-5} T + 1 \tag{4.14}$$

In the second option, one equation was proposed for the whole fire temperature range 20–1000°C. Fig. 4.36B shows that there is also good agreement between Eq. (4.14) and the test results. Eq. (4.13) provides more accurate predictions of postfire residual elastic modulus for S460. However, Eq. (4.14) is recommended if simpler predictive equation is needed.

4.5.3.2 Yield strength

Although four strain levels (0.2%, 0.5%, 1.5%, and 2%) were considered in determining the yield strength, only the residual factors based on the 0.2% proof stress method were used in deriving the empirical equations. It is because the yield strengths according to other strain levels have not been used widely.

Option 1:

$$20 \leq T \leq 800, \quad \frac{f_{yPT}}{f_{y20}} = -1.19 \times 10^{-9} T^3 + 1.03 \times 10^{-6} T^2 + 2.25 \times 10^{-4} T + 1.004 \tag{4.15a}$$

$$800 < \theta \leq 1000 \quad \frac{f_{yPT}}{f_{y20}} = 0.876 - \frac{(T-800)^{3.634}}{2.048 \times 10^6 T} \tag{4.15b}$$

Fig. 4.37A shows that good agreement exists between the predictions of Eq. (4.15) and test results. To develop a simpler equation, another option was worked out and compared with test results as shown in Fig. 4.37B. It can be seen that Eq. (4.15) provides a more accurate prediction, while Eq. (4.16) applies a simpler option. They can be chosen depending on different needs in practice.

Option 2:

$$20 \le T \le 1000 \quad \frac{f_{yPT}}{f_{y20}} = -3.24 \times 10^{-10} T^3 + 4.98 \times 10^{-8} T^2 + 4.52 \times 10^{-5} T + 0.998 \tag{4.16}$$

4.5.3.3 Ultimate strength

Option 1:

$$20 \le T \le 800, \quad \frac{f_{uPT}}{f_{u20}} = -1.19 \times 10^{-9} T^3 + 1.03 \times 10^{-6} T^2 + 2.25 \times 10^{-4} T + 1.004 \tag{4.17a}$$

$$800 < T \le 1000 \quad \frac{f_{uPT}}{f_{u20}} = 0.876 - \frac{(T - 800)^{3.634}}{2.048 \times 10^6 T} \tag{4.17b}$$

Fig. 4.38A shows that there is a good agreement between the predictions of Eq. (4.17) and test results. To develop a simpler equation, an alternative option was compared with test results as shown in Fig. 4.38B. It can be seen that Eq. (4.17) provides a more accurate prediction while Eq. (4.18) applies a simpler option.

Option 2:

$$20 \le T \le 1000 \quad \frac{f_{uPT}}{f_{u20}} = -2.79 \times 10^{-7} T^2 + 1.08 \times 10^{-4} T + 0.996 \tag{4.18}$$

4.6 Residual properties of high strength S690 steel

An experimental study was carried out by Qiang et al. (2012) to investigate the deterioration of mechanical properties of high strength S690 steel after

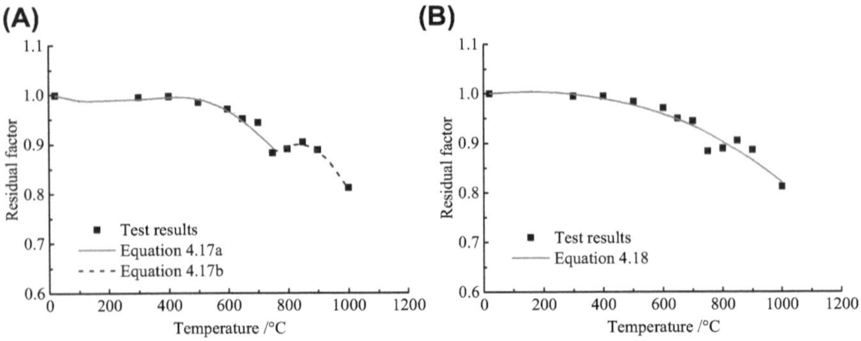

Figure 4.38 Comparison of predicted ultimate strength residual factor from equations with test results. (A) Eq. (4.17) and (B) Eq. (4.18).

fire. Tensile coupon tests were undertaken after cooling down from elevated temperatures up to 1000 °C. The post-fire elastic modulus, yield and ultimate strengths, ductility and stress–strain curves were obtained from this study. The results of the test are introduced below.

4.6.1 Test material and specimen

The test specimens were cut from S690QL steel sheet with nominal thicknesses of 5 mm S690QL is a high strength structural steel produced in compliance with EN 10025-6 (CEN, 2009). This material is heat-treated using the quenched and tempered process and has good bending and welding properties. S690QL is the grade designation abbreviation of this steel, where *S* means structural steel; 690 is its minimum nominal yield strength; *Q* means quenching and tempering; and *L* means low notch toughness testing temperature. The chemical composition of the tested HSS S690 is shown in Table 4.14. The dimensions of the specimens were prepared in accordance with EN 10002-5 (CEN, 2009) and ASTM standard E21-09 (2009). As shown in Fig. 4.29, one hole was provided at each end of the specimen to fix it to the grips of Gleeble 3800 using two steel bars.

Table 4.14 Chemical composition of S690QL material (%).

Chemical element	C	Si	Mn	P	S	Cr	Cu
	0.16	0.21	0.85	0.012	0.001	0.35	0.03
Chemical element	N	Nb	Ni	Ti	V	Al	Mo
	0.0026	0.025	0.05	0.006	0	0.093	0.2

4.6.2 Test equipment and procedure

The test setup and procedure for S690 steel are similar to those for S460 steel presented in Section 4.5.1.2.

4.6.3 Experimental results

The postfire stress–strain relationships obtained from this experimental study are plotted in Fig. 4.39. The postfire mechanical properties of S690 were investigated based on the stress–strain curves, as shown in this section. The results of S690 show that exposing to fire temperatures as low as 100 and 200°C almost has no effect on its mechanical properties, comparing with that at ambient temperature.

4.6.3.1 Elastic modulus

At elevated temperatures, the elastic modulus of steel is determined from the stress–strain curve, based on the initial slope. Similarly, the elastic modulus of postfire steel is also obtained from the stress–strain curve of steel after being exposed to certain elevated temperature. The deterioration of steel properties at elevated temperatures is represented by reduction factors at corresponding temperature, while the postfire material properties of steel are representative by residual factors. The residual factor is calculated as the ratio of the elastic modulus after cooling down from some elevated temperature (E_{PT}) to that at ambient temperature without being exposed to fire (E_{20}). The elastic modulus and residual factors of HSS after being exposed to fire obtained from this investigation are given in Table 4.15 Fig. 4.40 presents the variation of elastic modulus residual factors after cooling down from various elevated temperatures up to 1000°C for S690. As seen in Fig. 4.40, when S690 is cooled down after being exposed to elevated temperatures below 600°C, it can regain its elastic modulus. But beyond 600°C, there is a considerable degradation of its elastic modulus for S690, and it seems S690 loses its elastic modulus quickly. It is interesting to note that at least 75% of elastic modulus is regained after cooling from temperatures up to 800°C for S690 and at least 60% for temperatures up to 1000°C.

4.6.3.2 Yield strength

To determine the yield strength of steel at ambient temperature, the 0.2% proof stress method is normally used, while for elevated temperatures, the 0.5%, 1.5%, and 2.0% total strain methods are usually used as well (Outinen and Makelainen, 2004; Outinen, 2007; Chen et al., 2006). In this study, the yield strengths were determined using the 0.2% proof stress method as well

Table 4.15 Postfire elastic modulus and residual factors.

	Elastic modulus		Ultimate strength	
Temperature (°C)	E-modulus (MPa)	Residual factor	Ultimate strength (MPa)	Residual factor
20	202,619	1	640	1
100	—	—	—	—
200	—	—	—	—
300	202,300	0.998	638	0.996
400	198,870	0.981	637	0.995
500	194,910	0.962	631	0.985
600	191,320	0.944	621	0.97
650	190,000	0.938	608	0.95
700	186,140	0.919	605	0.945
750	176,184	0.87	566	0.884
800	173,660	0.857	571	0.892
850	170,780	0.843	580	0.906
900	163,980	0.809	568	0.887
1000	144,470	0.713	521	0.814

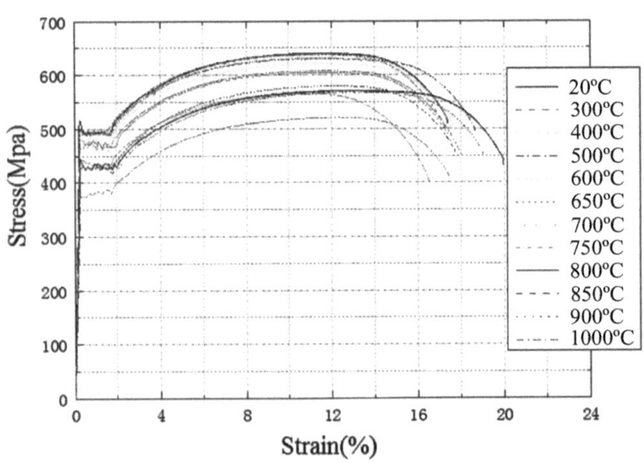

Figure 4.39 Stress–strain curves of S460 after cooling down from various fire temperatures.

as the 0.5%, 1.5%, and 2% total strain methods. The yield strength based on 0.2% proof stress method is measured from the intersection of the stress–strain curve and the proportional line offset by 0.2% strain level. The strengths at 0.5%, 1.5%, and 2.0% strain levels are measured from the intersection of stress–strain curve and the nonproportional vertical lines at given strain values.

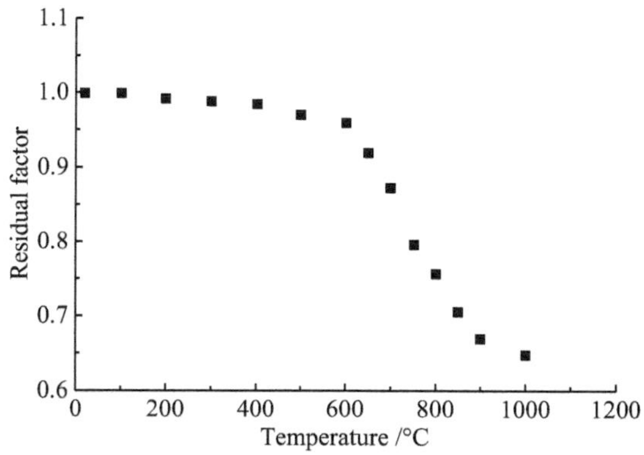

Figure 4.40 Elastic modulus residual factors.

After being exposed to various elevated temperatures, the postfire residual yield strengths of S690 are presented in Table 4.16 The yield strength residual factors after cooling down from elevated temperatures were calculated as the ratio of yield strength at elevated temperatures (f_{yPT}) to that at ambient temperature (f_{y20}), and the results based on different strain methods are given in Table 4.17 Fig. 4.41 shows the deterioration of residual yield strength with temperatures for S690. As seen in Fig. 4.41, the postfire yield strengths of S690 are almost not affected until they have achieved temperatures above 600°C. For S460, at least 75% of its yield strength can be preserved after cooling from temperatures up to 1000°C. For safe consideration, a conservative suggestion is recommended to use

Table 4.16 Postfire residual yield strengths (MPa).

Temperature (°C)	0.2%	0.5%	1.5%	2.0%
20	490	495	490	524
100	—	—	—	—
200	—	—	—	—
300	490	496	495	521
400	489	489	495	498
500	494	496	490	509
600	480	478	492	506
650	466	472	474	473
700	474	474	471	472
750	442	439	430	455
800	428	425	437	432
850	427	427	428	441
900	427	424	425	426
1000	374	375	387	400

Table 4.17 Postfire yield strength residual factors.

Temperature (°C)	0.2%	0.5%	1.5%	2.0%
20	1	1	1	1
100	—	—	—	—
200	—	—	—	—
300	1	1.002	1.01	0.995
400	0.997	0.988	1.011	0.951
500	1.007	1.002	1	0.972
600	0.98	0.966	1.005	0.965
650	0.95	0.954	0.968	0.903
700	0.968	0.959	0.96	0.901
750	0.901	0.887	0.876	0.868
800	0.874	0.858	0.891	0.824
850	0.871	0.863	0.873	0.843
900	0.871	0.858	0.867	0.813
1000	0.763	0.758	0.79	0.764

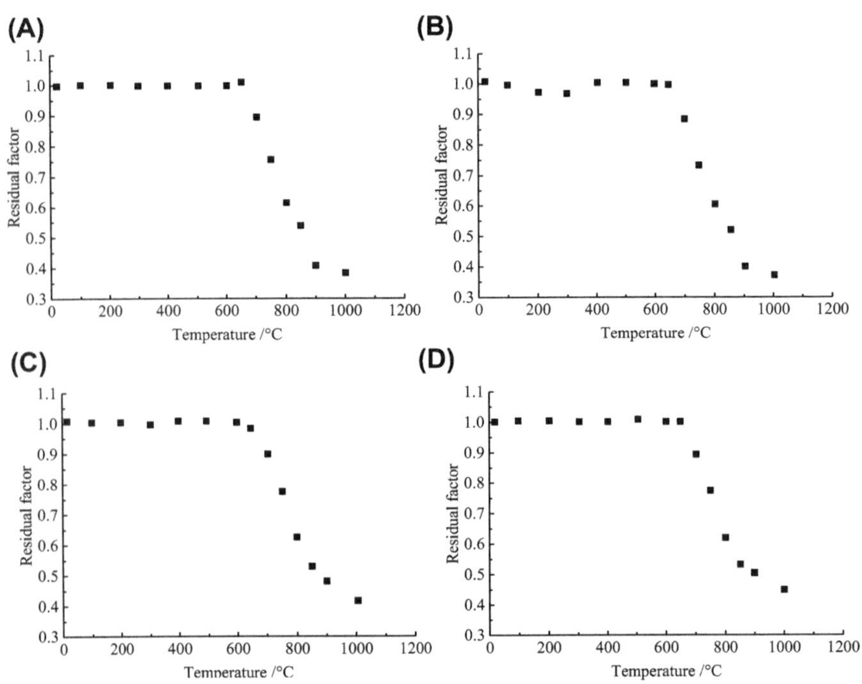

Figure 4.41 Yield strength residual factors based on various strain methods. (A) 0.2% proof stress method; (B) 0.5% total strain method; (C) 1.5% proof stress method and (D) 2.0% total strain method.

90% of the nominal yield strength for postfire S690 if it is exposed to fire temperature below 600°C. Therefore, if members made of S690 are exposed to fire temperature below 600°C and their distortions remain within the tolerances for straightness and shape, they are reusable after fire.

4.6.3.3 Ultimate strength

The tensile tests were carried out until specimen failed; therefore, the residual factors for ultimate strengths were calculated as the ratio of ultimate strength after cooling down from elevated temperature (f_{uPT}) to that at ambient temperature (f_{u20}). Table 4.15 and Fig. 4.42 present the residual ultimate strengths after being exposed to elevated temperatures up to 1000°C. The results show that, when S690 is exposed to elevated temperatures below 600°C, it can regain its nominal ultimate strength after cooling down. For S690, its residual ultimate strength is only about 55% when cooling down from 1000°C elevated temperature. It is also the result of quench and tempering, which S690 experienced in manufacture process.

4.6.3.4 Ductility

Ductility of steel is defined based on the deformation that steel can undergo before fracture. The postfire ductility of HSS is studied by comparing the strain values at fracture, as shown in Fig. 4.43. Fig. 4.44 shows the failure modes of HSS tensile test specimens conducted at ambient temperature after cooling down from elevated temperatures up to 1000°C; for comparison, the failure modes of HSS without being exposed to fire are also included. It is seen that every specimen shows ductile failure with necking, and no brittle failure is observed for S690 no matter how high temperature the specimen is exposed to. This observation also indicates that lack of

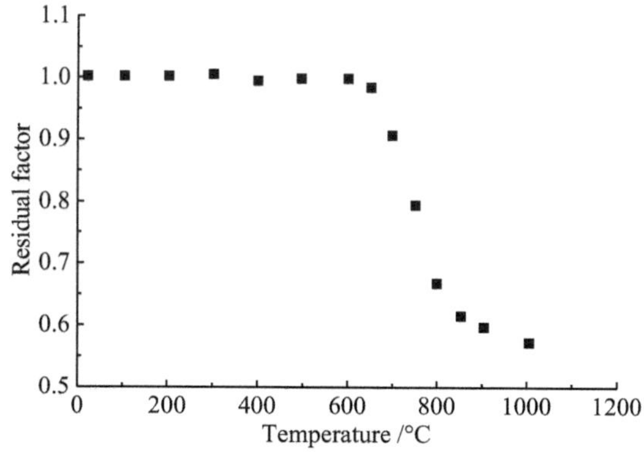

Figure 4.42 Postfire ultimate strength residual factors.

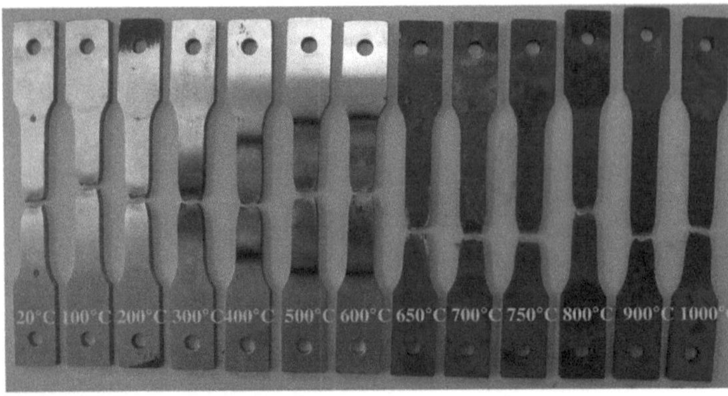

Figure 4.43 Failure modes of S690 tensile specimens after being exposed to various elevated temperatures.

ductility is not a concern for HSS S690 after being exposed to fire, which is very promising for the safety of steel structures with S690 going through fire disaster.

4.6.4 Predictive equations for residual properties

As temperature was the main reason causing the deterioration of material properties of steel, the equations were developed as a function of the highest temperature *T*, which specimens have been exposed to. Figs. 4.44 and 4.45 show that the proposed predictive equations agree well with the experimental results and therefore are able to accurately predict the postfire mechanical properties of S690.

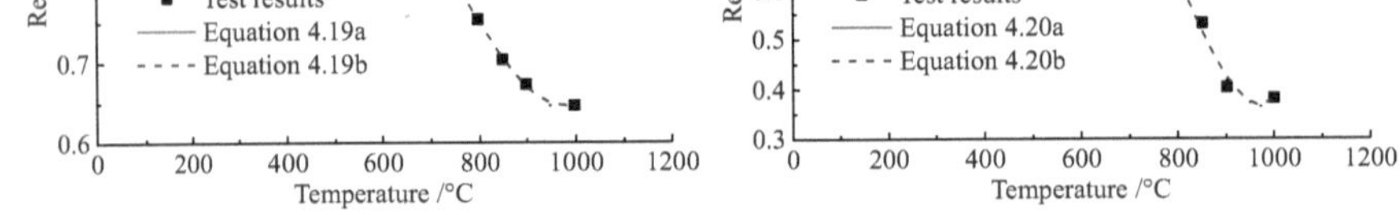

Figure 4.44 Comparison of predicted elastic modulus and yield strength residual factor from Eq. (4.19) and Eq. (4.20) with test results. (A) Elastic modulus and (B) Yield strength.

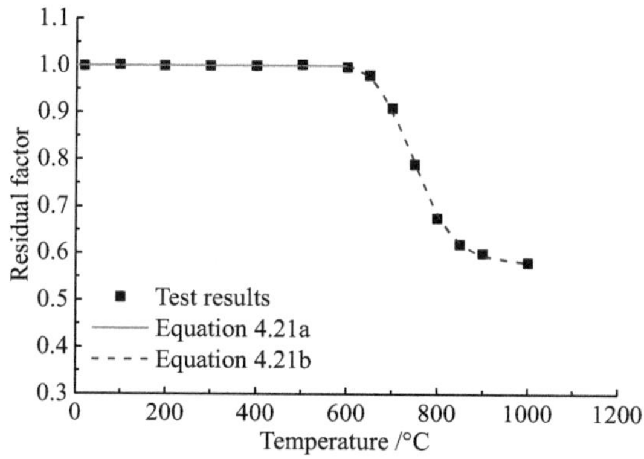

Figure 4.45 Comparison of predicted ultimate strength residual factor from Eq. (4.21) with test results.

4.6.4.1 Elastic modulus

For the residual elastic modulus of S690, the predictions from Eq. (4.19) are compared with test results obtained from this study in Fig. 4.44A. This figure shows that there is good agreement between the proposed equations and the test results, and Eq. (4.19) is not complicated in practical use. Therefore, it is recommended to use Eq. (4.19) to determine the postfire residual elastic modulus of S690 after being exposed to elevated temperatures up to 1000°C.

$$20 \leq T \leq 600, \quad \frac{E_{PT}}{E_{20}} = -1.52 \times 10^{-10}T^3 + 2.70 \times 10^{-8}T^2 - 3.35 \times 10^{-5}T + 1 \tag{4.19a}$$

$$600 < T \leq 1000 \quad \frac{E_{PT}}{E_{20}} = 6.27 \times 10^{-9}T^3 - 1.38 \times 10^{-5}T^2 + 8.59 \times 10^{-3}T - 0.806 \tag{4.19b}$$

4.6.4.2 Yield strength

Fig. 4.44B compares the predictions of Eq. (4.20) with test results on the postfire residual yield strength of S690. It is seen that the proposed equations agree well with the test results, and Eq. (4.20) is not complicated in practical use. So, it is recommended to determine the postfire residual yield strength of S690 after cooling down from elevated temperatures up to 1000°C.

$$20 \leq T < 650, \quad \frac{f_{yPT}}{f_{y20}} = 1 - \frac{(T-20)^{1.584}}{9957T} \tag{4.20a}$$

$$650 \leq T \leq 1000 \quad \frac{f_{yPT}}{f_{y20}} = 1.80 \times 10^{-8}T^3 - 4.03 \times 10^{-5}T^2 + 2.74 \times 10^{-2}T - 4.711 \tag{4.20b}$$

4.6.4.3 Ultimate strength

Fig. 4.45 compares the predictions of Eq. (4.21) for S690 with test results on its postfire residual ultimate strength. It seems that the proposed equations agree well with the test results. And Eq. (4.21) is not very complicated in practical use. Therefore, it is recommended to determine the postfire residual ultimate strength of S690 after cooling down from elevated temperatures up to 1000°C.

$$20 \leq T \leq 600, \quad \frac{f_{uPT}}{f_{u20}} = 1 \tag{4.21a}$$

$$600 < T \leq 1000 \quad \frac{f_{uPT}}{f_{u20}} = -1.24 \times 10^{-10}T^4 + 4.13 \times 10^{-7}T^3 - 5.077 \times 10^{-4}T^2 + 0.271T - 52.21 \tag{4.21b}$$

4.7 Residual properties of high strength S960 steel

In order to reveal post-fire material performance of very high strength steel S960, an experimental study was carried out by Qiang et al. (2013), which serves for the evaluation of post-fire performance of structures with components made of S960. Tensile tests were undertaken on specimens made of S960 after cooling down from temperatures up to 1000 °C. Its post-fire elastic modulus, yield strength, ultimate strength, and stress—strain curves were obtained.

4.7.1 Test specimen and equiment

The test specimens were cut from one S960QL steel sheet ordered for this study with nominal thicknesses of 5 mm S960QL is a very high strength structural steel produced in compliance with EN 10025-6 (CEN, 1992). This material is heat-treated using the quenched and tempered process and has good bending and welding properties. S960QL is the grade designation abbreviation of this steel, where S means structural steel; 960 is its minimum nominal yield strength; *Q* means quenching and tempering; and *L* means low notch toughness testing temperature. The chemical composition of the tested very HSS S960 is shown in Table 4.18 The dimensions of the specimens were similar to S460 steel presented in section 4.5.1.1. The test equipment and procedure are same as those for S460 steel introduced in Section 4.5.1.2.

4.7.2 Experimental results

4.7.2.1 Postfire stress–strain relationships

The postfire stress–strain relationships of VHSS S960 obtained from this experimental study are plotted in Fig. 4.46. Its postfire mechanical

Table 4.18 Chemical composition of high strength steel S690QL material provided by the supplier (%).

C	Si	Mn	P	S	Cr	Cu	Mo
0.19	0.39	1.43	0.008	0.0009	0.58	0.02	0.45
N	Nb	Ni	Ti	V	Al-g	B-g	
0.0047	0.022	0.05	0.004	0.03	0.044	0.0001	

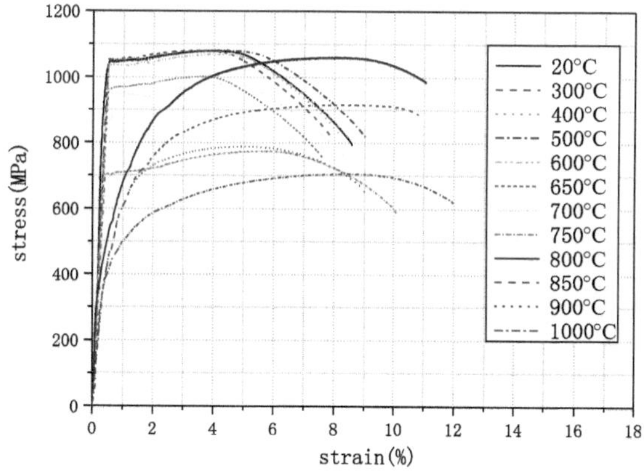

Figure 4.46 Stress–strain curves after cooling down from various fire temperatures.

properties were investigated based on the stress–strain curves, as described in this section. It is worthwhile to note that when cooled down from fire temperatures below 800°C, the postfire stress–strain curves of S960 exhibit yield plateau as for ambient temperature; however, beyond 800°C, the postfire stress–strain curves are of the gradual yielding type, and there is an obvious increase in the postfire strengths at 800°C and gradual decrease again afterward.

4.7.2.2 Elastic modulus

At elevated temperatures, the elastic modulus of steel is determined from the stress–strain curve, based on the initial slope. Similarly, the elastic modulus of postfire steel is also obtained according to the initial slope of the stress–strain curve for steel after exposed to certain elevated temperature. The postfire material performance of steels is representative by residual factor. The elastic modulus residual factor is calculated as the ratio of the elastic modulus after cooling down from some elevated temperature (E_{PT}) to that at ambient temperature without being exposed to fire (E_{20}). The elastic modulus and residual factors of VHSS S960 after being exposed to fire up to 1000°C obtained from this study are given in Table 4.19 As seen in Table 4.19, when S960 is cooled down after being exposed to elevated temperatures below 600°C, it is able to regain its elastic modulus. But beyond 600°C, there is a considerable degradation of its elastic modulus.

Table 4.19 Postfire elastic modulus and their residual factors of S960.

	Elastic modulus		Ultimate strength	
Temperature (°C)	E-modulus (MPa)	Residual factor	Ultimate strength (MPa)	Residual factor
20	213,586	1	1079	1
300	217,287	1.017	1080	1.001
400	208,889	0.978	1082	1.002
500	206,272	0.966	1074	0.995
600	201,870	0.945	1070	0.991
650	193,695	0.907	1002	0.928
700	186,649	0.874	983	0.911
750	171,418	0.803	776	0.719
800	160,587	0.752	1060	0.982
850	151,339	0.709	917	0.849
900	141,275	0.661	788	0.73
1000	138,648	0.649	695	0.644

It is interesting to note that at least 75% of elastic modulus is regained after cooling from temperatures up to 800°C and at least 60% for temperatures up to 1000°C.

4.7.2.3 Yield strength

To determine the yield strength of steel at ambient temperature, the 0.2% proof stress method is normally used, while for elevated temperatures, the 0.5%, 1.5%, and 2.0% total strain methods are usually used as well (Outinen and Makelainen, 2004; Outinen, 2007; Chen et al., 2006). In this study, the yield strengths were determined using the 0.2% proof stress method as well as the 0.5%, 1.5%, and 2% total strain methods. The typical method of determining yield strength based on 0.2% proof stress method and 0.5%, 1.5%, and 2.0% strain levels were used herein as same as that in previous research (Qiang et al., 2012). The yield strength based on 0.2% proof stress method is measured from the intersection of the stress—strain curve and the proportional line offset by 0.2% strain level. The strengths at 0.5%, 1.5%, and 2.0% strain levels are measured from the intersection of stress—strain curve and the nonproportional vertical lines at given strain values.

After being exposed to various elevated temperatures, the postfire residual yield strengths of S960 are presented in Table 4.20 The yield strength residual factors after cooling down from elevated temperatures are calculated as the ratio of yield strength at elevated temperatures (f_{yPT}) to that at ambient temperature (f_{y20}), and the results based on different strain methods are given in Table 4.20 as well. As shown in Table 4.20, S960 can regain its nominal yield strength if it is exposed to temperatures below 600°C, which is positive for the reuse of steel structures with VHSS S960 after fire. For safe consideration, a conservative suggestion is recommended to use 90% of the nominal yield strength for postfire S960 if steel structures are exposed to fire temperature below 600°C. Therefore, if members made of S960 are exposed to fire temperature below 600°C and their distortions remain within the tolerances for straightness and shape, they are reusable after fire. In addition, it is interesting to note that at strain levels of 1.5% and 2.0%, the postfire yield strength of S960 after cooling down from 800°C increases considerably, and the corresponding yield strengths after cooling down from temperatures higher than 800°C decrease gradually again.

4.7.2.4 Ultimate strength

The tensile tests were carried out until specimen failed; therefore, the residual factors for ultimate strengths are calculated as the ratio of ultimate

Table 4.20 Postfire yield strengths and their residual factors of S960.

	Steady-state test				Transient-state test				
	Residual factors					Residual factors			
Temperature(°C)	0.20%	0.50%	1.50%	2.00%	Temperature(°C)	0.20%	0.50%	1.50%	2.00%
20	1045	1041	1052	1058	1	1	1	1	20
300	1052	1054	1058	1068	1.006	1.012	1.006	1.01	300
400	1050	1054	1052	1058	1.004	1.012	1	1	400
500	1054	1054	1052	1058	1.008	1.012	1	1	500
600	1035	1039	1041	1045	0.99	0.998	0.99	0.988	600
650	967	965	977	981	0.925	0.926	0.929	0.927	650
700	755	753	807	807	0.722	0.723	0.767	0.763	700
750	701	705	716	722	0.671	0.677	0.68	0.682	750
800	629	581	826	909	0.601	0.558	0.785	0.859	800
850	529	512	786	869	0.506	0.492	0.747	0.822	850
900	388	400	570	601	0.371	0.385	0.542	0.568	900
1000	384	394	552	587	0.367	0.379	0.525	0.555	1000

strength after cooling down from elevated temperature (f_{uPT}) to that at ambient temperature (f_{u20}). Table 4.19 presents the residual ultimate strengths and their residual factors after being exposed to elevated temperatures up to 1000°C. The results show when S960 is exposed to elevated temperatures below 600°C, it is able to regain its nominal ultimate strength after cooling down. Even when it is exposed to temperatures as high as 1000°C, S960 can regain at least 60% of its ultimate strength. Similar to yield strength, the postfire ultimate strength increases after cooling down from fire temperature of 800°C and decreases after cooling down from higher fire temperatures.

4.7.2.5 Failure mode

Fig. 4.47 shows the failure modes of VHSS S960 tensile test specimens conducted at ambient temperature after cooling down from elevated temperatures up to 1000°C; for comparison, the failure mode of S960 without being exposed to fire is also included. It is found that every specimen shows ductile failure with necking, and no brittle failure is observed no matter how high temperature the specimen is exposed to. This observation is promising for the safety of steel members made of VHSS S960 going through fire disaster.

4.7.3 Predictive equations for residual properties

As fire temperature is the main reason causing the deterioration of material properties of steel, the predictive equations were developed as a function of the highest fire temperature T, to which specimens have been exposed to.

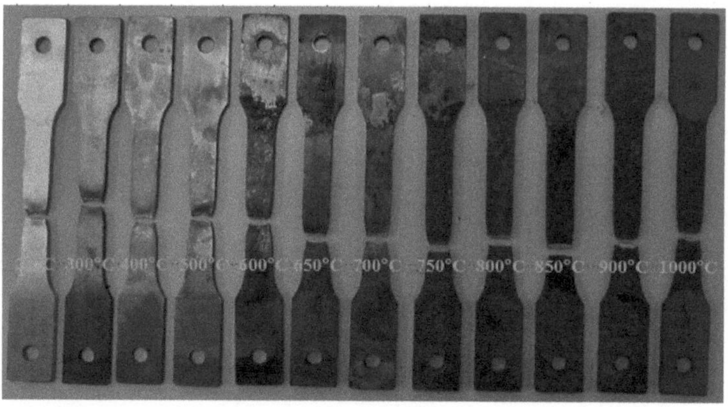

Figure 4.47 Failure modes of S960 tensile specimens after being exposed to various fire temperatures.

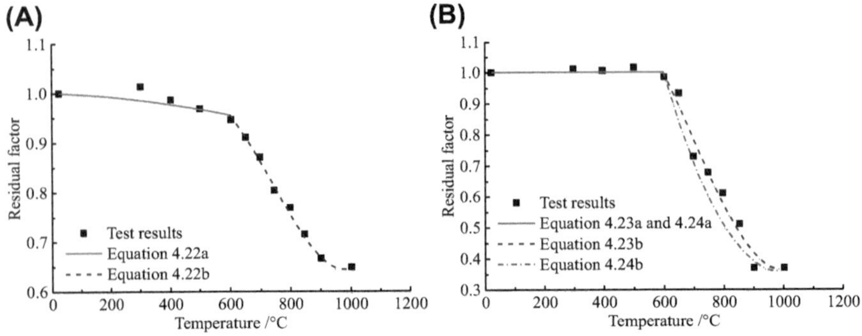

Figure 4.48 Comparison of predicted residual factor from equations with test results. (A) Elastic modulus and (B) yield strength.

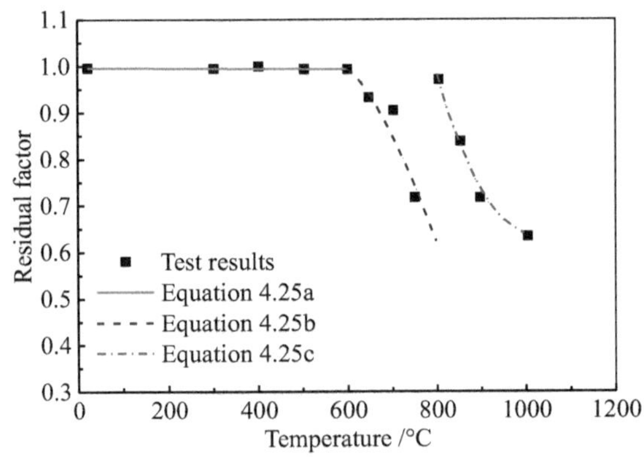

Figure 4.49 Comparison of predicted ultimate strength residual factor from Eq. (4.25) with test results.

Figs. 4.48 and 4.49 show that the proposed predictive equations agree well with the experimental results and therefore are able to accurately predict the postfire performance of VHSS S960.

4.7.3.1 Elastic modulus

As the elastic modulus residual factors of S960 are very similar to that of S690, the predictive equations (Eq. 4.22) proposed previously for evaluating residual elastic modulus of S690 (Qiang et al., 2012) were compared with the test results from this study as shown in Fig. 4.48A. This figure shows that there is a good agreement between the proposed equations and the test results of S960. Eq. (4.22) is not complicated in practical use. Therefore, it is recommended to use Eq. (4.22) to determine the postfire residual elastic modulus of VHSS S960 after being exposed to fire temperatures up to 1000°C.

$$20 \le \theta \le 600, \quad \frac{E_{P\theta}}{E_{20}} = -1.52 \times 10^{-10}\theta^3 + 2.70 \times 10^{-8}\theta^2 - 3.35 \times 10^{-5}\theta + 1 \tag{4.22a}$$

$$600 < \theta \le 1000 \quad \frac{E_{P\theta}}{E_{20}} = 6.27 \times 10^{-9}\theta^3 - 1.38 \times 10^{-5}\theta^2 + 8.59 \times 10^{-3}\theta - 0.806 \tag{4.22b}$$

4.7.3.2 Yield strength

Although four strain levels (0.2%, 0.5%, 1.5%, and 2%) were considered in determining the yield strength, only the residual factors based on the 0.2% proof stress method were used in deriving the empirical equations. It is because the yield strengths according to other strain levels have not been used widely.

Two sets of predictive equations were worked out for the postfire residual yield strength of S960 and compared with test results in Fig. 4.48B. Eq. (4.23) is more accurate according to all test data but sometimes not conservative. For conservative consideration, Eq. (4.24) is recommended for practical use, to determine the postfire residual yield strength of S960 after cooling down from fire temperatures up to 1000°C.

$$20 \le T < 600, \quad \frac{f_{yPT}}{f_{y20}} = 1 \tag{4.23a}$$

$$600 \le T \le 1000 \quad \frac{f_{yPT}}{f_{y20}} = 8.157 \times 10^{-9}T^3 - 1.685 \times 10^{-5}T^2 + 9.388 \times 10^{-3}T - 0.333 \tag{4.23b}$$

$$20 \le T < 600, \quad \frac{f_{yPT}}{f_{y20}} = 1 \tag{4.24a}$$

$$600 \le T \le 1000 \quad \frac{f_{yPT}}{f_{y20}} = 4.4 \times 10^{-6}T^2 - 8.637 \times 10^{-3}T + 4.596 \tag{4.24b}$$

4.7.3.3 Ultimate strength

$$20 \leq T \leq 600, \quad \frac{f_{uPT}}{f_{u20}} = 1 \tag{4.25a}$$

$$600 < T < 800 \quad \frac{f_{uPT}}{f_{u20}} = 1.006 - \frac{(T-600)^{1.158}}{9.567 \times 10^5 T} \tag{4.25b}$$

$$800 \leq T \leq 1000, \quad \frac{f_{uPT}}{f_{u20}} = 7.762 \times 10^{-6} T^2 - 1.568 \times 10^{-2} T + 8.564 \tag{4.25c}$$

Because of the abrupt increase of the postfire ultimate strength of S960 around 800°C caused by the aforementioned phase transformations in microstructures, three equations were proposed instead of two to predict the postfire ultimate strength of S960 after cooling down from fire temperatures up to 1000°C. Fig. 4.49 shows that a good agreement exists between the predictions of Eq. (4.25) and test results of the postfire residual ultimate strength.

4.8 Residual properties of high strength A572 steel

4.8.1 Test equipment

The specimens after heating in a furnace and cooling to RT are tested in an MTS-810 strength machine to evaluate residual strength of steel. This test setup is composed of a universal testing machine and a data collection system. The universal testing machine has a loading capacity of 250 kN, and the axial displacement of the specimen in tension is measured through an MTS extensometer. The data are recorded using an advanced data acquisition system. The test setup for residual strength test is shown in Fig. 4.50.

4.8.2 Test procedure

The residual tensile tests were conducted on steel coupons after exposure to temperatures of 20, 400, 500, 600, 700, 800, and 1000°C. For residual tensile tests, the steel coupon was placed in the furnace and heated to a target temperature at a heating rate of 10°C/min. Then, the specimen was maintained at the target temperature for 2 h. After that, heating in the furnace was turned off, and the coupon was allowed to cool down to RT. Two methods of cooling, namely air cooling and water quenching, were adopted to evaluate the effect of cooling type on strength recovery. In the first method, the coupons were left in the furnace after heating, to naturally

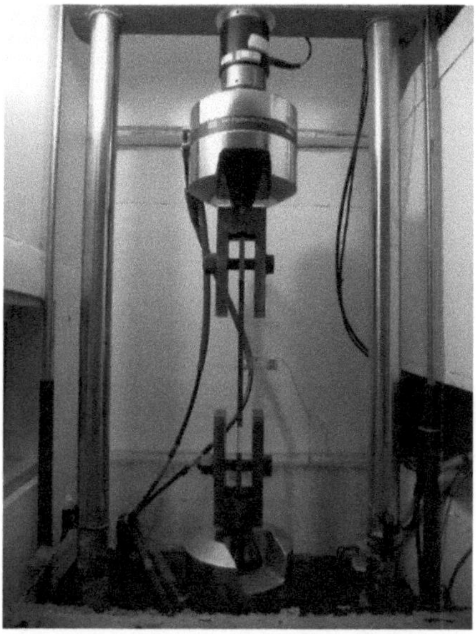

Figure 4.50 Test setup for residual tensile strength tests.

cool down to *RT*, while in the second method, the coupons were quenched in water to cool down to *RT*. No load was applied on the coupons during heating or cooling phase. Following cooling to *RT*, residual tensile strength tests were carried out on the coupons as per ASTM E8 provisions (ASTM, 2005) using MTS-810 machine test.

The placement of coupon for residual tensile strength test setup is shown in Fig. 4.50 Displacement control loading was applied at a displacement rate of 0.002 mm/s up to 3 mm displacement (yielding), and then the rate of displacement rate was raised to 0.021 mm/s up to rapture of the steel coupon.

4.8.3 Residual strength

The load–displacement data recorded during uniaxial residual tensile strength tests are utilized to generate residual stress–strain curves after cooling down from various target temperatures. The residual stress–stain response of A572 specimens following either air cooling or water quenching is plotted in Fig. 4.51A,B, respectively. The residual yield strength is evaluated at 0.2% strain offset, while maximum stress attained during the test is taken as the ultimate stress. The significant points on residual stress–strain curve in air-cooled and water-quenched specimens, as obtained from uniaxial tensile strength tests, are summarized in Table 4.21,

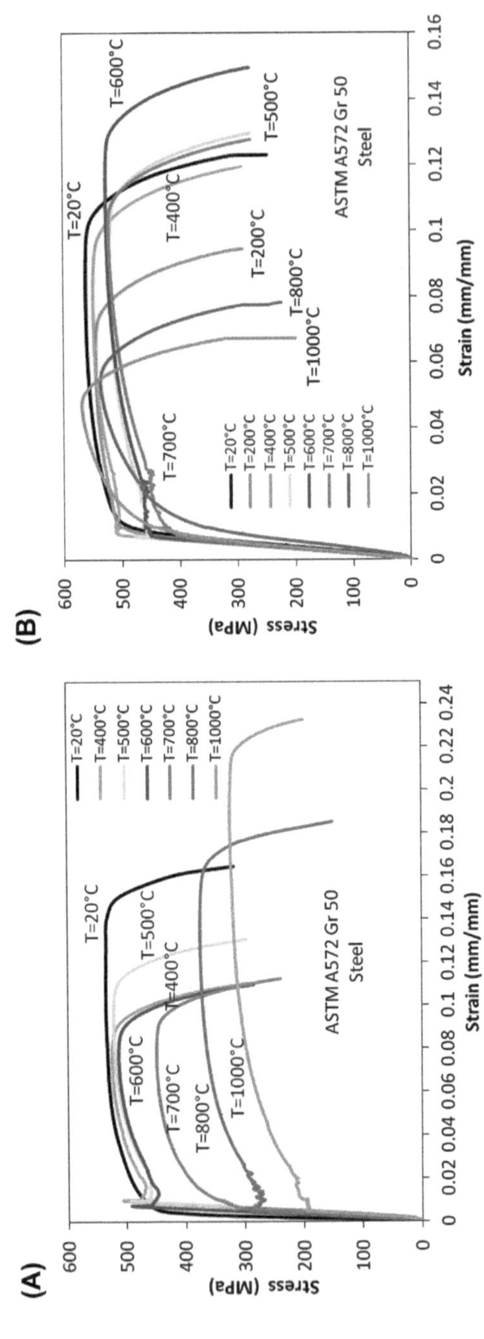

Figure 4.51 Residual stress–strain response of A572 Gr. 50 steel after heating. (A) Air cooling and (B) quenching with water.

Table 4.21 Summary of high-temperature residual tensile strength tests on A572 Gr. 50 steel after air cooling.

Temperature (°C)	$f_{y,R,0.2\%}$ (MPa)	$f_{u,R}$ (MPa)	Rupture strain	Elongation (%)	Reduction in area (%)	$E_{,R}/E_{20}$	$f_{y\ R}/f_{y20}$	$f_{u,R}/f_{u20}$
20	452.5	534.7	0.164	8.2	74.4	1.00	1.00	1.00
400	477.9	507.8	0.110	5.6	68.6	1.20	1.06	0.95
500	465.9	522.3	0.130	6.5	69.6	1.17	1.03	0.98
600	452.3	513.5	0.110	5.7	72.5	1.32	1.00	0.96
700	325.5	449.1	0.112	6.1	75.6	1.29	0.72	0.84
800	272.8	374.2	0.185	9.3	77.4	1.26	0.60	0.70
1000	204.0	324.5	0.232	11.3	78.3	1.13	0.45	0.61

respectively. The strength and stiffness factors given in these tables are calculated as a ratio of residual yield strength to that of yield strength at *RT*.

The residual strength factors for heated A572 steel coupons after air cooling and water quenching are plotted in Fig. 4.52 It can be seen that A572 steel recovers almost all (100%) of its *RT* strength (both yield and ultimate) when it is heated to a temperature below 600°C and then cooled down to *RT*, regardless of the method of cooling (air cooling or water quenching). But, A572 steel loses up to 40% of its *RT* yield strength and up to 30% of its *RT* ultimate strength when it is heated to 800°C and then air cooled to ambient temperature. However, when the same steel is heated to 800°C and then cooled through quenching in water, a 75% of its *RT* yield strength and almost 95% of its *RT* ultimate strength are recovered.

Results from residual tensile strength tests tabulated in Table 4.21 indicate that the residual modulus of elasticity of A572 steel is fully reversible when heated to temperature 700°C or below and cooled down to *RT* regardless the method of cooling. When the steel is heated to 800°C and then cooled through water quenching, 45% reduction in residual elastic modulus is observed. In addition, heating A572 steel and cooling to *RT* through water quenching result in significant ductility (elongation) reduction as compared with heating and air cooling, which experiences lower reduction in ductility. This can be clearly seen by comparing strain level at rupture in Fig. 4.52A,B for both air cooling and water quenching test coupons. It can be seen that the rapture strain decreases when test coupons are heated to temperature up to 700°C and then air cooled to *RT*; however, this trend alters for heating temperature beyond 700°C, in which the rapture strain increases for temperatures of 800 and 1000°C. On the

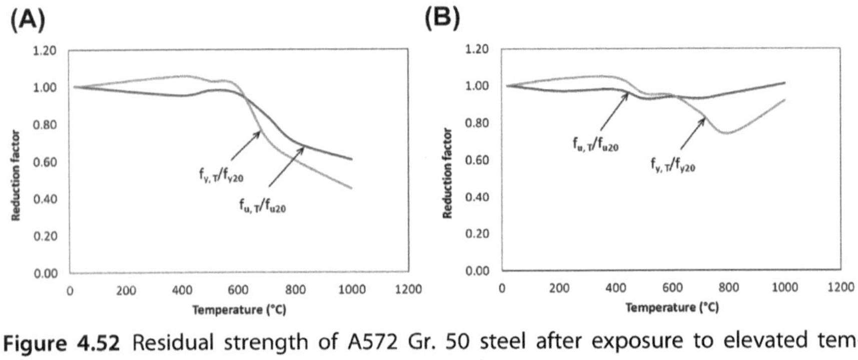

Figure 4.52 Residual strength of A572 Gr. 50 steel after exposure to elevated temperature. (A) Air cooling and (B) quenching with water.

other hand, significant reduction in rapture strain (ductility) is observed when test coupons are heated to 800 and 1000°C and then cooled through water quenching.

4.8.4 Temperature-dependent mechanical properties of A572 steel

For deriving high-temperature properties specific to A572 steel, a nonlinear regression analysis, with least sum of squares of errors, is carried out on the data generated from high-temperature tensile and residual strength tests on A572 Gr. 50 steel coupons. Using the results from this analysis, the following relations are proposed for strength and stiffness reduction factors at elevated temperature, as well as residual strength reduction factors of A572 Gr. 50 steel after cooling to RT under different cooling regimes. The proposed relations can be utilized for undertaking fire resistance analysis or to evaluate the residual capacity of fire-exposed steel structural members made of A572 steel.

4.8.5 Proposed reduction factors for residual strength properties

For evaluating residual strength of A572 Gr. 50 steel after heating and cooling to RT through air cooling or water quenching, the following relations are proposed:

(a) Air cooling after heating:

For $20°C \leq T < 400°C$,

$$\frac{f_{ypR}}{f_{y20}} = 1.0 \tag{4.26}$$

For $400°C \leq T_s \leq 1000°C$,

$$\frac{f_{ypR}}{f_{y20}} = (1.0e-8)T_s^3 - (2.12e-5)T_s^2 + (1.31e-2)T_s - 1.43 \tag{4.27}$$

$$\frac{f_{upR}}{f_{u20}} = (7.64e-9)T_s^3 - (1.68e-5)T_s^2 + (1.1e-2)T_s - 1.26 \tag{4.28}$$

(b) Water quenching after heating:

For $20°C \leq T_s \leq 1000°C$,

$$\frac{f_{ypR}}{f_{y20}} = (5.94e - 12)T_S^4 - (9.71e - 9)T_S^3 + (4.04e - 6)T_S^2 - (3.6e - 4)T_S + 1.0 \quad (4.29)$$

$$\frac{f_{upR}}{f_{u20}} = 0.95 \quad (4.30)$$

Predictions from the proposed relations (Eqs. 4.26–4.30), namely, residual strength for both air cooling and water quenching with corresponding test data, are compared with published data for high strength S690 steel, as well as conventional A992 steel, as shown in Figs. 4.53 and 4.54 (Qiang et al., 2012; Lee et al., 2012). From Fig. 4.53A,B, conventional A992 steel can recover up to 80% of its RT yield strength and 90% of its RT

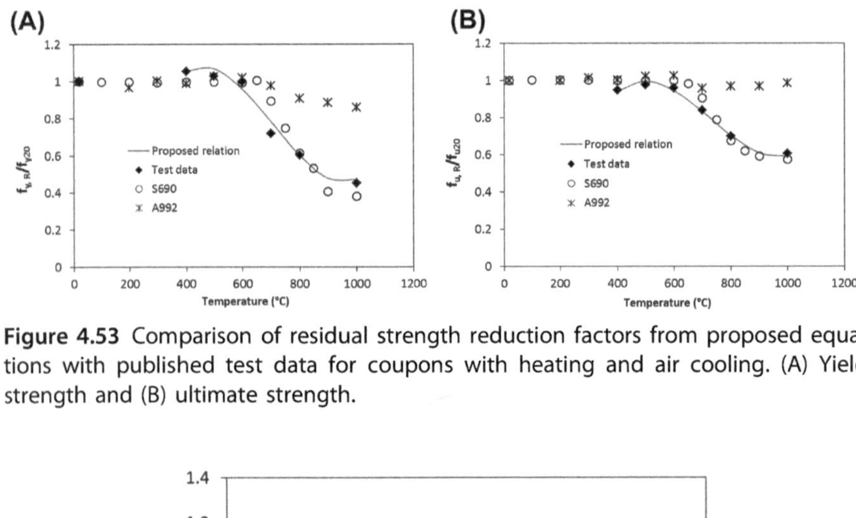

Figure 4.53 Comparison of residual strength reduction factors from proposed equations with published test data for coupons with heating and air cooling. (A) Yield strength and (B) ultimate strength.

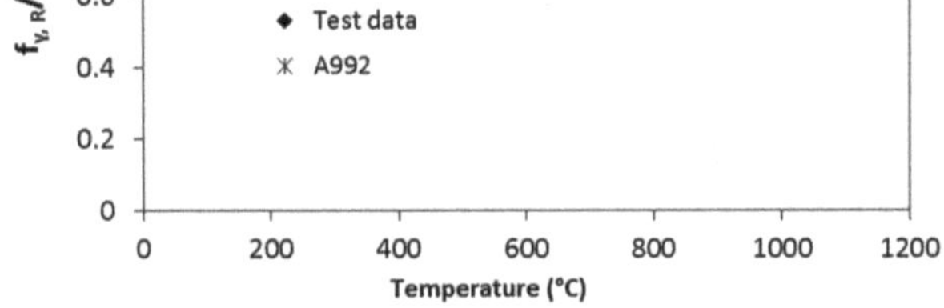

Figure 4.54 Comparison of residual strength reduction factors from proposed equations with published test data for coupons with heating and water quenching.

tensile strength (ultimate strength) when heated to temperatures up to 1000°C and cooled down to *RT*. This is as compared with 40% yield strength and 60% ultimate strength recovery for high-strength, low-alloy A572 steel. On the other hand, Fig. 4.54 shows less difference in residual strength reduction factors between conventional A992 and A572 steel, when water quenching is used to cool down heated (fire-exposed) steel.

4.9 Residual properties of high strength RQT-S690 high strength steel

Heat treated high strength steel (HSS) has attracted more and more applications in both on-shore and off-shore structures. According to the classical theories of metallurgy, the hardening effects in the microstructures that grant HSS its high strength may be weakened or eliminated by heat. Structures built using such materials may behave differently after they are heated and cooled down (e.g. during and after a fire situation). An experimental study was carried out by Chiew et al. (2014) to investigate the mechanical properties of reheated, quenched and tempered (RQT) HSS plate manufactured to grade S690 at elevated temperatures as well as its residual strength after heating and cooling down.

4.9.1 The test device

The heating device attached to the tensile test machine is a split tube furnace that can heat the specimen at a rate of 30°C/min up to 1050°C. Inside the furnace, there were three temperature detectors monitoring the inner temperatures from top to bottom. In addition, a thermocouple was attached to the specimen directly to measure its real-time temperature profile. The configuration of the setup is shown in Fig. 4.55.

To ensure consistence in specimen dimension and test environment, in the postfire residual strength test, the postfire specimens were tested in the same tensile test machine with the furnace turned off. The laboratory heating furnace Nabertherm LH216 with a maximum heating capacity of 1200°C and robust refractory bricks inside was employed to simulate the heating and cooling in fire condition (Fig. 4.56). The internal dimensions (width × depth × height) of the oven were 500 × 500 × 700 mm. As the thermal inertia of these specimens is very small compared with that of the furnace and they are located so as not to impede the airflow, a homogenous atmosphere has been achieved. To structurally isolate the members during heat treatment, specimens were simply supported on ceramic roller supports to allow flexibility of the specimens during thermal expansion as maintained.

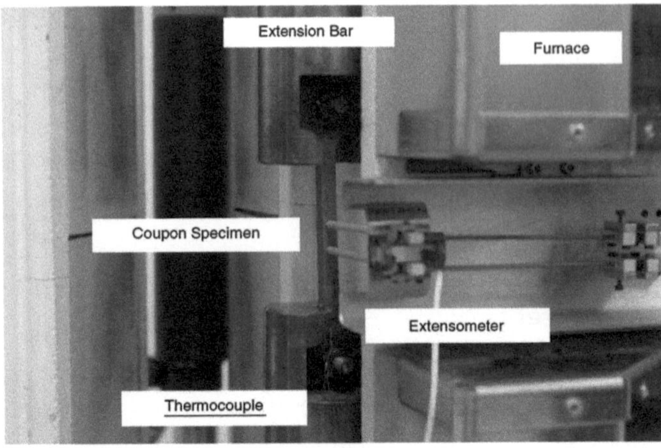

Figure 4.55 The high-temperature tensile test machine with heating chamber.

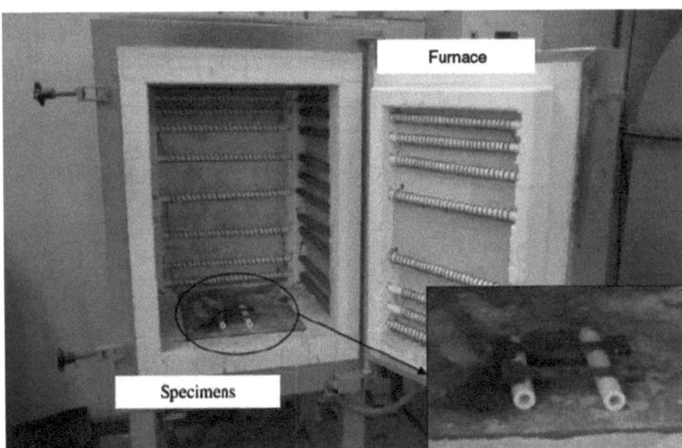

Figure 4.56 The furnace employed for heating the specimens.

4.9.2 Tested materials and specimens

RQT steel plates exhibit better homogeneity in through-thickness mechanical properties compared with traditional directly quenched and tempered steel plates. The materials and specimens studied in this chapter have a specified nominal yield strength of 690 N/mm^2, tensile strength between 790 and 930 N/mm^2, and elongation capacity around 15%. These RQT HSS plates comply with the EN 10025-6 grade S690 specification (BSI, 2004), which is approximately equivalent to the ASTM A514 steel (ASTM, 2009).

Table 4.22 Characteristic strengths of RQT-S690 high strength steel at room temperature (MPa).

Strengths	$f_{0.2}$	$f_{0.5}$	$f_{1.5}$	f_2	f_u	E	Elongation (%)
RQT-S690 HSS	769.0	772.6	810.2	817.1	849.8	201.3	14.7

The mechanical properties of RQT-S690 HSS at *RT* acquired from coupon tensile test are shown in Table 4.22. The chemical composition of the tested RQT-S690 HSS with thickness between 8 and 30 mm is shown in Table 4.23, in which both the maximum values and the typical values are provided. The calculated carbon equivalent (CE) value corresponding to the maximum element content state (second row of Table 4.23) is 0.43, while that corresponding to the typical values (last row of Table 4.23) is 0.4. Note that the CE value corresponding to the typical value is within the suggested welding limit of 0.40 by the AWS (2008), and this grants RQT-S690 HSS better weldability compared with other high-strength, high-alloy steels.

The specimens were cut from 8 mm thick RQT-S690 HSS plates in the longitudinal direction. The configuration was designed according to BS EN 10002-5, as shown in Fig. 4.57. Two holes were drilled at both ends to fix the specimen into the load cell during the elevated temperature test. Both the elevated temperature and the postfire residual strength tests adopted the same specimen configuration. During the heat treatment for postfire test specimens, it was found that oxidized layer would form on the surface if the coupons were heated to temperatures above 600°C. Hence, those specimens were polished and remeasured before the first postfire tensile test.

4.9.3 Test results

4.9.3.1 General descriptions

The stress—elongation curves of RQT-S690 HSS after heating and cooling down once are shown in Fig. 4.58. The temperatures in the legends refer to the highest temperature to which the specimens were once heated by the furnace. In general, it is observed that RQT-S690 HSS obtained better ductility and higher tensile to yield ratio but a lower strength after the heat treatment. The mechanical properties were stable until 400°C after which the strengths kept decreasing but ductility tended to increase. 900°C was literally the end of the property change. The difference between

Table 4.23 Chemical composition of RQT-S690 high strength steel (%).

Elements	C	Si	Mn	S	P	Cr	Mo	Nb	Ti	V	Ni	Cu	B	Al	CE
Maximum	0.20	0.50	1.60	0.010	0.025	0.25	0.20	0.060	0.040	0.08	0.70	0.20	0.004	0.060	**0.43**
Typical	0.14	0.40	1.35	0.003	0.012	0.01	0.12	0.035	0.025	0.05	0.01	0.01	0.002	0.035	**0.40**

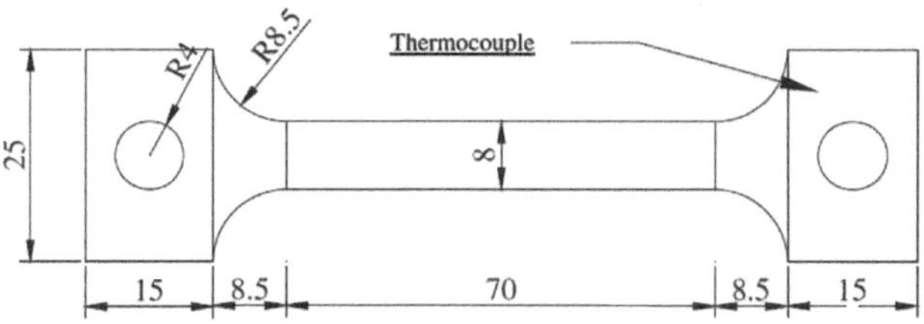

Figure 4.57 The dimensions of the specimen.

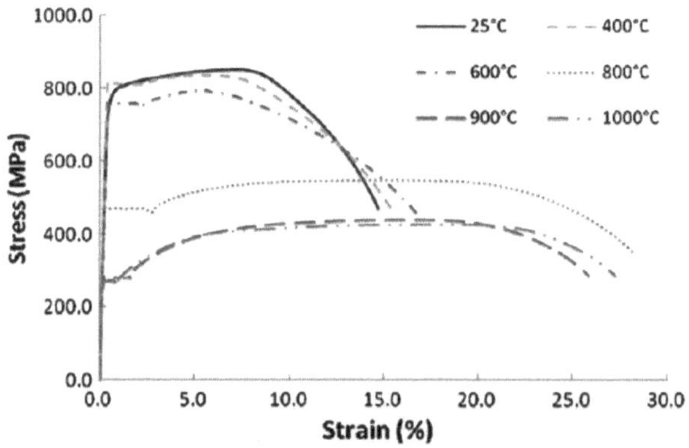

Figure 4.58 Stress–strain curves of RQT-S690 after heated to different temperatures and cooled down once.

stress–strain curve of 900 and 1000°C is only a small amount increase in ductility for the 1000°C curve, which corresponds to the fact of austenite transformation in the microstructure around 900°C (Bhadeshia and Honeycombe, 2006).

The transformation in strength and ductility also happened to S355J2H CFS (Outinen, 2007) and the quenched and tempered steel S690QL HSS (Qiang et al., 2012). After heated to 1000°C and cooled down, S690QL HSS showed yield strength of 320 MPa, tensile ratio of 1.5, and ductility of 21%, while the corresponding values for RQT-S690 HSS are 265 MPa, 1.61%, and 27.3%, respectively. Accordingly, it was confirmed again that heat treatment/work-hardened steels would "return back" into their "parent" steel before hardening if heated properly. Furthermore, assuming the values just quoted were exactly the raw data of the "parent" steel before quenching and tempering, the hardening extent for each material could be calculated as the ratio between nominal strength at *RT* before heating and the strength after heated to 1000°C and cooled down once. If yield strength could be used in the calculation to replace the nominal strength at *RT*, the hardening extent of S690QL HSS and RQT-S690 HSS would be 247%

and 290%, respectively. Thus, this apparently explains why the deterioration of RQT-S690 HSS was more serious than other materials at elevated temperatures.

4.9.3.2 Analysis of the deterioration of mechanical properties

To further analyze the dependence of strength to temperature history, the characteristic strength reduction factor method was employed again. The selected parameters included the yield strength, 2.0% strain strength, ultimate strength, and elastic modulus, as presented in Table 4.24. Strengths corresponding to 0.5% and 1.5% strains were not adopted in the analysis because of the uncertainty in the length of the yield stage, i.e., sometimes they were still within the yield stage and sometimes do not.

As can be seen in Fig. 4.59A, despite the fact that the curve of ultimate strength was higher than the other two at 900 and 1000°C, the difference among those curves was not big. The reduction of 0.2% and 2.0% strength almost coincided, while the less deterioration of ultimate strength explained the growing tensile ratio. The strengths did not change remarkably unless the heating temperature went above 600°C (less than 10% at 600°C), which was the same for S690QL HSS (Qiang et al., 2012). If the heating temperature was between 600 and 900°C, the strengths decreased rapidly. Among all the reduction factor curves, the 2.0% proof strength curve is the lowest, thus it is recommended to represent the strength—heating temperature relationships.

Qiang et al. (2012) reported the strength reduction of S690QL HSS after being heated and cooled down, including 0.2%, 0.5%, 1.5%, and 2.0%

Table 4.24 Reduction factors of characteristic strengths and elastic modulus.

Temperature (°C)	$f_{pT}/f_{p,n}$	$f_{0.5,t}/f_{0.5,n}$	$f_{1.5,t}/f_{1.5,n}$	$f_{2.0,t}/f_{2.0,n}$	$f_{u,t}/f_{u,n}$	E_t/E_n
25	1.00	1.00	1.00	1.00	1.00	1.00
100	0.99	1.00	0.96	0.96	0.96	1.01
200	0.94	0.98	0.98	0.97	0.95	1.02
300	0.89	0.95	0.98	0.98	0.97	0.96
400	0.85	0.89	0.87	0.87	0.84	1.01
450	0.61	0.66	0.68	0.67	0.64	0.91
500	0.40	0.51	0.52	0.51	0.50	0.77
600	0.14	0.15	0.16	0.16	0.19	0.66
700	0.05	0.06	0.07	0.07	0.10	0.38
800	0.05	0.06	0.07	0.07	0.07	0.29

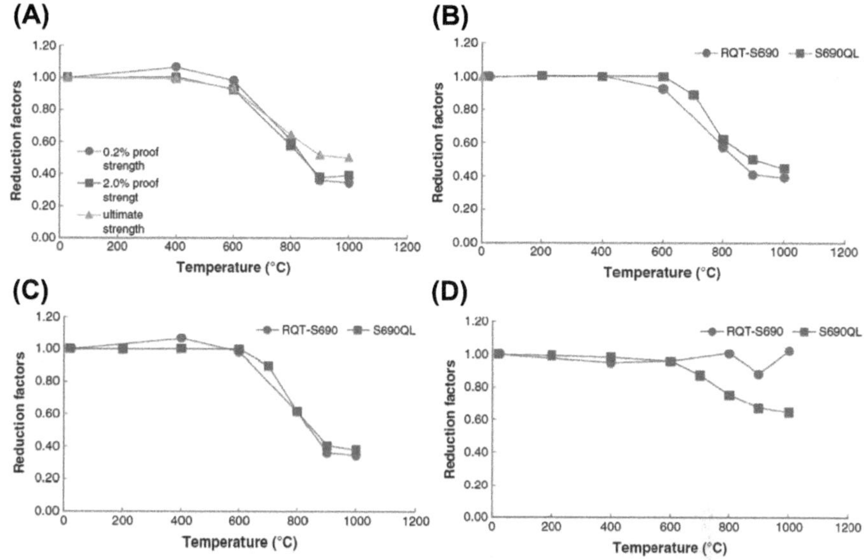

Figure 4.59 Residual factors of RQT-S690 high strength steel (HSS) after heated and cooled down once. (A) Different strengths of the RQT-S690 HSS; (B) lower yield strength; (C) 2.0% strain strength; and (D) elastic modulus.

strain strengths. This study only took the lower yield strength, 2.0% strain strength, and the ultimate strength into consideration because 0.5% and 1.5% strains were sometimes still within the yield stage, as can be seen in Fig. 4.58. The comparisons between RQT-S690 HSS and S690QL HSS regarding postfire residual yield strength (lower yield strength for RQT-S690 HSS and the 0.2% strain strength for S690QL) and 2.0% strain strength are shown in Fig. 4.59B,C, respectively. For yield strength, both materials were stable below 600°C. However, the strength reduction of RQT-S690 HSS occurred rapidly for temperatures between 600 and 900°C. Accordingly, the final residual strength of RQT-S690 HSS was lower than that of S690QL HSS. As for the 2.0% strain strengths, the deterioration starting temperature for RQT-S690 HSS shifted to be lower than 600°C, so did the ultimate strength. This change revealed that the deterioration of mechanical properties would start from tensile ratio and strength at large strain state. In short, the heating temperature–strength relationships of both RQT-S690 HSS and S690QL HSS were quite similar, though the deterioration of RQTS690 HSS seemed to be slightly more serious between 600 and 900°C. The largest difference between the two lay in the elastic modulus. As shown in Fig. 4.59D, after heated once and cooled down to *RT*, the elastic modulus of RQT-S690 HSS is generally unchanged except for slight fluctuation when the reheating temperature is 900°C, while a gradual decrease could be observed in the reduction curve of S690QL HSS.

4.9.3.3 The influence of repeated heating

According to the classic theory of metallurgy, the effects of heating to quenched and tempered steel is highly dependent on the maximum heating temperature and time. Longer time or higher temperature may lead to the same result. Therefore, to verify the stability of mechanical properties after heating and investigate the influence of repeated heating and cooling, another group of specimens were heated and cooled twice before tensile test at *RT*. The heating procedures in the second run were the same as those in the first run. The reduction factors ($f_{y,t}/f_{0.2,n}$, $f_{2.0,t}/f_{2.0,n}$, $f_{u,t}/f_{u,n}$, E_t/E_n) of RQT-S690 HSS after heated and cooled down twice are presented in Fig. 4.60A–D and are compared with those only heated once. Similar reduction regulation in strength and elastic modulus was observed.

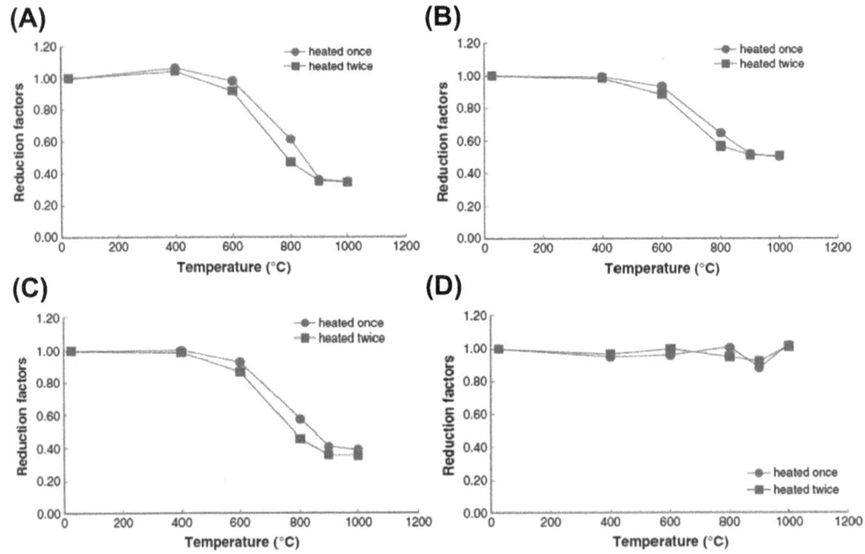

Figure 4.60 Comparison of the heating the specimen once and twice. (A) Lower yield strength; (B) 2.0% proof strength; (C) ultimate strength; and (D) elastic modulus.

However, when heated and cooled twice, the temperature associated with structural deformation was lowered to 400°C and thus assuming a critical temperature of 600°C was not conservative for long time heating. The influence of repeated heating almost existed only in the descent part, i.e., 400–900°C. However, the strengths before the deterioration started and the residual strength after the deterioration remained the same. As the strengths decreased further in this stage, it was not surprising to find out that the ductility at corresponding temperatures improved accordingly. In addition, it was worth noting that the gap between these two series of strength reduction curves was getting closer from the lower yield strength to the ultimate strength. As a result, the tensile ratio was further improved after repeated heating. However, such beneficial changes only took place from 400 to 900°C, accompanied with a huge loss in strength.

4.9.4 Test specimens

The chemical compositions of Grade 1200 and Grade 800 steels considered in this chapter and their manufacturing process are similar to those presented in Ref (Azhari et al., 2015). The test specimens were sectioned from the ultra-high strength steel (UHSS) and HSS tubes using high-pressure water jet cutting with nominal external diameter of 76.1 mm and wall thickness of 3.2 mm. As shown in Fig. 4.61, the shape and dimensions of the specimens are determined in accordance with ASTM E8 (2011) The specimens are cut from 2 strips located at right angles (90 degrees) to the tube weld line as indicated in Fig. 4.62. To fit the samples into the grips for the tensile tests, they are mechanically flattened based on the instructions given in AS1391 (1991). Each test is repeated until the results obtained from two similar tests have a difference of less than 3%. The cross-sectional area of the gauge length of the test specimens is determined according to AS1391 (1991).

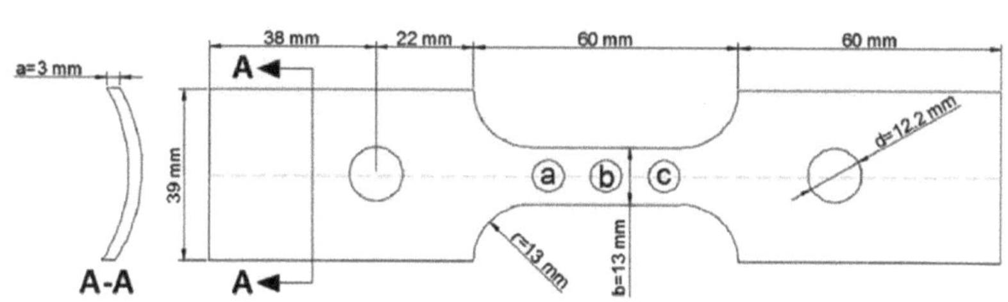

Figure 4.61 Dimensions of test specimens.

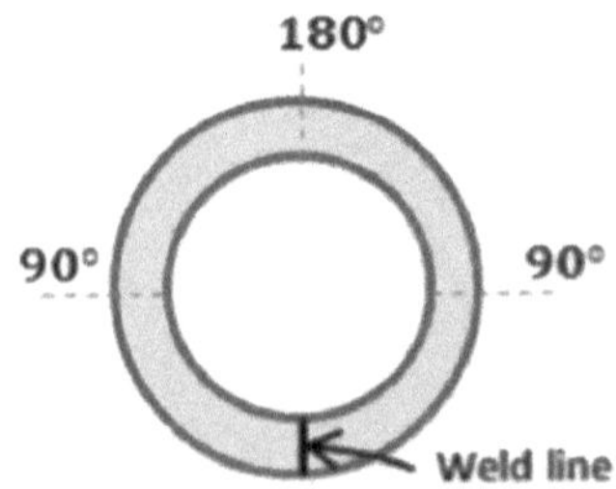

Figure 4.62 The tube section.

4.9.5 Testing method

4.9.5.1 Heat-up tests

For the heat-up tests, the SF-16 split furnace shown in Fig. 4.63 is installed on the Instron 5982 100 kN testing machine. In this furnace, a resistance wire is wound on to a recrystallized alumina tube in three independent zones to form the furnace element. This three-zone format allows the user to modify the furnace temperature gradient, creating a uniform central zone. The furnace can heat the specimen up to 900°C at a rate of ~20°C/min.

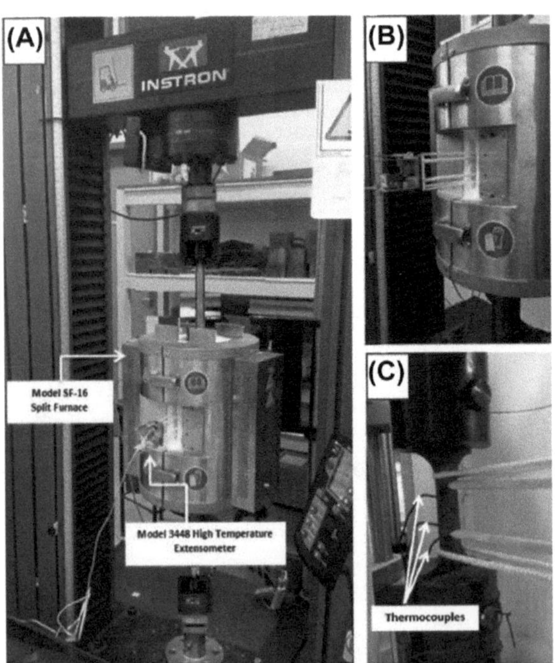

Figure 4.63 (A) and (B) Test setup from different views; (C) inside the furnace.

As illustrated in Fig. 4.63C, three thermocouples labeled as T_1, T_2, and T_3 are attached to three points (a, b, and c) on the gauge length of the tensile coupon shown in Fig. 4.61 to record its real-time temperature profile. Nine different target temperatures are considered in this study, i.e., *RT*, 300, 470, 540, 600, 650, 700, 750, and 800°C. It is worth mentioning that the tests under fire temperatures below 600°C have been already conducted by the authors (Azhari et al., 2015) for the same UHSS tube material. However, the device employed in the previous study to heat up the specimens was an environmental chamber that uses a different heating mechanism from the furnace used in the present study. The environmental chamber uses forced convection heat transfer method to provide an optimum temperature distribution. Forced convection stands for the transfer of heat by motion of a fluid that is instigated by application of an external motive force (Narang, 2005). In an environmental chamber, the fluid transferring the heat is the air and the external force moving the air is generated by the chamber's fan. However, in model SF-16 split furnace, the specimen gripped inside this furnace is heated through a combination of convection and radiation, dependent on the test temperature. Thermal radiation is the transfer of heat by means of the emission of photons in electromagnetic waves that carry the heat energy away from the source to the surrounding objects.

It is well-known that temperature measurements in heating devices that use radiation to transfer the heat to the specimen can be affected by radiation errors. From the literature (Brohez et al., 2004; Harada et al., 1997), it was concluded that the main factor controlling the radiation errors is the thermal inertia of the furnace wall lining material. Harada et al. (1996) investigated the reproducibility of the fire tests data for two commonly used wall lining materials, i.e., fire brick and ceramic fiber. According to their results, because of small thermal inertia of ceramic fiber, the range of data scatter in the ceramic fiber wall furnaces was considerably lower than that in the fire brick wall furnaces. In the furnace used for this study, high-performance ceramic fiber insulation is used to reduce heat losses and provide fast heating rates. To obtain a uniform set of data, all fire temperatures ranging from 300 to 800°C are considered in this study. Because both UHSS and HSS tube specimens lose ~97% of their strength at temperatures above 800°C, the maximum temperature studied in this research does not exceed 800°C. When the difference between the temperature T_i of each thermocouple and the target temperature T_t is less than 2%, i.e., $|T_i - T_t| \leq 2\%$, it is assumed that the thermal profile is uniform throughout the specimen and the quasi-static tensile test can be started. Hence, by using the Instron 5982 100 kN testing machine, a strain-controlled tensile test with an applied strain rate of $0.005 \pm 0.002\text{min}^{-1}$ (ASTM, 2011) is carried out on the specimen until failure. As for strain measurement of the specimen during the tensile test, the MODEL 3448 high-temperature contact extensometer with approximate gauge length of 25 mm is used. As shown in Fig. 4.63B, the front cutout of the furnace allows the extensometer to enter inside the furnace and be in contact with the specimen gripped inside the furnace.

4.9.5.2 Cooling tests

To evaluate the postfire mechanical response of the UHSS and HSS tubes after cooling from fire temperatures of up to 800°C, the test specimens are heated up to the target temperature as explained in the previous section. Once the temperatures shown by the three thermocouples attached to the specimen are stabilized at the target temperature (T_t), the temperature is held constant for 20 min. The specimen is then allowed to air cool at its own rate by turning off the furnace and opening the door as shown in Fig. 4.64. The strain-controlled tensile test with an applied strain rate of $0.005 \pm 0.002\text{min}^{-1}$ is then conducted on the test specimen once it reaches *RT*.

Figure 4.64 The cooling test setup for the postfire stage ($T_t = 800°\text{C}$).

Finally, to help interpret the results obtained from the cooling tests, microstructural examination of the UHSS materials cooled from some very high fire temperatures is conducted using optical and scanning electron microscopy.

4.9.6 Results and discussion

4.9.6.1 General descriptions

To investigate the effect of high-temperature exposure on mechanical properties of the tested steels, the stress–strain curves, characteristic tensile strengths, strain values, and energy absorption are determined. As the stress–strain curves do not exhibit a pronounced yield point, the tensile strength at different strain levels including 0.2%, 0.5%, 1.5%, and 2.0% strains and also the ultimate tensile strength are reported as characteristic tensile strengths. The values of these parameters are illustrated in a typical stress–strain curve shown in Fig. 4.65 As indicated in Fig. 4.65, the values of 0.2% proof stress ($f_{0.2}$) and 0.5% stress ($f_{0.5}$) are determined by finding the intersection of the stress–strain curve with a line starting from the 0.2% and 0.5% strains, respectively, parallel to the elastic part of the curve. The strain corresponding to the 0.2% proof stress denotes the proof strain (ε_y). It is worth noting that in most research works, the value of $f_{0.5}$ is calculated by simply finding the stress at 0.5% stress level (Chiew et al., 2014; Gunalan

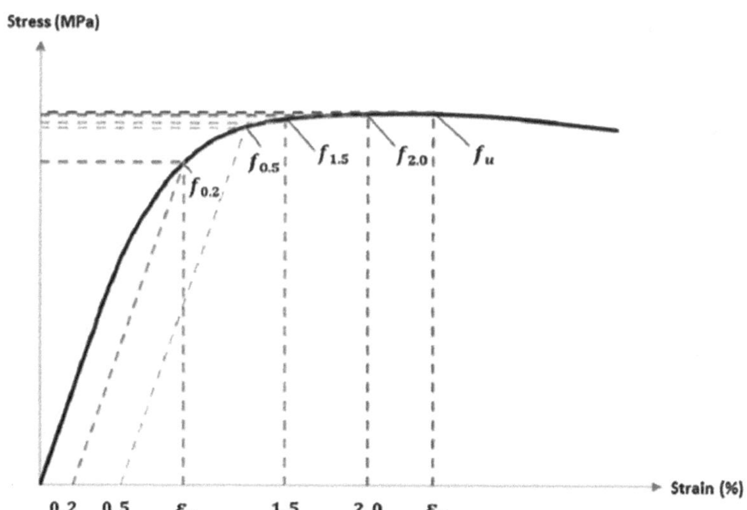

Figure 4.65 Definition of characteristic strengths and strains on a typical stress–strain curve.

et al, 2014). However, as in the stress strain curves obtained in this research, the 0.5% strain is usually within the yield stage, and the $f_{0.5}$ value is found by the aforementioned method. As shown in Fig. 4.65, to find the strength values at 1.5% and 2.0% strains, i.e., $f_{1.5}$ and $f_{2.0}$, the conventional method is applied. The maximum stress value in the engineering stress—engineering strain curve denotes the ultimate tensile strength (f_u) and the corresponding strain value is the uniform elongation (ε_u). Finally, the area under the stress—strain curve starting from zero strain to the uniform elongation (ε_u) represents the energy absorption (U^*).

4.9.6.2 Stress—strain curves

Fig. 4.66 shows the elevated temperature stress—strain curves resulting from the heat-up tests. Because of the changes that occur in the cross section of the specimens during necking and the fact that the cross-sectional geometry during necking depends on specimen size (Ling, 1996), the strains are no longer uniform once necking starts such that readings of the strain values are not accurate. In spite of that, the tensile tests are not terminated and the full engineering stress—strain curves are plotted up to the failure of the material. However, the onset of necking is signified by a black cross sign (X) on the plotted curves. In this study, the UHSS specimen tested at elevated temperature T is shown as UHSS-HT, where H represents the Heat-up test. For example, the stress—strain curve of the UHSS specimen tested at 800°C

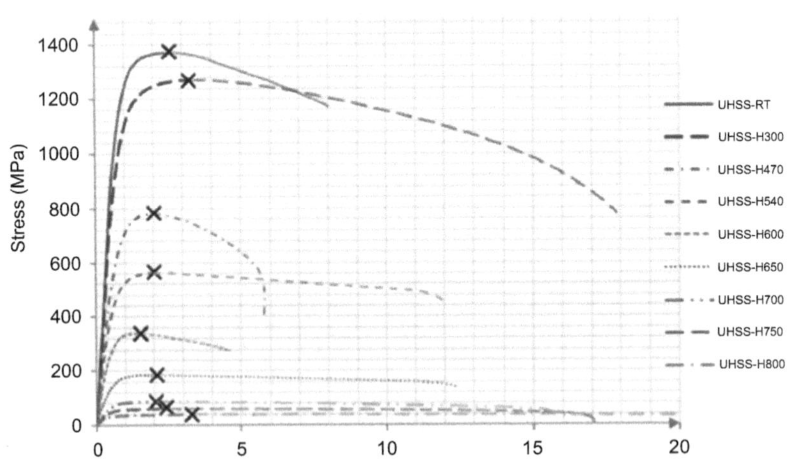

Figure 4.66 Stress—strain curves for ultra-high strength steel specimens at elevated temperatures ranging from 300 to 800°C.

is labeled as UHSS-H800. Also, the UHSS-RT represents the stress—strain curve of the virgin UHSS at *RT*. As illustrated in Fig. 4.66, the strength of UHSS is reduced as the maximum temperature increases. In addition, a considerable drop in strength is observed when the temperature exceeds 600°C.

The stress—strain curves of the UHSS specimens under strain-controlled tensile test at *RT* after being cooled from different fire temperatures are presented in Fig. 4.67 Similar to the curve labeling system explained for the heat-up tests, the stress—strain curve of the UHSS specimens tested at *RT* after cooling from the elevated temperature *T* is labeled as UHSS-CT. For example, UHSS-C800 represents the stress—strain curve of the UHSS specimen cooled from 800°C to *RT*. It is evident from the curves shown in Fig. 4.67 that the strength of UHSS is not regained after cooled from temperatures at or above 470°C. However, according to (Qiang et al., 2012, 2013), the postfire mechanical properties of S460, S690, and S960 steels are not affected until they are exposed to fire temperatures above 600°C. It is clear that the thermal response of the UHSS is fundamentally different to these previously studies steels. It can be also interpreted that the residual strength of UHSS is significantly reduced after being cooled from fire temperatures above 700°C. More detailed explanations about this phenomenon are provided in the microstructure characterization section of this paper. In addition, it is curious that the UHSS-C750 and UHSS-C800

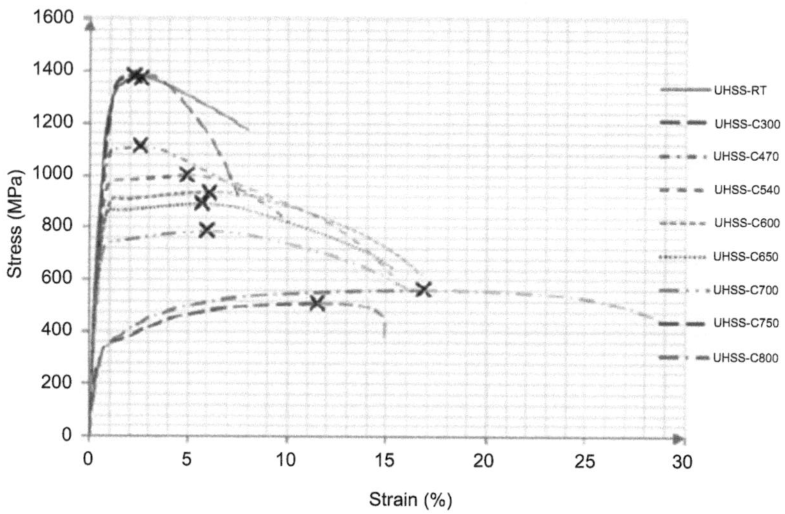

Figure 4.67 Stress—strain curves of ultra-high strength steel specimens at room temperature after cooling from the elevated temperatures ranging from 300 to 800°C.

curves are almost the same, and seemingly the variation of the residual strength of the UHSS with temperature is stabilized when subjected to fire temperatures above 750°C. The reason behind this phenomenon is also explained in the microstructure characterization section.

It is evident that higher the temperature the UHSS specimens have experienced, the more residual uniform elongation they experience when tested after being cooled to *RT*. The time–temperature curves of the UHSS specimens after being cooled from different fire temperatures are summarized in Fig. 4.68 It can be seen that in most of the cooling tests, the approximate time at which the specimen reached its maximum temperature is around 90 min. This is consistent with real fire conditions in which during a certain time, different parts of the structure are heated up to different fire temperatures.

4.9.6.3 Strength

To discuss the variation of characteristic strengths with temperature of the steel specimens in both heat-up and cooling tests, reduction factors have been calculated as the ratio of the strength values at elevated temperatures and after cooling to *RT* to the corresponding strength values of virgin material at *RT*. The characteristic strength reduction factors for the heat-up tests and cooling tests are presented in Tables 4.25 and 4.26. In these Tables, *T* and *CT* indices correspond to the strength values for UHSS specimens tested at elevated temperature *T* and after cooling from temperature *T* to

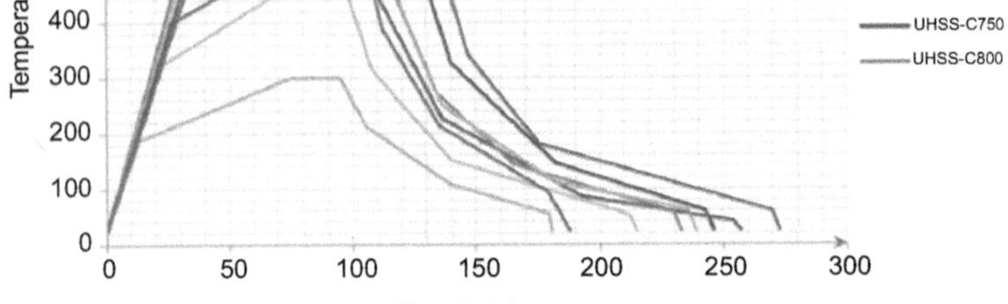

Figure 4.68 Temperature–time profile of ultra-high strength steel specimens cooled from the elevated temperature *T* to room temperature.

Table 4.25 Reduction factors of characteristic strengths for ultra-high strength steel under various fire temperatures.

Temperature	300°C	470°C	540°C	600°C	650°C	700°C	750°C	800°C
$\frac{f_{0.2,T}}{f_{0.2,RT}}$	0.889	0.496	0.368	0.222	0.091	0.041	0.032	0.025
$\frac{f_{0.5,T}}{f_{0.5,RT}}$	0.897	0.550	0.397	0.248	0.123	0.057	0.043	0.026
$\frac{f_{1.5,T}}{f_{1.5,RT}}$	0.896	0.570	0.411	0.252				
$\frac{f_{2,T}}{f_{2,RT}}$	0.912	0.569	0.409					
$\frac{f_{u,T}}{f_{u,RT}}$	0.925	0.568	0.410	0.247	0.134	0.064	0.047	0.030

Table 4.26 Reduction factors of characteristic strengths for ultra-high strength steel after cooling from various fire temperatures.

Temperature	300°C	470°C	540°C	600°C	650°C	700°C	750°C	800°C
$\frac{f_{0.2,T}}{f_{0.2,RT}}$	1.00	0.850	0.774	0.650	0.665	0.575	0.264	0.258
$\frac{f_{0.5,T}}{f_{0.5,RT}}$	1.00	0.840	0.750	0.687	0.664	0.565	0.271	0.269
$\frac{f_{1.5,T}}{f_{1.5,RT}}$	1.02	0.818	0.730	0.676	0.650	0.558	0.273	0.285
$\frac{f_{2,T}}{f_{2,RT}}$	1.02	0.808	0.723	0.667	0.645	0.553	0.281	0.301
$\frac{f_{u,T}}{f_{u,RT}}$	1.03	0.809	0.729	0.681	0.653	0.574	0.372	0.411

RT, respectively. In addition, the *RT* index refers to the strength values of the virgin material tested at *RT*. For some heat-up tests, as the 1.5% and 2.0% strains are within the postnecking part of the stress—strain curve of the specimen, the reduction factors of their corresponding strengths, i.e., $f_{1.5,T}/f_{1.5,RT}$, and $f_{2.0,T}/f_{2.0,RT}$, are not reported in Table 4.25. Based on the data shown in Table 4.25, the UHSS specimens tested at temperatures above 700°C lose more than 94% of their strength. The reduction of the values of $f_{0.2}$, $f_{0.5}$, and f_u occurs to the same extent.

As can be seen from the results shown in Table 4.26, the characteristic strength reduction factors obtained from the cooling tests have a decreasing trend with increasing temperature. There is a significant drop in the reduction factors of the specimens cooled from 750°C and above. Because the strength reduction factors of UHSS specimens after cooling from 750 to 800°C are almost the same, it can be concluded that they have been stabilized such that they will not experience any higher reduction in their strength when they are cooled from higher fire temperatures. This phenomenon is rationalized in the microstructure characterization section based on the plots of the thermodynamic stability of the phases present in the UHSS microstructure. Elaborating deeper into the results shown in Table 4.26, it can be seen that for specimens cooled from 750–800°C to *RT*, the values of ultimate tensile strength are not reduced to the same extent as other characteristic strengths. The higher values of the ultimate tensile strength for the UHSS specimens cooled from these two temperatures show that the postyield part of the UHSS-C750 and UHSS-C800 stress–strain curves has a more round shape compared with that of other curves as shown in Fig. 4.67 This is a signature of a major microstructural change in the steel leading to a structure capable of exhibiting significantly more strain hardening capacity than the virgin material.

To illustrate the effect of cooling phase of a fire on the UHSS's strength, the reduction factors of the 0.2% proof stress ($f_{0.2}/f_{0.2,RT}$) and the ultimate tensile strength ($f_u/f_{u,RT}$) for both heat-up and cooling tests are plotted in Figs. 4.69 and 4.70 with respect to the maximum temperature the specimens have experienced. For the purposes of comparison, the values provided by the AS4100 (2005) and AISC (2005) standards for the 0.2% proof strength and the ultimate tensile strength reduction factors of mild steel at elevated temperatures are presented. The vertical distance between the curves obtained from the heat-up and cooling tests represents the strength recovered (compared with the elevated temperature strength) during the cooling phase of a fire.

As shown in Figs. 4.69 and 4.70, using the values given by the standards to predict the tensile mechanical behavior of UHSS structural members at

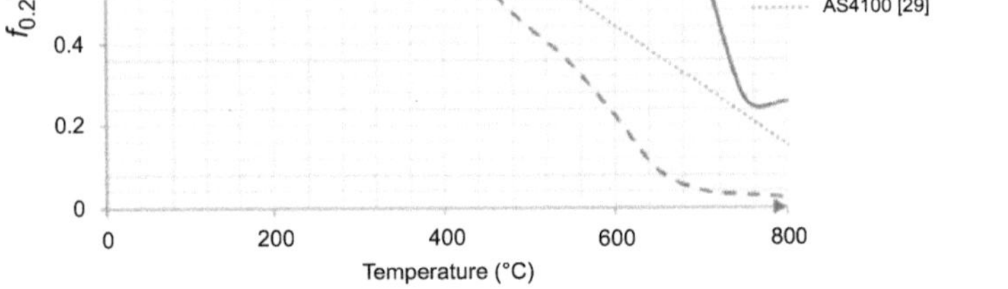

Figure 4.69 Variation of 0.2% proof stress reduction factor for ultra-high strength steel under in-fire and postfire conditions.

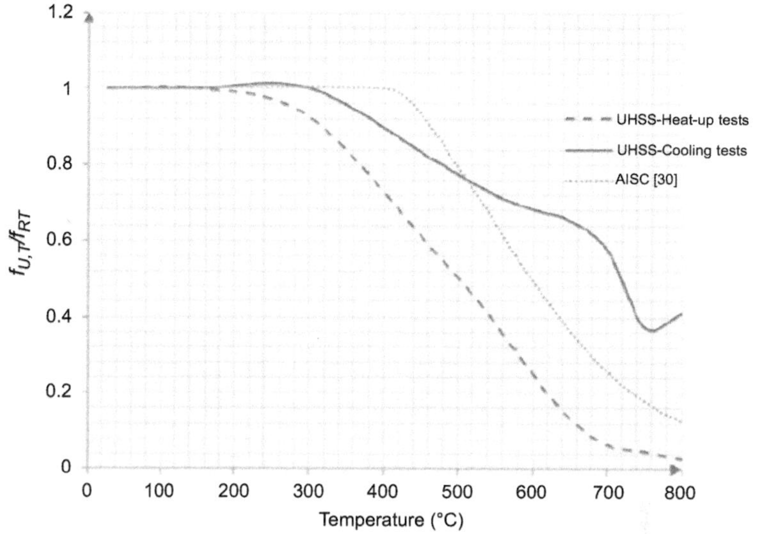

Figure 4.70 Variation of ultimate tensile strength reduction factors for ultra-high strength steel under in-fire and postfire conditions.

fire temperatures may lead to an unsafe design. On the other hand, there is no provision in the building codes for the UHSS materials under cooling phase of a fire. To predict the in-fire and postfire behavior of the UHSS structural members, a new set of strength design formulations must be proposed.

4.9.6.4 Ductility

Several parameters can be calculated to discuss the ductility of a material under fire conditions. In this chapter, the first parameter used to indicate the in-fire/postfire ductility of the UHSS is the ratio of the uniform elongation (ε_u) of the UHSS specimens obtained from both heat-up and cooling tests to that of the virgin UHSS tested at *RT* ($\varepsilon_{u,RT}$). In Fig. 4.71, the variation of $\varepsilon_u/\varepsilon_{u,RT}$ ratios with respect to the maximum temperature the material has experienced are plotted. The results show that the variation of uniform elongation of the UHSS specimens tested at elevated temperatures do not share the same trend with those of the UHSS specimens tested after being cooled to *RT*. While the values of $\varepsilon_u/\varepsilon_{u,RT}$ ratio for the UHSS specimens under cooling tests increase up to 6.5 as the temperature reaches 800°C, their maximum value for the UHSS specimens under heat-up tests is only 1.27. The elongation of a material depends strongly on the strain

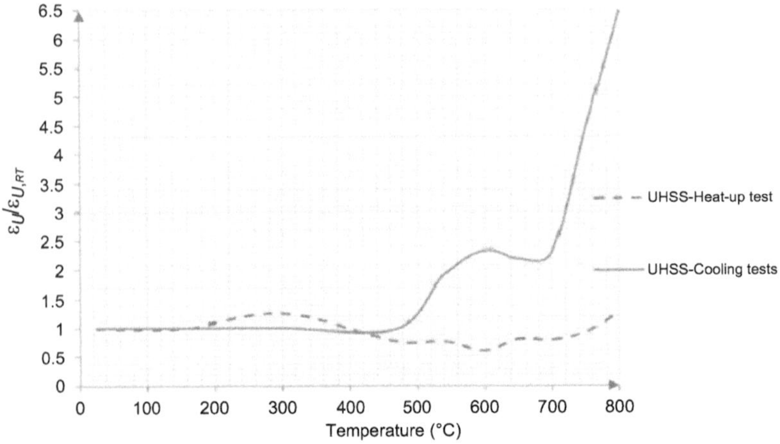

Figure 4.71 Variation of uniform elongation reduction factors for ultra-high strength steel under in-fire and postfire conditions.

hardening capacity of its structure, and the rapid increases in the elongation after cooling from temperatures above 700°C again signifies a major change in the structure of these steels when heated above ~700°C.

Another parameter characterizing the ductility or the energy absorption of the material is the strain energy U^* absorbed by the material. Because the engineering stress and strain values are calculated based on the original dimensions of the tested specimen, they do not take into account the changes in the cross-sectional area of the specimen during the tensile test (ASM, 2002). Therefore, evaluation of the exact energy absorption up to the fracture of the tested specimens cannot be conducted based on the engineering stress—strain curves. Considering that the accurate determination of the true stress—strain curve during necking of the specimens is still under investigation (Fonseca and Dario, 2014; Wang and Tong, 2015; Zhao et al., 2016) and the fact that response before necking is size-independent, in this chapter, the energy absorption is calculated up to the onset of necking just as an indicator for the specimen's capability to absorb energy, which is defined by

$$U^* = \int_0^{\varepsilon u} \sigma d\varepsilon \tag{4.31}$$

where σ and ε are the engineering stress and strain of the material, respectively. The values of U^* obtained from the heat-up tests and cooling tests are normalized with respect to that of the virgin UHSS tested at RT

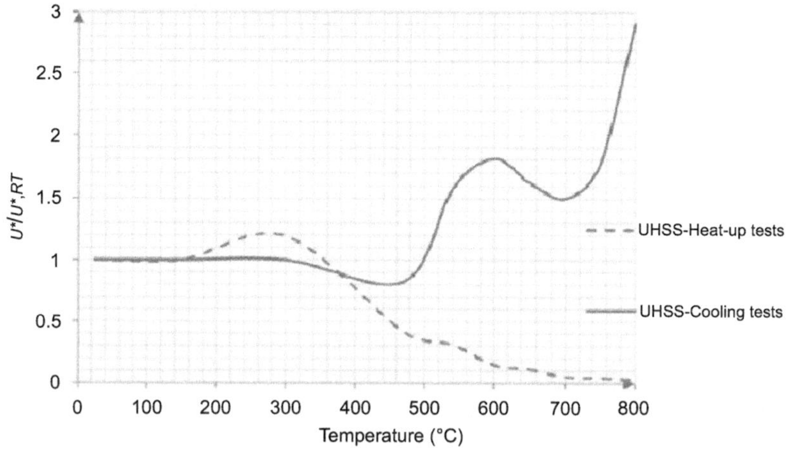

Figure 4.72 Variation of energy absorption reduction factors for ultra-high strength steel under in-fire and postfire conditions.

($U^*_{,RT}$), and the variation of $U^*/U^*_{,RT}$ ratios with respect to the maximum temperature of the UHSS specimens are shown in Fig. 4.72. It can be seen that for the UHSS specimens under cooling tests, $U^*/U^*_{,RT}$ ratio generally increases as the maximum temperature increases (although this shows a local maximum in the temperature range 600–700°C). However, for the UHSS specimens under heat-up tests, the ratio of $U^*/U^*_{,RT}$ considerably decreases as the maximum temperature increases. The reason behind this phenomenon is that in addition to the values of strain, the strength has also a considerable contribution to the magnitude of energy absorption. Therefore, the calculated energy absorption for the UHSS tested at 800°C is decreased to ~3% of that of the virgin UHSS tested at *RT*.

The ductility of a structure is an important parameter to determine whether or not sufficient plastic deformation can be developed. Among a number of requirements recommended by the codes of practice such as AS4100 (1998), AS/NZS 4600 (2005), and Eurocode 3 (2010) to determine whether the plastic analysis can be applied, the ratios of $f_u/f_{0.2}$, which has been addressed by all the aforementioned standards, and $\varepsilon_u/\varepsilon_y$ recommended by Eurocode 3 (2010) are discussed in this study. According to AS4100 (1998) and Eurocode 3 (2010), the ratio of $f_u/f_{0.2}$ must be equal to or greater than 1.2, while AS/NZS 4600 (2005) recommends a limit of 1.08 for this ratio. Moreover, $\varepsilon_u/\varepsilon_y$ must be equal to or greater than 20 as suggested by Eurocode 3 (2010).

Tables 4.27 and 4.28 present the variations of $f_u/f_{0.2}$ and $\varepsilon_u/\varepsilon_y$ for the UHSS specimens tested at elevated temperature and after cooling from different fire temperatures. It can be seen that while the values obtained for the ratio of $f_u/f_{0.2}$ from the heat-up tests for temperatures equal to or greater than 300°C are within the acceptable limits recommended by the standards, those obtained for the UHSS specimens cooled from fire temperatures to *RT* do not meet the requirements of AS4100 (1998) and Eurocode 3 (2010) for all temperatures. However, they are still within the acceptable limit suggested by AS/NZS 4600 (2005). In addition, according to the results shown in the second row of Tables 4.27 and 4.28, the values of $\varepsilon_u/\varepsilon_y$ do not meet the requirement of Eurocode 3 (2010) for all temperature exposures, meaning that it may have to be modified to take into account the higher strength steel grades, the fire temperature exposure, and the cooling phase.

4.9.6.5 Comparison with grade 800 high strength steel

To investigate the effect of steel grade on the changes occurring in the mechanical behavior of steel under fire conditions, the stress–strain curves of the HSS (Grade 800) tube specimens subjected to strain-controlled quasi-static tensile test at high fire temperatures and after cooling to *RT* are summarized in Figs. 4.73 and 4.74, respectively. According to the results shown by the authors (Azhari et al., 2015) for the same HSS material, the strength of the HSS tube specimens cooled from fire temperatures below 600°C to *RT* is fully recovered. Therefore, in this study, to investigate the effect of steel grade on the residual mechanical properties of steel, only temperatures above 700°C are considered. In Figs. 4.73 and 4.74, similar to the labeling system used for the UHSS specimens, the HSS-HT and HSS-CT curves represent the stress–strain curves of the HSS specimens tested at temperature *T* and at *RT* after being cooled from temperature *T*, respectively. Also, the stress–strain curve of the virgin HSS specimen tested at *RT* is labeled as HSS-RT. As shown in Fig. 4.73, the strength and ductility of the HSS specimens tested at temperatures above 700°C are considerably reduced. In addition, it can be seen from Fig. 4.74 that while the strength of HSS is almost fully regained when cooled from 700°C to *RT*, there is a significant reduction in the strength of the HSS tube specimens cooled from temperatures above 700°C.

To better compare the effect of cooling phase of a fire on the strength and ductility of UHSS and HSS, the variations of the ultimate tensile strength and the uniform elongation with the maximum fire temperature

Table 4.27 Variation of the ratios of $f_u/f_{0.2}$ and $\varepsilon_u/\varepsilon_y$ at various fire temperatures.

Temperature	Room temperature	300°C	470°C	540°C	600°C	650°C	700°C	750°C	800°C
$f_u/f_{0.2}$	1.174	1.221	1.345	1.309	1.306	1.724	1.827	1.743	1.398
$\varepsilon_u/\varepsilon_y$	3.106	3.892	3.153	3.514	3.255	6.551	7.767	9.231	12.565

Table 4.28 Variation of the ratios of $f_u/f_{0.2}$ and $\varepsilon_u/\varepsilon_y$ after cooling down from various fire temperatures.

Temperature	Room temperature	300°C	470°C	540°C	600°C	650°C	700°C	750°C	800°C
$f_u/f_{0.2}$	1.174	1.170	1.117	1.106	1.231	1.152	1.171	1.653	1.867
$\varepsilon_u/\varepsilon_y$	3.106	3.100	3.341	6.806	9.939	8.963	10.391	20.613	30.963

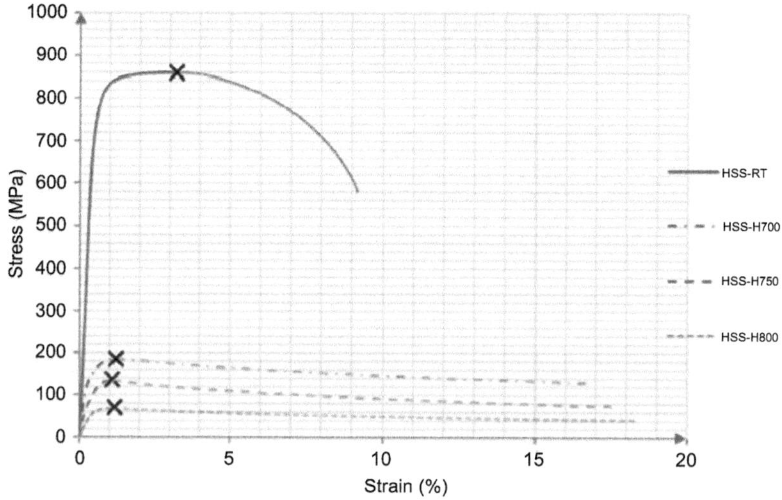

Figure 4.73 Stress–strain curves for the high strength steel specimens at high fire temperatures.

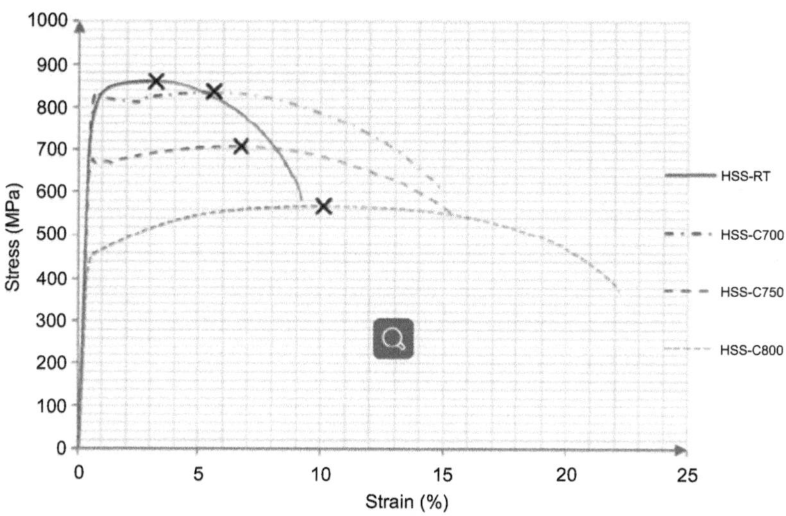

Figure 4.74 Stress–strain curves for the high strength steel specimens at room temperature after cooling from high fire temperatures.

are plotted in Figs. 4.75 and 4.76. It can be seen that while the ultimate strength of the HSS specimen cooled from 800°C to *RT* is reduced to ~66% of that of the virgin HSS specimen at *RT*, the corresponding reduction factor for the UHSS (Grade 1200) is 41% (Fig. 4.75). It may be noted that the ultimate stresses of the UHSS and the HSS, after cooling

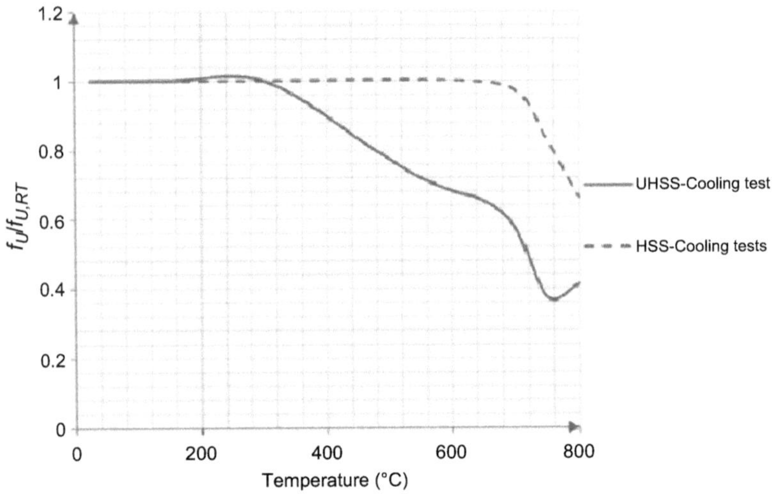

Figure 4.75 Variation of ultimate tensile strength of high strength steel and ultra-high strength steel specimens after cooling from various fire temperatures.

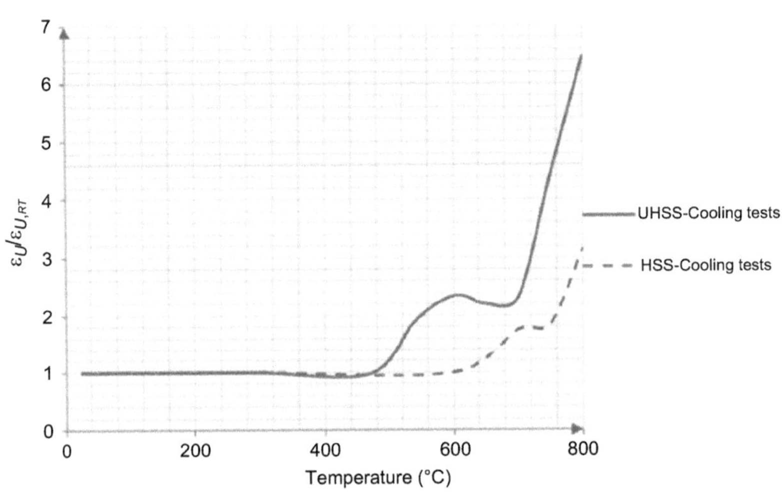

Figure 4.76 Variation of uniform elongation of high strength steel and ultra-high strength steel specimens after cooling from various fire temperatures.

from 800°C, are basically the same ~500 MPa. This indicates that the special processing that is performed to give these two steels their *RT* properties is completely undone by heating to 800°C and the structure generated on cooling back to *RT* is similar, and much weaker, in both steels. In addition, the ductility of the HSS specimens tested after being cooled from 800°C to RT is three times its value for the virgin HSS tested at RT; however, the UHSS specimen cooled from 800°C experiences a uniform elongation of 6.5 times its original value (Fig. 4.76). Thus, the UHSS loses its strength quicker than HSS while the rate of ductility increase in HSS subject to cooling phase of a fire is lower than that of UHSS. These results indicate the higher sensitivity of the UHSS to the temperature history of the material.

4.9.7 Predictive equations

In this section, the aim is to develop empirical equations for the stress–strain response of the UHSS material cooled from different fire temperatures to RT. According to Fig. 4.68, the postfire stress–strain curves obtained for the UHSS tube materials do not exhibit a pronounced yield point. The rounded and extended elastoplastic transition observed for the UHSS is typical of advanced HSSs and can also be observed in the stress–strain curves obtained for stainless steel specimens (Arrayago et al., 2015; Mirambell and Real, 2000; Rasmussen, 2003). The sharp yield point is a feature especially characteristic of mild steels. In this study, the model used to predict the stress–strain response of the UHSS tube cooled from fire temperatures T to RT is based on the compound Ramberg–Osgood material model (Mirambell and Real, 2000; Rasmussen, 2003) such that

$$\varepsilon = \frac{\sigma}{E_{CT}} + 0.002\left(\frac{\sigma}{f_{0.2,CT}}\right)^{n_{CT}} \quad \text{for } \sigma \le f_{0.2,CT}$$

$$\varepsilon = \frac{\sigma - f_{0.2,CT}}{E_{0.2,CT}} + \left(\varepsilon_{u,CT} - \varepsilon_{y,CT} - \frac{f_{u,CT} - f_{0.2,CT}}{E_{0.2,CT}}\right)\left(\frac{\sigma - f_{0.2,CT}}{f_{u,CT} - f_{0.2,CT}}\right)^{n'_{CT}} + \varepsilon_{y,CT} \quad \text{for } f_{0.2,CT} \le \sigma \le f_{u,CT} \tag{4.32}$$

where

$$E_{0.2,CT} = \frac{E_{CT}}{1 + 0.002 n_{CT} \dfrac{E_{CT}}{f_{0.2,CT}}} \tag{4.33}$$

In the relations given in Eqs. (4.32) and (4.33), ε and σ are the engineering strain and stress, respectively, and the CT index is used for the properties of UHSS tube specimens tested after cooling from temperature T to RT. Also, n and n' are the exponential coefficients to take the material nonlinearity into consideration while E and $E_{0.2}$ are the material's Young's modulus and the tangent modulus at 0.2% proof stress, respectively. The definition of the remaining parameters used in Eq. (4.32) are illustrated in the schematic stress—strain curve shown in Fig. 4.65.

By using the experimental results presented in the previous sections and finding the polynomials that have the best fit to the series of data points, the following predictive equations are obtained for the values of $f_{0.2,CT}$, $f_{u,CT}$, and $\varepsilon_{u,CT}$ in terms of the maximum temperature T the UHSS specimens have experienced and also the corresponding parameter of the virgin material at RT.

$$\frac{f_{0.2,CT}}{f_{0.2,RT}} = \begin{cases} 1 & 25 \le T \le 300 \\ 118.151 - 1.48514T + 7.64380 \times 10^{-3}T^2 - 2.04374 \times 10^{-5}T^3 & \\ +2.99481 \times 10^{-8}T^4 - 2.28328 \times 10^{-11}T^5 + 7.08474 \times 10^{-15}T^6 & 300 < T \le 800 \end{cases} \tag{4.34}$$

$$\frac{f_{u,CT}}{f_{u,RT}} = \begin{cases} 1 & 25 \le T \le 300 \\ 73.04820 - 9.05365 \times 10^{-1}T + 4.63958 \times 10^{-3}T^2 - 1.24018 \times 10^{-5}T^3 & \\ +1.82209 \times 10^{-8}T^4 - 1.39579 \times 10^{-11}T^5 + 4.35813 \times 10^{-15}T^6 & 300 < T \le 800 \end{cases} \tag{4.35}$$

$$\frac{\varepsilon_{u,CT}}{\varepsilon_{u,RT}} = \begin{cases} 1 & 25 \le T \le 300 \\ 33.3125 - 2.49832 \times 10^{-1}T + 6.27862 \times 10^{-4}T^2 - 4.49286 \times 10^{-7}T^3 & \\ -3.33432 \times 10^{-10}T^4 + 4.2094 \times 10^{-13}T^5 & 300 < T \le 800 \end{cases} \tag{4.36}$$

where the RT index represents the virgin UHSS tube material tested at RT. It is worth noting that E and $E_{0.2}$ are independent of the temperature history of the UHSS tube specimens and are assumed to be constant, i.e., $E_{RT} = E_{CT}$ and $E_{0.2,RT} = E_{0.2,CT}$.

In addition, according to Fig. 4.65, $\varepsilon_{y,CT}$ can be determined from

$$\varepsilon_{y,CT} = \frac{f_{0.2,CT}}{E_{RT}} + 0.002 \tag{4.37}$$

To find the values of the exponential coefficients n_{CT} and n'_{CT} in terms of the maximum fire temperature (T) the specimens have experienced, a nonlinear regression analysis is performed using the stress—strain curves shown in Fig. 4.67 and Eq. (4.32). Hence, the following equations are derived:

$$n_{CT} = \begin{cases} 6.4 & 25 \le T \le 300 \\ -125.431 + 1.09094T - 3.21568 \times 10^{-3}T^2 & \\ +4.02685 \times 10^{-6}T^3 - 1.82703 \times 10^{-9}T^4 & 300 < T \le 800 \end{cases} \tag{4.38}$$

$$n_{CT} = \begin{cases} 13.3 & 25 \le T \le 300 \\ -138.733 + 9.71289 \times 10^{-1}T - 1.88696 \times 10^{-3}T^2 & \\ +1.12958 \times 10^{-6}T^3 & 300 < T \le 800 \end{cases} \tag{4.39}$$

To validate the proposed material model shown in Eq. (4.32), the stress–strain curves obtained from the model and the experimental tests for the UHSS tube specimens cooled from different fire temperatures are depicted in Fig. 4.77. Similar to the labeling system used in Fig. 4.67, the stress–strain curves of the UHSS cooled from fire temperature T to RT obtained from the experiment and the proposed model are shown as UHSS-CT and UHSS-CT model, respectively. In addition, the stress–strain curves of the virgin UHSS material at RT obtained from the experiment and the proposed model are displayed as UHSS-RT and UHSS-RT model, respectively. Fig. 4.77 indicates that a good accuracy is

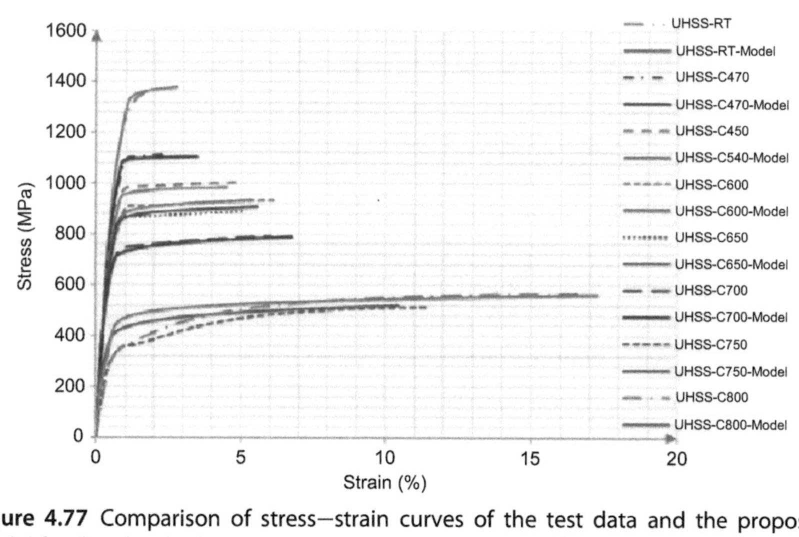

Figure 4.77 Comparison of stress–strain curves of the test data and the proposed model for the ultra-high strength steel tube specimens cooled from different elevated temperatures to room temperature.

resulted from the stress−strain curves predicted by the proposed material model for the UHSS throughout the entire range of fire temperatures. It is worth noting that the main advantage of the model is that it is only taking the basic parameters of the UHSS at RTs, i.e., E_{RT}, $f_{0.2,RT}$, $f_{u,RT}$, $\varepsilon_{u,RT}$, and the maximum fire temperature T as input data. These parameters are usually known by the engineers and are commonly provided in the material data sheet by the steel manufacturing companies. Therefore, although using only $f_{0.2,RT}$ and $f_{u,RT}$ as input characteristic strength values have led to a gap between the 0.2% proof stress and the flat plastic region of UHSS-C800 and UHSS-C800 model curves, the accuracy of the proposed model is still good enough to be applied for the engineering problems.

4.10 Residual properties of GLG460, GLG550, GLG650, and GLG835 steel

An experimental investigation was conducted by Chen et al. (2016) to reveal the post-fire mechanical properties of four widely used high-strength steel tie rod grades, namely, GLG460, GLG550, GLG650, and GLG835. Specimens were heated to 13 various preselected temperatures up to 1000 °C and subsequently cooled down to ambient temperature via two different methods: air cooling and water cooling. Tensile coupon tests were conducted, and the post-fire stress−strain curves, elastic modulus, yield strengths, ultimate strengths, and ductility were obtained.

4.10.1 Test material and specimens

The test specimens were cut from products of steel tie rod GLG460, GLG550, GLG650, and GLG835 ordered for this study with the nominal diameter of 25 mm. In this section, the naming method of steel tie rods suggested in GB/T 20934−2007 (2007) is adopted, where GLG is the abbreviation of steel tie rods (in Chinese Pinyin); 460, 550, 650, and 835 means their minimum nominal yield strengths are 460, 550, 650, and 835 N/mm^2, respectively. All of these four grades of steel tie rods were manufactured with alloy structural steel 35CrMo using different quenching and tempering methods. 35CrMo is the grade designation abbreviation of this alloy structural steel, where 35 means that the average content of carbon element is 0.35%, and CrMo means that its major alloy elements are Cr and Mo. Table 4.29 shows the chemical composition of 35CrMo, which is in accordance with GB/T 3077−1999 (1999). Table 4.30 presents

Table 4.29 Chemical compositions of alloy structural steel 35CrMo material (%).

Chemical element	C	Si	Mn	P	S	Cr	Ni	Cu	Ti	Mo	V	B	Al
	0.35	0.20	0.56	0.014	0.011	0.94	0.02	0.01	0.007	0.164	0.004	0.0003	0.014

Table 4.30 Nominal mechanical properties of steel tie rods.

Grade	Yield strength (MPa)	Ultimate strength (MPa)	Elongation after fracture (%)
GLG460	≥460	≥610	≥19
GLG550	≥550	≥750	≥17
GLG650	≥650	≥850	≥15
GLG835	≥835	≥1030	≥13

the nominal mechanical properties of these four grades of steel tie rods specified in GB/T 20934–2007 (2007).

The shapes and sizes of the specimens tested in this study are in accordance with GB/T 228–2010 (2010) and GB/T 4338–2010 (2010) (Fig. 4.78). The diameter of each specimen was measured at three points within the gauge length using a vernier caliper. The average values of the measured dimensions were used to calculate the mechanical properties of the steel tie rods.

4.10.2 Test equipment and procedure

The entire procedure of the test mainly contains two steps. In the first step, the specimens were initially heated to the preselected elevated temperatures and subsequently cooled down to ambient temperature. In the second step, the tensile coupon test was conducted on the specimens at ambient temperature. The specimens were heated by a temperature-controlled electrical furnace (Fig. 4.79). The thermocouple located inside the furnace measured the air temperature of the furnace and fed back to the control system,

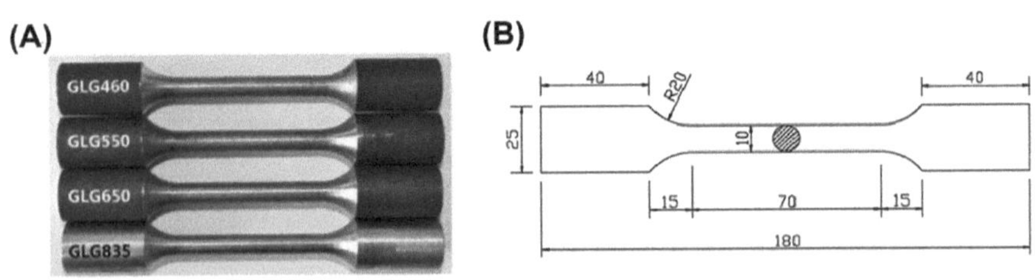

Figure 4.78 Tensile test specimens and dimensions. (A) Test specimens and (B) dimensions (mm).

Figure 4.79 Temperature-controlled electrical furnace.

which thereby adjusted the heating rate; thus, a closed control loop was formed. In this test, the 13 preselected temperatures were 100, 200, 300, 400, 500, 600, 650, 700, 750, 800, 850, 900, and 1000°C. Temperatures higher than 1000°C were not considered because steel tie rods become very soft at such temperatures. In the heating process, the furnace temperature was initially raised at 15°C/min to a temperature of 30°C less than the target temperature and held for 10 min, after which the furnace temperature was set to the target temperature and held for another 20 min. Because the influence of heating duration can be ignored according to previous literature (Smith et al., 1981) such heating process could be adopted to ensure the uniform temperature distribution of the specimens as well as to avoid the exceeding of actual temperature from the target temperature. Subsequently, the specimens were removed from the furnace and cooled down to ambient temperature. Both air and water cooling methods were considered. For air cooling method, the specimens were placed in the air and cooled down at their own rate to simulate the situation that fire goes out for itself. For water cooling method, the specimens were cooled down by water spraying using a water jet to simulate the situation that fire is extinguished by fire guns. The water volume used in this experiment was determined according to the principle that the water volume sprayed on each unit area of the specimen surface is proportional to the water volume sprayed on the members in an actual fire extinguishment. Given the fire-fighting parameters and flux of the water jet, the time of spraying can be calculated by Eq. (4.31), as follows:

$$T_2 = \frac{3.6 Q_1 T_1 A}{Q_2 \pi R^2} \tag{4.40}$$

where T_1 refers to the fire extinguishing time in actual fire events (h); Q_1 refers to the water flux in fire extinguishment (L/s), R refers to cover radius of the fire guns (m); T_2 refers to the time of water spraying in this test (s); Q_2 refers to the flux of water jet (mL/s); and A refers to the surface area of a specimen (mm^2). The firefighting parameters employ the values suggested in GB 50045-95 (2005), where $T_1 = 2$ h, $Q_1 = 15$ L/s, and $R = 15$ m. Q_2 was kept constant in the test as 20 mL/s.

Tensile test was conducted using a universal material testing machine, which can exert as much as 600 kN force in tension. The specimen was connected to the top and bottom grips. The bottom end was fixed, whereas the top end was free to move upwards (Fig. 4.80). An electronic extensometer was attached to the specimen to measure the strain precisely until fracture occurred. The tensile load was applied at a constant strain rate of 0.003/min, which satisfied the requirement of GB/T 4338−2010 (2010). All the data obtained from the experiment, such as force, stroke, stress,

Figure 4.80 Tensile test setup.

strain, and stress–strain curves, were recorded by a computer that is compatible with the testing machine. For comparison, tensile tests were also conducted at ambient temperature on specimens that had not been exposed to elevated temperatures. To minimize the accidental errors, three specimens, which were cooled down from the same elevated temperature by the same cooling method, were tested and regarded as a group. A total of 108 groups with 324 specimens (including four groups without exposure to elevated temperatures) were tested. The average value of the three specimens in each group was adopted to analyze the mechanical properties of steel tie rods after fire.

Additional tests were also conducted to explore the possible degeneration in the mechanical properties of steel tie rods caused by multiple heating and cooling cycles. The specimens were heated to preselected elevated temperatures and cooled down to ambient temperature repeatedly up to three times. After conducting the tensile coupon tests to obtain their mechanical properties, the effects of cyclic heating and cooling were analyzed. Specimens of low-grade GLG460 and high-grade GLG835 were tested. Because the main purpose of this additional test is to investigate the influence of multiple heating–cooling cycles rather than other factors such as cooling methods and exposure temperatures on the mechanical properties of steel tie rods, only air cooling method and three elevated temperatures, namely, 300, 600°C, and 900°C, were considered.

4.10.3 Experimental results

4.10.3.1 Thermochromism of steel tie rods

The specimens of all four grades of steel tie rods showed obvious thermochromism after cooling down from elevated temperatures (Fig. 4.81). Nearly no difference was observed between the specimens cooled down by air and water spraying, except that the water cooling specimens showed more rust on surface. Thermochromism is mainly closely related to the highest exposed temperatures, and all four grades of steel tie rods showed similar trend with respect to the exposed temperature. The color of the specimen surface turned obviously yellow from the original silvery white after exposure to temperatures within the range of 300–400°C; subsequently, the color became darker as the exposure temperature increased and presented gray black at 650–750°C. After exposure to temperatures above 750°C, flaking was observed in the form of oxide; furthermore, the color of the specimen surface turned lighter gradually and showed light gray when the exposed temperature was 1000°C. After cooling from 750 to 850°C,

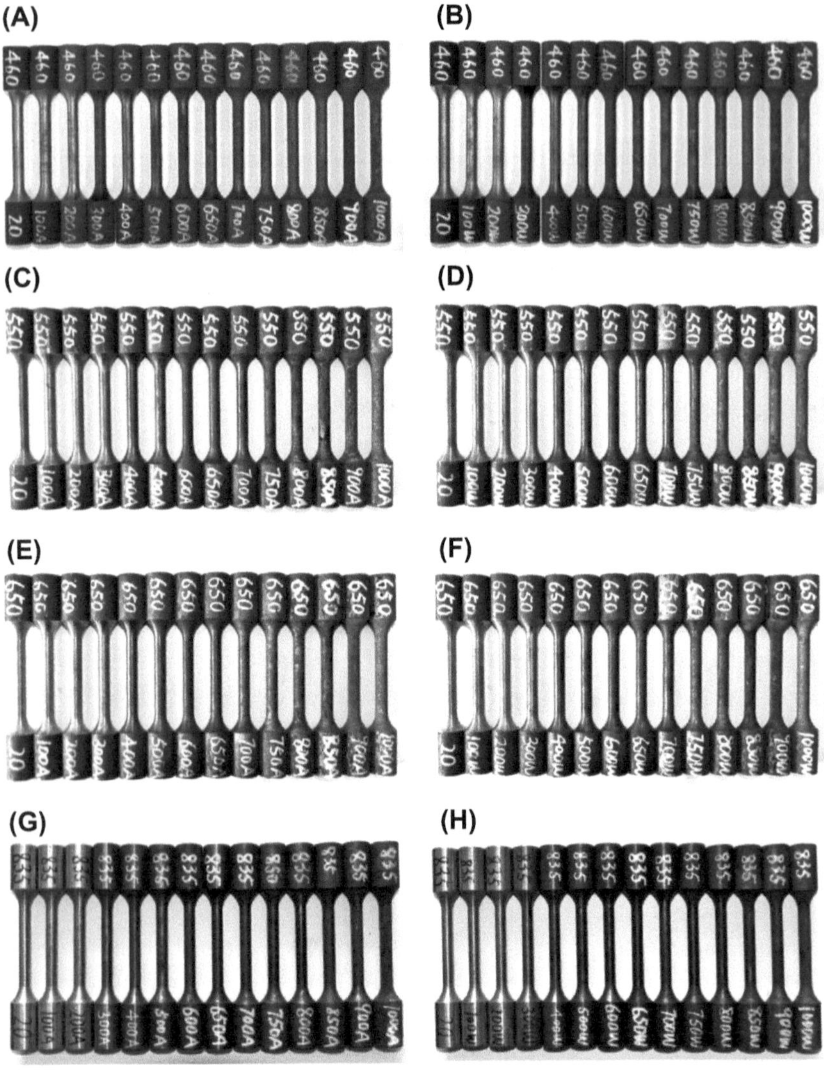

Figure 4.81 Failure modes of test specimens. (A) GLG460 specimens after air cooling; (B) GLG460 specimens after water cooling; (C) GLG550 specimens after air cooling; (D) GLG550 specimens after water cooling; (E) GLG650 specimens after air cooling; (F) GLG650 specimens after water cooling; (G) GLG835 specimens after air cooling; and (H) GLG835 specimens after water cooling.

GLG460, GLG550, and GLG650 showed evident rust, and GLG835 also showed relatively apparent rust after exposure to 850°C, which can be considered as an obvious feature to distinguish water-cooled and air-cooled steel tie rods. The thermochromism shown in Fig. 4.81 is useful in estimating the highest exposure fire temperature of steel tie rods.

4.10.3.2 Failure modes of test specimens

Tensile coupon tests were conducted on the specimens of steel tie rods cooled down from elevated temperatures until fracture. The failure modes of the specimens were then obtained (Fig. 4.82). For comparison, the failure modes of the specimens without exposure to elevated temperatures were also included. Generally, the failure modes of all four grades of steel tie rods showed similar trend with respect to the exposure temperature though lower grade steel tie rods presented more obvious necking, which means higher ductility. And the influences of different cooling methods are also significant. With the increasing exposure temperature, all grades of steel tie rods cooled in air showed different degrees of ductile failures with necking, no matter how high temperature the specimens had been exposed to. However, for the water-cooled specimens, more brittle failures with nearly no necking were observed when the specimens were cooled from temperatures above 850°C; this observation indicates that considerable attention should be given to prevent the occurrence of brittle fractures in the reuse of steel tie rods extinguished by fire guns.

4.10.3.3 Stress—strain relationships

The postfire stress—strain relationships obtained from this experiment are plotted in Fig. 4.83 (each curve plotted is one of the three curves obtained from the specimens in a group, and for ease to distinguish and identify the curves, only the postfire stress—strain curves for 20, 200, 400, 600, 700, 750 800, 850 900, and 1000°C were plotted). All of the postfire mechanical properties of these four grades of steel tie rods were studied based on these curves, as discussed in Sections 3.2—3.6. The postfire stress—strain curves differ considerably with respect to the exposure temperatures and steel tie rod grades. The curves of GLG460 and GLG550 show obvious yield plateau at ambient temperature and after exposure to temperatures up to 800°C; meanwhile, high-grade GLG650 and GLG835 present yield plateau only after exposure to temperatures within the range of 700—800°C. For GLG460, GLG550, and GLG650, the stress—strain curves of specimens after exposure to temperatures $<650°C$ are similar to that of ambient temperature specimens though slight variations in ductility can be observed. When the exposure temperature continued to grow, significant changes in strength, ductility, and shape of the stress—strain curves occurred. By contrast, the high-grade GLG835 was more sensitive to the effects of heating and cooling, and the significant changes in stress—strain curves occurred when the exposure temperature exceeded 500°C.

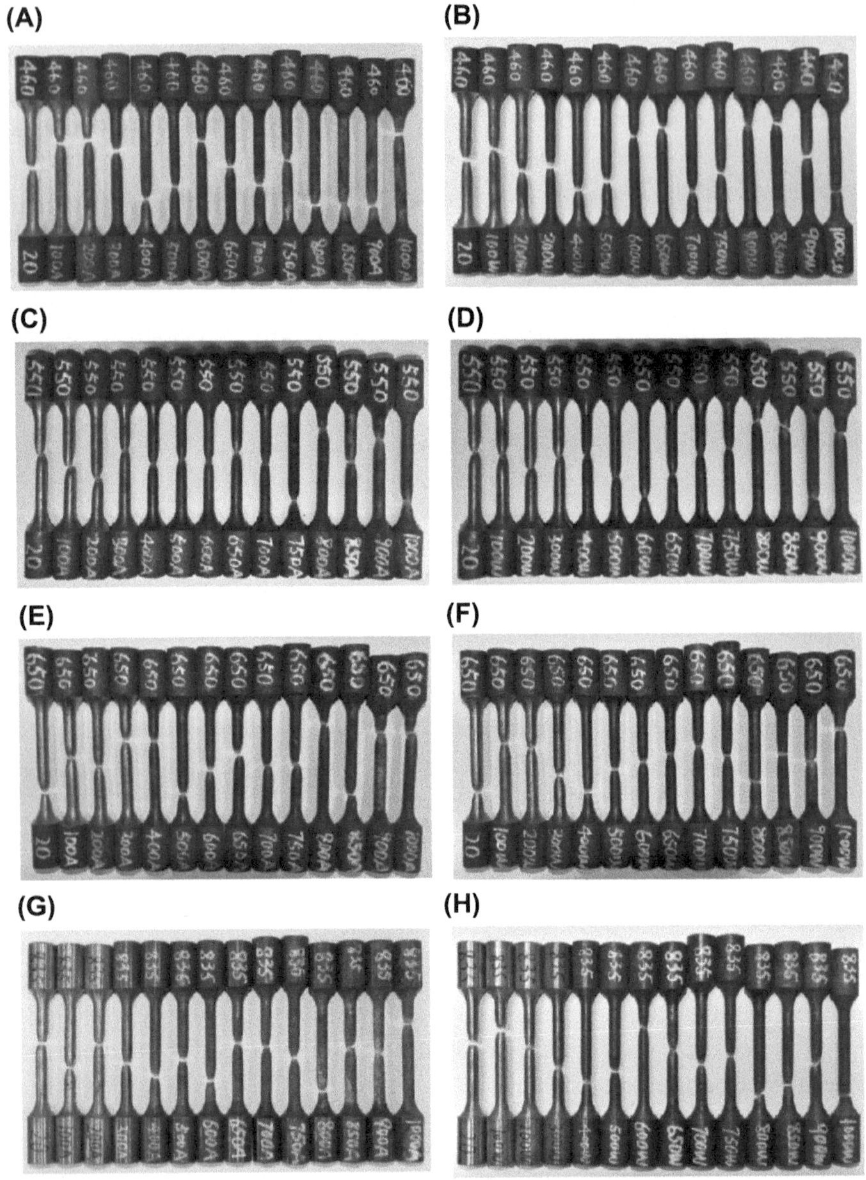

Figure 4.82 Failure modes of test specimens. (A) GLG460 specimens after air cooling; (B) GLG460 specimens after water cooling; (C) GLG550 specimens after air cooling; (D) GLG550 specimens after water cooling; (E) GLG650 specimens after air cooling; (F) GLG650 specimens after water cooling; (G) GLG835 specimens after air cooling; and (H) GLG835 specimens after water cooling.

In addition, the influence of different cooling methods also played an important role in the change of postfire mechanical properties. After exposure to temperatures exceeding 800°C, the steel tie rods cooled by water spraying showed higher strength but lower ductility than those cooled in air.

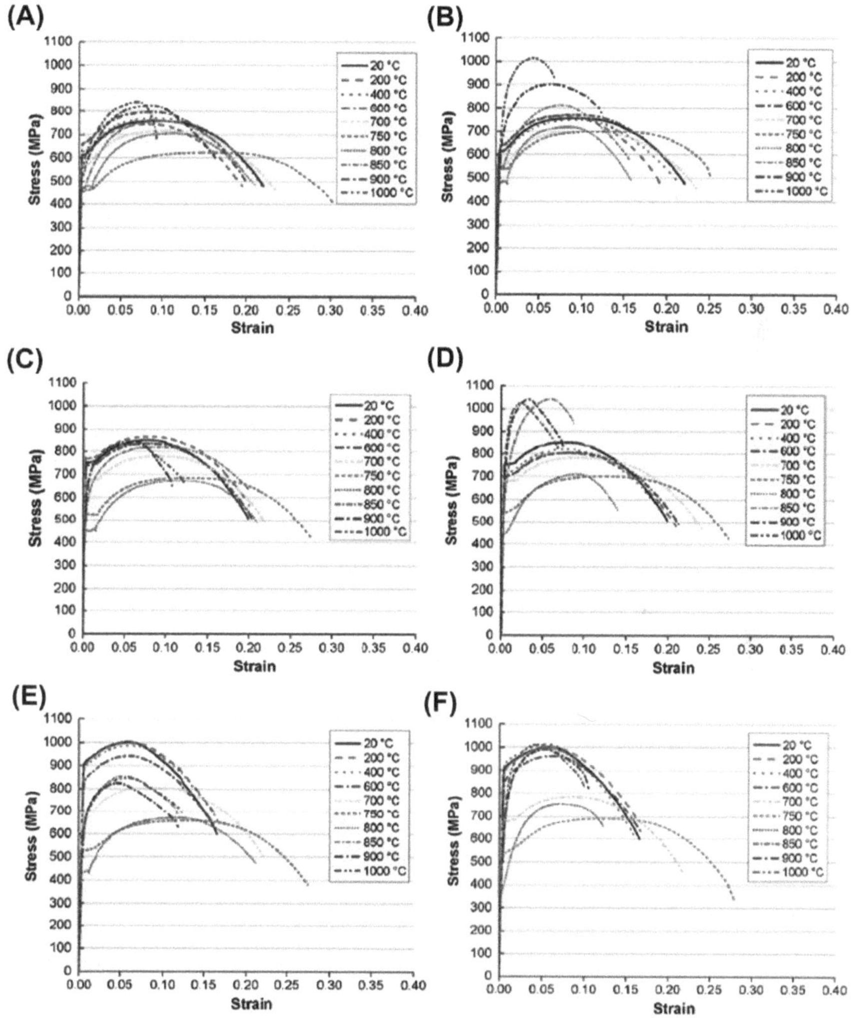

Figure 4.83 Stress–strain curves after exposure to various fire temperatures. (A) GLG460 specimens after air cooling; (B) GLG460 specimens after water cooling; (C) GLG550 specimens after air cooling; D) GLG550 specimens after water cooling; (E) GLG650 specimens after air cooling; (F) GLG650 specimens after water cooling; (G) GLG835 specimens after air cooling; and (H) GLG835 specimens after water cooling.

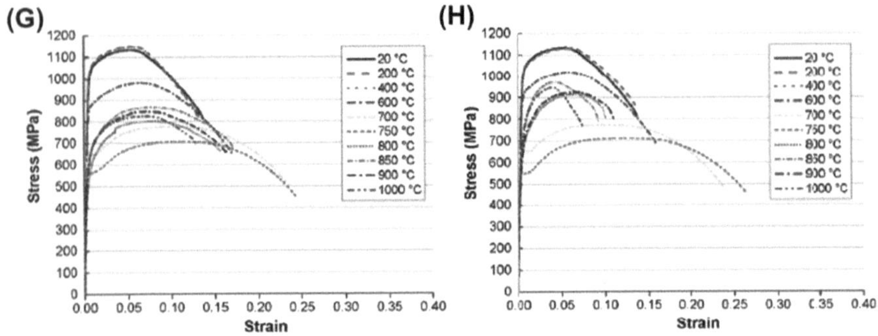

Figure 4.83 Cont'd

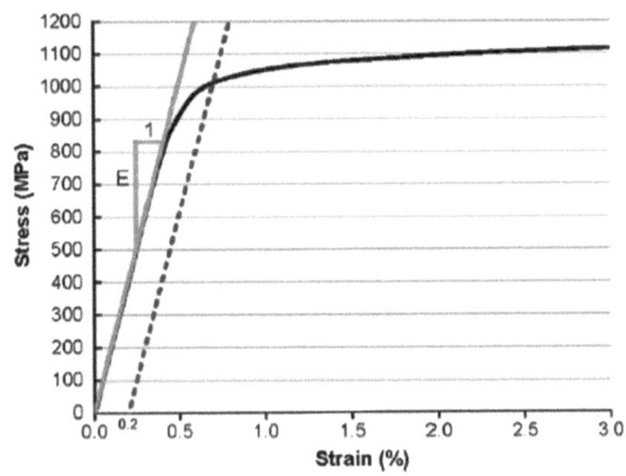

Figure 4.84 Elastic modulus and yield strength determination.

4.10.3.4 Elastic modulus

The postfire elastic moduli of steel tie rods tested in this experiment are defined as the initial slope of their stress–strain curves (Fig. 4.84). The method adopted to calculate the initial slope was in accordance with GB/T 22315-2008 (2008). The residual factor of elastic modulus is defined as the ratio of elastic modulus after cooling down from elevated temperatures (E_{PT}) to that at ambient temperature (E) to describe the deterioration in the elastic moduli of steel tie rods after fire. Tables 4.31 and 4.32 provide the list of the postfire elastic moduli and corresponding residual factors obtained from this experiment for all four grades of steel tie rods. The residual factors are also plotted as the function of exposure temperature in Fig. 4.85 (each value listed and plotted is the average value of the three specimens in a group).

All four grades of steel tie rods can regain their elastic moduli after exposure to temperatures up to 600°C. Subsequently, an obvious reduction

Table 4.31 Postfire elastic moduli (GPa).

	GLG460		GLG550		GLG650		GLG835	
Temperature (°C)	Air cooling	Water cooling	Air cooling	Water cooling	Air cooling	Water cooling	Air cooling	Water cooling
20	205.4	205.4	204.3	204.3	196.2	196.2	211.0	211.0
100	206.4	208.7	203.8	203.6	195.8	197.5	212.0	209.2
200	206.5	205.5	203.8	204.3	195.2	195.6	206.7	202.2
300	202.4	207.6	205.5	204.7	189.0	187.7	207.7	203.2
400	200.2	206.3	203.5	203.2	195.3	193.4	208.9	200.1
500	204.8	202.7	200.7	199.0	189.2	193.7	207.8	203.0
600	205.5	201.9	199.6	198.8	191.4	189.0	203.6	203.1
650	200.4	200.8	193.3	192.0	190.3	188.7	204.6	204.7
700	194.5	193.5	194.1	190.2	190.6	192.2	195.9	195.5
750	188.2	190.9	189.9	190.1	183.2	180.1	198.3	197.7
800	191.0	190.8	188.5	179.0	177.9	175.6	181.3	192.6
850	192.5	186.3	173.6	175.6	176.9	160.6	182.0	181.5
900	182.6	178.0	172.5	167.3	159.3	158.1	176.9	183.2
1000	175.1	167.2	169.7	165.1	151.6	148.4	163.0	169.0

Table 4.32 Postfire elastic modulus residual factors.

Temperature (°C)	GLG460		GLG550		GLG650		GLG835	
	Air cooling	Water cooling	Air cooling	Water cooling	Air cooling	Water cooling	Air cooling	Water cooling
20	1.000	1.000	1.000	1.000	1.000	1.000	1.000	1.000
100	1.005	1.016	0.998	0.997	0.998	1.007	1.005	0.991
200	1.005	1.000	0.998	1.000	0.995	0.997	0.980	0.958
300	0.985	1.011	1.006	1.002	0.963	0.957	0.984	0.963
400	0.975	1.004	0.996	0.995	0.995	0.986	0.990	0.948
500	0.997	0.987	0.982	0.974	0.964	0.987	0.985	0.962
600	1.000	0.983	0.977	0.973	0.976	0.963	0.965	0.963
650	0.976	0.978	0.946	0.940	0.970	0.962	0.970	0.970
700	0.947	0.942	0.950	0.931	0.971	0.980	0.928	0.927
750	0.916	0.929	0.930	0.930	0.934	0.918	0.940	0.937
800	0.930	0.929	0.923	0.876	0.907	0.896	0.859	0.913
850	0.937	0.907	0.850	0.860	0.902	0.819	0.863	0.860
900	0.889	0.867	0.844	0.819	0.812	0.806	0.838	0.868
1000	0.852	0.814	0.831	0.808	0.773	0.756	0.773	0.801

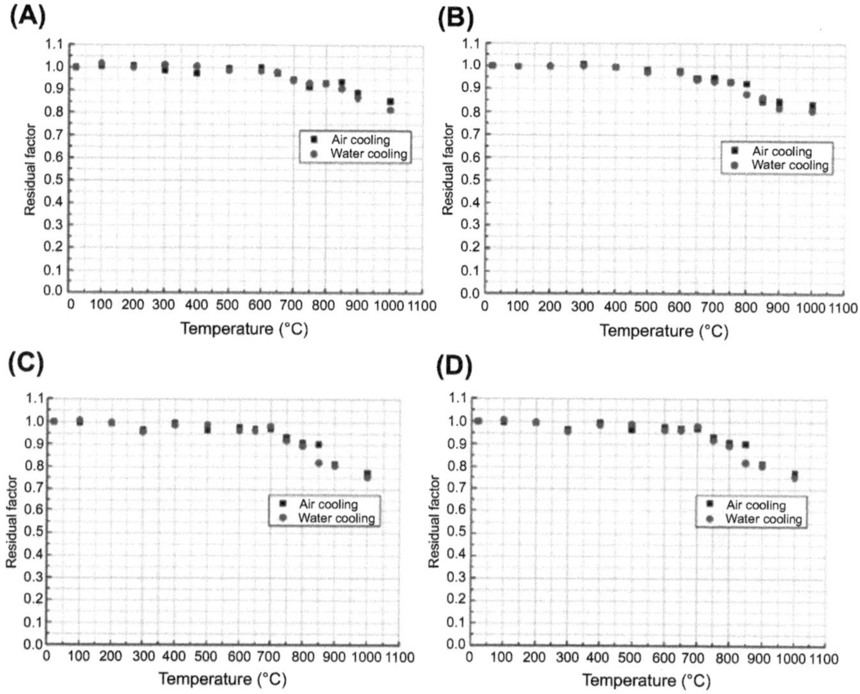

Figure 4.85 Postfire elastic modulus residual factors.(A) GLG460; (B) GLG550; (C) GLG650; and (D) GLG835.

in the elastic moduli of all of these four grades of steel tie rods was observed. However, the difference in residual elastic modulus caused by two different cooling methods is not significant. For GLG460 and GLG550, at least 80% of elastic modulus was regained after cooling from elevated temperatures up to 1000°C and at least 75% for GLG650 and GLG835. This result is positive for the reuse of steel tie rods after fire.

4.10.3.5 Yield strength

For comparison, the 0.2% proof stress method was used to define the yield strengths of steel tie rods studied in this experiment, no matter whether they have yield plateau or not. The 0.2% proof stress f0.2 was determined from the intersection of the stress–strain curve and proportional line offset by 0.2% strain level (Fig. 4.84). The residual factor of yield strength is defined as the ratio of the yield strength after cooling down from elevated temperatures (f_{yPT}) to that at ambient temperature without the heating and cooling (f_y) processes. Tables 4.33 and 4.34 provide the list of the yield strengths and residual factors of all four grades of steel tie rods obtained from

Table 4.33 Postfire yield strengths (MPa).

Temperature (°C)	GLG460		GLG550		GLG650		GLG835	
	Air cooling	Water cooling	Air cooling	Water cooling	Air cooling	Water cooling	Air cooling	Water cooling
20	613	613	753	753	885	885	1019	1019
100	611	628	743	720	885	863	1007	1021
200	602	627	767	695	902	892	1027	1010
300	601	650	737	718	887	894	1020	1033
400	622	608	735	723	878	930	1030	1020
500	625	642	727	724	917	901	1030	1032
600	623	625	740	700	834	868	878	912
650	633	611	732	739	871	772	873	859
700	565	556	674	680	693	674	650	655
750	450	541	531	568	530	526	573	569
800	472	494	452	448	446	386	502	605
850	481	567	537	607	608	634	575	641
900	528	608	620	747	613	642	585	630
1000	566	701	617	816	608	740	600	678

Table 4.34 Postfire yield strength residual factors.

Temperature (°C)	GLG460		GLG550		GLG650		GLG835	
	Air cooling	Water cooling	Air cooling	Water cooling	Air cooling	Water cooling	Air cooling	Water cooling
20	1.000	1.000	1.000	1.000	1.000	1.000	1.000	1.000
100	0.997	1.024	0.987	0.956	1.000	0.975	0.988	1.002
200	0.982	1.023	1.019	0.923	1.019	1.008	1.008	0.991
300	0.980	1.060	0.979	0.954	1.002	1.010	1.001	1.014
400	1.015	0.992	0.976	0.960	0.992	1.051	1.011	1.001
500	1.020	1.047	0.965	0.961	1.036	1.018	1.011	1.013
600	1.016	1.020	0.983	0.930	0.942	0.981	0.862	0.895
650	1.033	0.997	0.972	0.981	0.984	0.872	0.857	0.843
700	0.922	0.907	0.895	0.903	0.783	0.762	0.638	0.643
750	0.734	0.883	0.705	0.754	0.599	0.594	0.562	0.558
800	0.770	0.806	0.600	0.595	0.504	0.436	0.493	0.594
850	0.785	0.925	0.713	0.806	0.687	0.716	0.564	0.629
900	0.861	0.992	0.823	0.992	0.693	0.725	0.574	0.618
1000	0.923	1.144	0.819	1.084	0.687	0.836	0.589	0.665

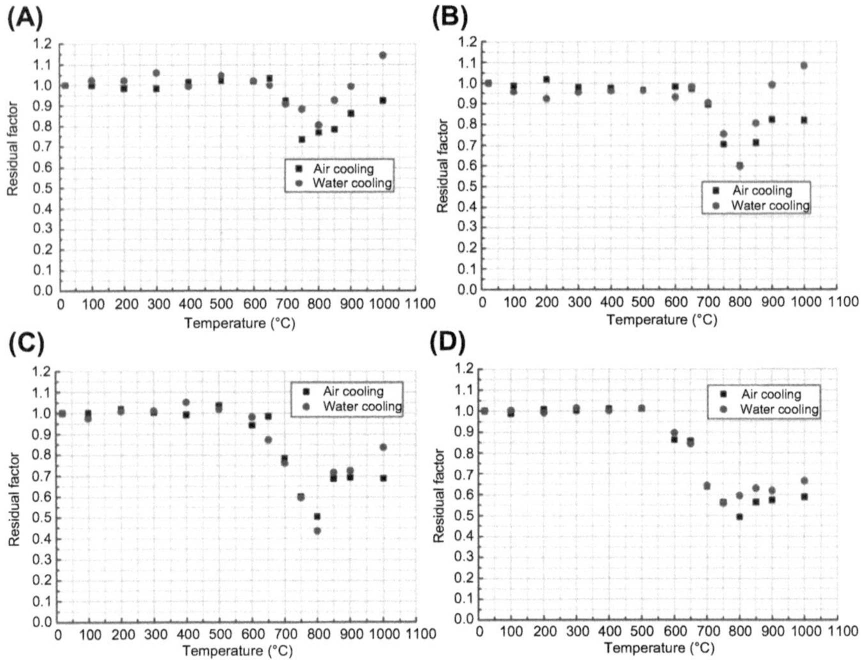

Figure 4.86 Postfire elastic modulus residual factors. (A) GLG460; (B) GLG550; (C) GLG650; and (D) GLG835

this experiment, respectively. The residual factors are also plotted as the function of exposure temperature in Fig. 4.86 (each value listed and plotted is the average value of the three specimens in a group). The yield strength of all four grades of steel tie rods began to decrease significantly after exposure to certain elevated temperatures, while the reduction characteristics of different grades of steel tie rods are different, which are similar to the results obtained from previous experimental study on hot-rolled HSSs (Qiang et al, 2012, 2013) and cold-formed steels (Gunalan and Mahendran, 2014). Moreover, the influences of different cooling methods are significant. For relatively low-grade GLG460, GLG550, and GLG650, the yield strengths were almost not affected until 650°C (except for the water cooling specimens of GLG650) and thereafter decreased at a rapid rate. The yield strength reached the lowest value after exposure to temperature within the range of 750–800°C, with a reduction of approximately 30%, 40%, and 55% for GLG460, GLG550, and GLG650, respectively. It indicates that high-grade steel tie rods regained less of their original yield strength than low-grade steel tie rods did. For high-grade GLG835, its yield strength began to decrease after exposure to temperatures exceeding 500°C; therefore, this grade is more sensitive to the heating and cooling effects; its yield strength also decreased rapidly and reached the lowest value within the range of 750–800°C, with a reduction of approximately 50%. Notably, an obvious rebound in yield strength was observed for all steel tie rod grades when the exposure temperature exceeded approximately 800°C, which was attributed to the change in the crystal structure of the alloy steel that was used to manufacture the steel tie rods.

The influences of different cooling methods on the residual yield strength were significant, particularly in the rebound stage (i.e., the exposure temperature range exceeding approximately 800°C). After exposure to elevated temperatures within this range, the water-cooled specimens of all steel tie rod grades regained more of their original yield strength than those of air-cooled specimens; this observation was attributed to the strengthening effect caused by water cooling, which is equivalent to quenching. Moreover, the strengthening effect was more evident on low-grade steel tie rods than on high-grade steel tie rods. The strengthening effects of water cooling indicates that water cooling method is beneficial for steel tie rods to preserve more of their yield strengths after exposure to relatively high fire temperatures.

4.10.3.6 Ultimate strength

Tensile coupon tests were conducted on the specimens until fracture, thus the ultimate strengths were obtained. The residual ultimate strength factor is defined as the ratio of the ultimate strength after cooling down from elevated temperatures (f_{uPT}) to that at ambient temperature without exposure to fire (f_u). Tables 4.35 and 4.36 present the ultimate strengths obtained from this experiment and their corresponding residual factors, respectively. The residual factors are also plotted in Fig. 4.87 as the function of exposure temperatures (each value listed and plotted is the average value of the three specimens in a group).

The degradation trend of the ultimate strength for all four grades of steel tie rod is similar to that of the yield strength. The ultimate strength remained nearly unchanged after cooling from elevated temperatures up to 650°C for GLG460, GLG550, and GLG650 and up to 500°C for GLG835. Thereafter, the strength decreased rapidly and reached the lowest value when the exposure temperature was within the range of 750°C–800°C. The reduction in the ultimate strength was less than the reduction in yield strength for all these four steel tie rod grades. A rebound of ultimate strength was also observed after cooling from temperatures exceeding 800°C, which was caused by the aforementioned change in the crystal structure of the alloy structural steel.

Table 4.35 Postfire ultimate strengths (MPa).

Temperature (°C)	GLG460		GLG550		GLG650		GLG835	
	Air cooling	Water cooling	Air cooling	Water cooling	Air cooling	Water cooling	Air cooling	Water cooling
20	758	758	853	853	996	996	1140	1140
100	755	767	845	830	976	984	1134	1137
200	752	769	869	805	999	999	1148	1131
300	748	792	844	830	988	1006	1135	1149
400	770	752	837	820	978	1012	1146	1142
500	768	776	832	829	999	998	1144	1142
600	752	766	840	805	942	967	996	1024
650	770	750	836	834	964	895	986	972
700	707	702	777	781	805	784	775	776
750	625	692	685	700	658	692	710	717
800	714	719	674	716	682	743	800	915
850	765	815	820	1043	848	995	865	971
900	820	888	848	1041	824	991	847	931
1000	838	1015	838	1023	848	1010	830	947

Table 4.36 Postfire ultimate strength residual factors.

Temperature (°C)	GLG460		GLG550		GLG650		GLG835	
	Air cooling	Water cooling	Air cooling	Water cooling	Air cooling	Water cooling	Air cooling	Water cooling
20	1.000	1.000	1.000	1.000	1.000	1.000	1.000	1.000
100	0.996	1.012	0.991	0.973	0.980	0.988	0.995	0.997
200	0.992	1.015	1.019	0.944	1.003	1.003	1.007	0.992
300	0.987	1.045	0.989	0.973	0.992	1.010	0.996	1.008
400	1.016	0.992	0.981	0.961	0.982	1.016	1.005	1.002
500	1.013	1.024	0.975	0.972	1.003	1.002	1.004	1.002
600	0.992	1.011	0.985	0.944	0.946	0.971	0.874	0.898
650	1.016	0.989	0.980	0.978	0.968	0.899	0.865	0.853
700	0.933	0.926	0.911	0.916	0.808	0.787	0.680	0.681
750	0.825	0.913	0.803	0.821	0.661	0.695	0.623	0.629
800	0.942	0.949	0.790	0.839	0.685	0.746	0.702	0.803
850	1.009	1.075	0.961	1.223	0.851	0.999	0.759	0.852
900	1.082	1.172	0.994	1.220	0.827	0.995	0.743	0.817
1000	1.106	1.339	0.982	1.199	0.851	1.014	0.728	0.831

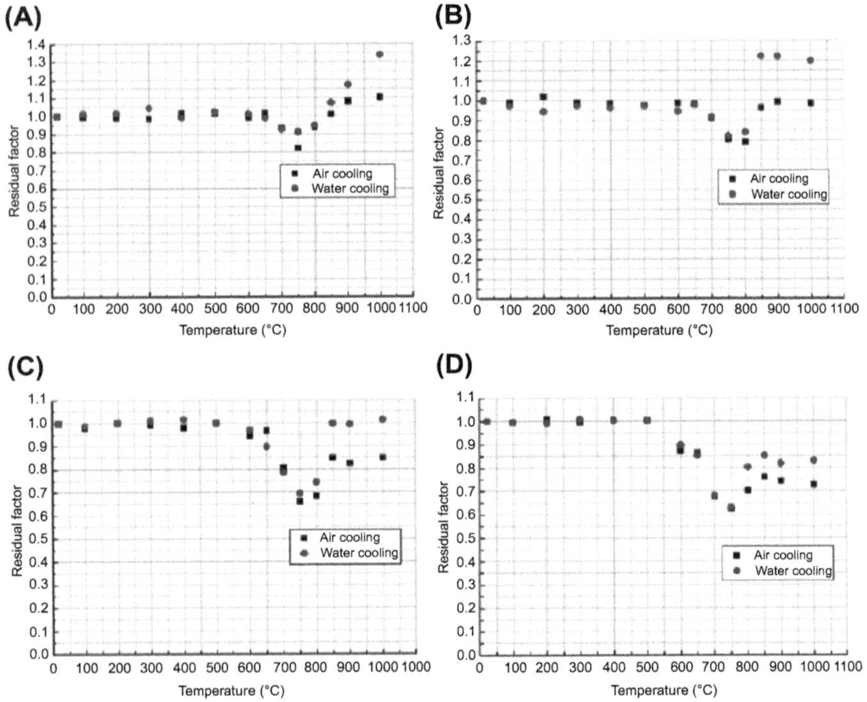

Figure 4.87 Postfire ultimate strength residual factors. (A) GLG460; (B) GLG550; (C) GLG650; and (D) GLG835.

The influence of cooling methods on the postfire ultimate strength is also similar to that on the postfire yield strength. After exposure to elevated temperatures within the range of the rebound stage, the water-cooled specimens presented higher ultimate strength than air cooling specimens because of the aforementioned strengthening effect of quenching. The strengthening effect was also more evident on low-grade steel tie rods. This result indicates that water cooling method has positive effects on the postfire ultimate strength of steel tie rods.

4.10.3.7 Ductility

The ductility of the steel tie rod is the indicator reflecting its plastic deformation capacity, which is defined based on the deformation that the steel tie rod specimens can undergo before fracture. The influence of exposure temperatures, steel tie rod grades, and cooling methods on the postfire ductility of steel tie rods can be obtained by comparing the fracture strain values (Fig. 4.83). The postfire ductility residual factor is defined as the ratio of the fracture strains after cooling down from elevated

Table 4.37 Postfire fracture strains (%).

	GLG460		GLG550		GLG650		GLG835	
Temperature (°C)	**Air cooling**	**Water cooling**	**Air cooling**	**Water cooling**	**Air cooling**	**Water cooling**	**Air cooling**	**Water cooling**
20	21.2	21.2	19.9	19.9	16.6	16.6	13.5	13.5
100	19.8	18.6	20.5	20.5	17.0	15.1	14.5	12.8
200	19.2	19.1	20.1	21.0	16.4	16.7	13.2	13.4
300	21.1	21.3	19.4	20.1	17.1	15.2	14.3	14.5
400	21.8	21.5	20.2	20.4	16.7	15.9	13.3	13.8
500	20.3	20.4	20.2	20.4	15.2	16.3	14.5	13.0
600	19.9	20.7	20.3	21.3	16.5	16.9	17.0	15.8
650	21.8	23.3	20.2	20.9	17.5	15.8	16.5	17.7
700	23.8	23.4	22.1	24.3	22.0	21.8	23.7	24.0
750	29.5	25.6	27.3	27.3	27.5	27.2	24.4	26.0
800	21.0	16.0	21.1	14.0	21.2	12.2	15.7	10.0
850	19.6	15.5	20.1	8.7	11.6	9.5	16.9	9.8
900	16.2	11.8	10.8	7.4	11.7	10.3	15.3	11.0
1000	9.3	6.8	12.1	6.9	12.2	9.9	12.7	7.5

Table 4.38 Postfire ductility residual factors.

Temperature (°C)	GLG460		GLG550		GLG650		GLG835	
	Air cooling	Water cooling	Air cooling	Water cooling	Air cooling	Water cooling	Air cooling	Water cooling
20	1.000	1.000	1.000	1.000	1.000	1.000	1.000	1.000
100	0.934	0.877	1.030	1.030	1.024	0.910	1.074	0.948
200	0.906	0.901	1.009	1.055	0.988	1.006	0.978	0.993
300	0.995	1.005	0.975	1.010	1.030	0.916	1.059	1.074
400	1.028	1.014	1.015	1.025	1.006	0.958	0.985	1.022
500	0.958	0.960	1.015	1.025	0.916	0.982	1.074	0.963
600	0.939	0.976	1.020	1.070	0.994	1.018	1.259	1.170
650	1.028	1.099	1.015	1.050	1.054	0.952	1.222	1.311
700	1.123	1.104	1.111	1.221	1.325	1.313	1.756	1.778
750	1.392	1.208	1.372	1.372	1.657	1.639	1.807	1.926
800	0.991	0.755	1.060	0.704	1.277	0.735	1.163	0.741
850	0.925	0.731	1.010	0.436	0.699	0.572	1.252	0.726
900	0.764	0.557	0.543	0.373	0.705	0.620	1.133	0.815
1000	0.439	0.321	0.606	0.347	0.735	0.596	0.941	0.552

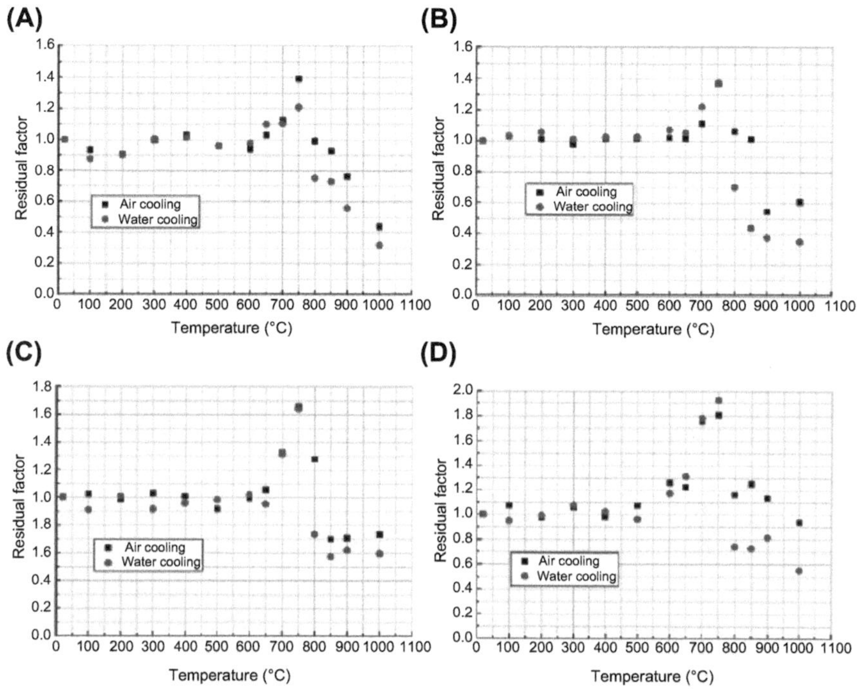

Figure 4.88 Postfire ductility residual factors. (A) GLG460; (B) GLG550; (C) GLG650; and (D) GLG835.

temperatures (δ_{uPT}) to that at ambient temperature without exposure to fire (δ_u). Tables 4.37 and 4.38 present the postfire fracture strains obtained from this experiment and corresponding residual factors, respectively. The residual factors are also plotted in Fig. 4.88 as the function of exposure temperatures (each value listed and plotted is the average value of three specimens in a group).

Low-grade steel tie rods showed higher ductility than high-grade steel tie rods, whereas the general trends of residual factors with respect to exposure temperatures were similar. Ductility was not affected when exposure temperatures were below 650°C for GLG460, GLG550, and GLG650 and 500°C for GLG835. Subsequently, a remarkable increase in ductility emerged up to 750°C for all these four grades of steel tie rods, which is positive to their reuse after fire events. Furthermore, high-grade steel tie rod (GLG650 and GLG835) showed more obvious growth in ductility than low-grade specimens (GLG460 and GLG550). However, after cooling from temperatures above 750°C, all the steel tie rods presented sharp decrease in ductility though high-grade specimens showed less reduction than low-grade ones.

The influence of different cooling methods on postfire ductility of steel tie rods was significant, particularly when the exposure temperature exceeded 750°C. Within this range of exposure temperatures, the specimens cooled by water spraying showed much lower ductility than the specimens cooled in air, which is negative to the reuse of the steel tie rods after exposure to a fire event extinguished by fire guns.

4.10.3.8 Effects of cyclic heating and cooling

Additional tensile coupon tests were conducted on specimens of GLG460 and GLG835 after multiple heating and cooling cycles to explore if possible degeneration in mechanical properties of steel tie rods exists; elevated temperatures 300, 600, and 900°C were considered. Related mechanical properties, such as yield strength, ultimate strength, elastic modulus, and ductility after cyclic heating and cooling, were obtained. Degeneration factors are defined as the ratio of the mechanical properties after multiple heating and cooling cycles (E_{MPT}, f_{yMPT}, f_{uMPT}, and δ_{uMPT}) to that after one heating and cooling cycle (E_{PT}, f_{yPT}, f_{uPT}, and δ_{uPT}). Tables 4.39–4.41 present the mechanical properties after multiple heating and cooling cycles obtained in this experiment and their corresponding degeneration factors. The degeneration factors are also plotted in Fig. 4.89 as the function of heating and cooling cycle index.

All the elastic modulus, yield strength, ultimate strength, and ductility of both GLG460 and GLG835 after multiple heating and cooling cycles were kept nearly the same as those after only one heating and cooling cycle, regardless of how high the exposure temperature was. This indicates that the effects of cyclic heating and cooling on the mechanical properties of both low- and high-grade steel tie rods are insignificant. Hence, they can be ignored in the assessment of the postfire mechanical properties of steel tie rods. They were also not considered in the predictive equations proposed in Section 4.10.4.

4.10.4 Predictive equations for the postfire residual mechanical properties

The experimental results described above demonstrate that the mechanical properties of steel tie rods have markedly changed after exposure to elevated temperatures. Although the characteristic of these changes can be estimated approximately according to the data and figures presented, to propose a series of predictive equations that are sufficiently accurate for practical use and can easily be mastered by engineers to evaluate the postfire

Table 4.39 Mechanical properties and degeneration factors after multiple heating and cooling circles (300°C).

Cycle index	E_{MPT} (GPa)	$\frac{E_{MPT}}{E_{PT}}$	$f_{y,MPT}$ (MPa)	$\frac{f_{y,MPT}}{f_{yPT}}$	$f_{u,MPT}$ (MPa)	$\frac{f_{u,MPT}}{f_{uPT}}$	$\delta_{u,MPT}$ (%)	$\frac{\delta_{u,MPT}}{\delta_{uPT}}$
GLG460								
1	202.4	1.000	601	1.000	748	1.000	21.1	1.000
2	203.4	1.005	592	0.985	737	0.985	20.7	0.981
3	201.8	0.997	590	0.982	742	0.992	21.5	1.019
GLG835								
1	207.7	1.000	1020	1.000	1135	1.000	14.3	1.000
2	206.8	0.996	1005	0.985	1121	0.988	14.6	1.021
3	208.6	1.004	1019	0.999	1135	1.000	14.8	1.035

Table 4.40 Mechanical properties and degeneration factors after multiple heating and cooling circles (600°C).

Cycle index	E_{MPT} (GPa)	$\frac{E_{MPT}}{E_{PT}}$	$f_{y,MPT}$ (MPa)	$\frac{f_{y,MPT}}{f_{yPT}}$	$f_{u,MPT}$ (MPa)	$\frac{f_{u,MPT}}{f_{uPT}}$	$\delta_{u,MPT}$ (%)	$\frac{\delta_{u,MPT}}{\delta_{uPT}}$
GLG460								
1	205.5	1.000	623	1.000	752	1.000	19.9	1.000
2	204.3	0.994	611	0.981	756	1.005	21.3	1.070
3	202.8	0.987	614	0.986	756	1.005	21.4	1.075
GLG835								
1	203.6	1.000	878	1.000	996	1.000	17.0	1.000
2	204.8	1.006	892	1.016	1007	1.011	16.4	0.965
3	202.2	0.993	831	0.946	920	0.924	17.8	1.047

Table 4.41 Mechanical properties and degeneration factors after multiple heating and cooling circles (900°C).

Cycle index	E_{MPT} (GPa)	$\frac{E_{MPT}}{E_{PT}}$	$f_{y,MPT}$ (MPa)	$\frac{f_{y,MPT}}{f_{yPT}}$	$f_{u,MPT}$ (MPa)	$\frac{f_{u,MPT}}{f_{uPT}}$	$\delta_{u,MPT}$ (%)	$\frac{\delta_{u,MPT}}{\delta_{uPT}}$
GLG460								
1	182.6	1.000	528	1.000	820	1.000	16.2	1.000
2	180.9	0.991	502	0.951	783	0.955	16.8	1.037
3	182.4	0.999	502	0.951	775	0.945	16.7	1.031
GLG835								
1	176.9	1.000	585	1.000	847	1.000	15.3	1.000
2	179.6	1.015	583	0.997	872	1.030	14.4	0.941
3	177.5	1.003	570	0.974	842	0.994	16.4	1.072

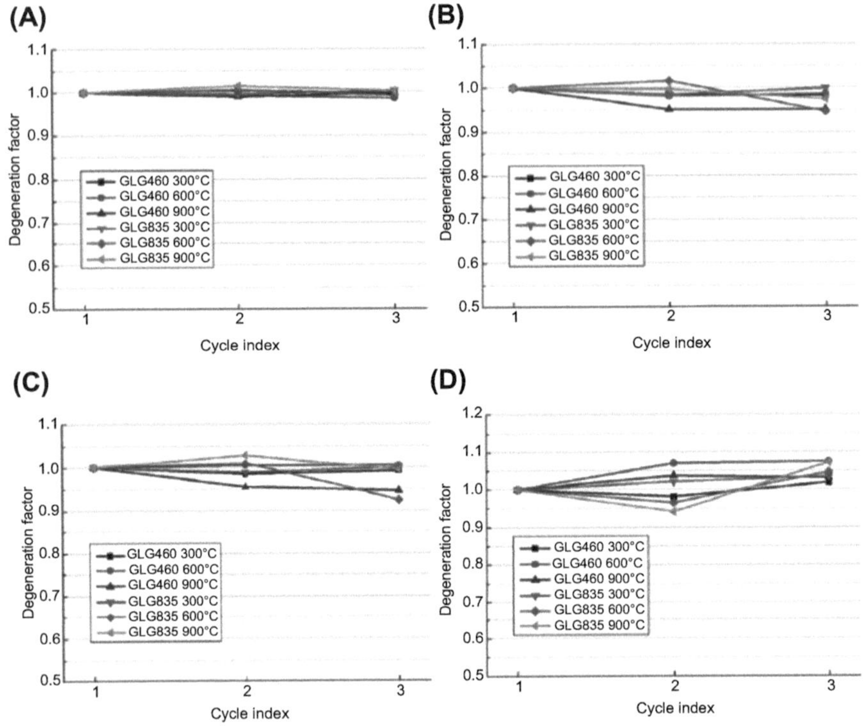

Figure 4.89 Degeneration factors of steel tie rods after multiple heating and cooling cycles. (A) Elastic modulus; (B) yield strength; (C) ultimate strength; and (D) ductility.

performance of steel tie rods is significant. However, the available equations proposed for commonly used structural steels (hot-rolled or cold-formed) are not suitable to predict the postfire mechanical properties of steel tie rods because of considerable differences in chemical composition and manufacturing process between them. The influence of cooling methods is also significant according to the results of this experiment, but it has been barely considered even in the study on structural steels, let alone steel tie rods. Thus, a new set of predictive equations, which are simple in form yet have considerable precision, were proposed to evaluate the postfire mechanical properties of steel tie rods in varying grades based on the experimental results. As the exposure temperature is the main factor causing the deterioration of mechanical properties, the predictive equations were developed as a function of the highest exposure temperature, where the influence of different cooling methods was also incorporated.

4.10.4.1 Residual elastic modulus

According to the experimental results presented in Section 3.3, the following predictive equations were proposed to present the reduction in the elastic modulus of steel tie rods after cooling from elevated temperatures. The influence of different cooling methods was not considered because nearly no difference was observed between them.

For GLG460,

$$20^\circ C \le T \le 600^\circ C, \frac{E_{PT}}{E} = 1.006 - 2.76 \times 10^{-5}T \tag{4.41a}$$

$$600^\circ C < T \le 1000^\circ C, \frac{E_{PT}}{E} = 1.050 + 2.39 \times 10^{-5}T - 2.45 \times 10^{-7}T^2 \tag{4.41b}$$

For GLG550,

$$20^\circ C \le T \le 600^\circ C, \frac{E_{PT}}{E} = 1.006 - 4.23 \times 10^{-5}T \tag{4.42a}$$

$$600^\circ C < T \le 1000^\circ C, \frac{E_{PT}}{E} = -0.445 + 6.16 \times 10^{-3}T - 8.45 \times 10^{-6}T + 3.54 \times 10^{-9}T^3 \tag{4.42b}$$

For GLG650,

$$20^\circ C \le T \le 700^\circ C, \frac{E_{PT}}{E} = 1.000 - 4.67 \times 10^{-5}T \tag{4.43a}$$

$$700^\circ C < T \le 1000^\circ C, \frac{E_{PT}}{E} = 1.91 - 1.77 \times 10^{-3}T + 6.23 \times 10^{-7}T^2 \tag{4.43b}$$

For GLG835,

$$20^\circ C \le T \le 650^\circ C, \frac{E_{PT}}{E} = 0.994 - 1.47 \times 10^{-4}T + 1.47 \times 10^{-7}T^2 \tag{4.44a}$$

$$650^\circ C < T \le 1000^\circ C, \frac{E_{PT}}{E} = 1.295 - 5.03 \times 10^{-4}T \tag{4.44b}$$

Fig. 4.90 shows a comparison of the elastic modulus residual factors predicted by Eqs. (4.41a)–(4.44b) with the experimental results, and a good agreement can be observed between them.

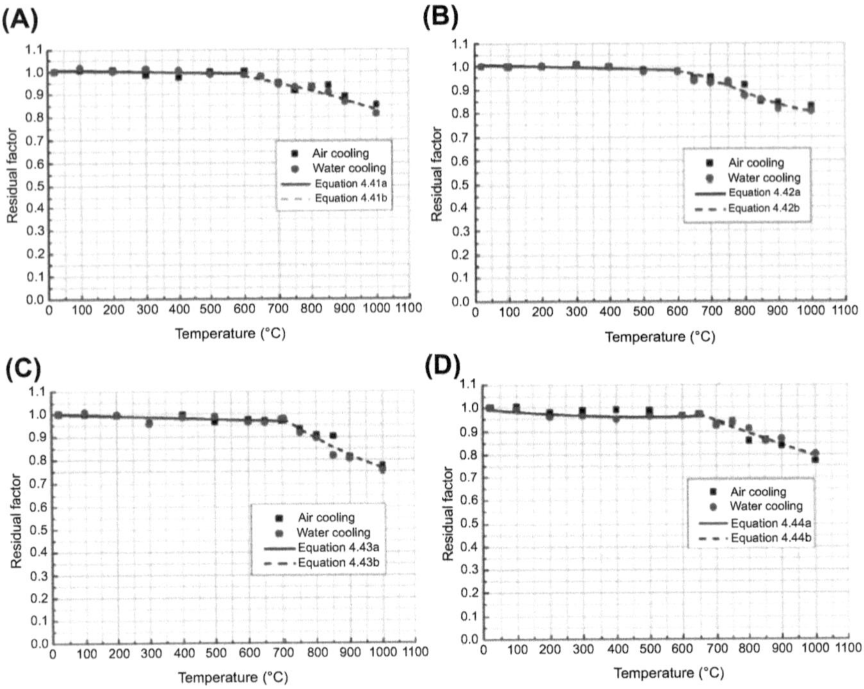

Figure 4.90 Comparison of predicted elastic moduli residual factors with the experimental results. (A) GLG460; (B) GLG550; (C) GLG650; and (D) GLG835.

4.10.4.2 Residual yield strength

The residual yield strength factors of all four steel tie rod grades presented three main regions with respect to exposure temperature because of the appearance of yield strength rebound. Hence, predictive equations containing three equations instead of two were developed to ensure accuracy, where the influence of different cooling methods was considered.

GLG460:

For air cooling,

$$20^{\circ}\text{C} \leq T \leq 650^{\circ}\text{C}, \frac{f_{yPT}}{f_y} = 0.999 - 7.71 \times 10^{-5}T + 2.01 \times 10^{-7}T^2 \tag{4.45a}$$

$$650^{\circ}\text{C} < T \leq 750^{\circ}\text{C}, \frac{f_{yPT}}{f_y} = -4.457 + 1.83 \times 10^{-2}T - 1.52 \times 10^{-5}T^2 \tag{4.45b}$$

$$750^{\circ}\text{C} < T \leq 1000^{\circ}\text{C}, \frac{f_{yPT}}{f_y} = 0.408 + 1.69 \times 10^{-4}T + 3.49 \times 10^{-7}T^2 \tag{4.45c}$$

For water cooling,

$$20^\circ\mathrm{C} \leq T \leq 650^\circ\mathrm{C}, \frac{f_{yPT}}{f_y} = 1.000 + 2.08 \times 10^{-4}T - 3.12 \times 10^{-7}T^2 \tag{4.46a}$$

$$650^\circ\mathrm{C} < T \leq 800^\circ\mathrm{C}, \frac{f_{yPT}}{f_y} = 2.446 - 3.10 \times 10^{-3}T + 1.31 \times 10^{-6}T^2 \tag{4.46b}$$

$$800^\circ\mathrm{C} < T \leq 1000^\circ\mathrm{C}, \frac{f_{yPT}}{f_y} = -2.520 + 6.20 \times 10^{-3}T - 2.51 \times 10^{-6}T^2 \tag{4.46c}$$

Fig. 4.91A compares the postfire yield strength residual factors of GLG460 predicted by Eqs. (4.45a)–(4.45c) and (4.46a)–(4.46c) with the experimental results. A good agreement is observed between the predictions and test results.

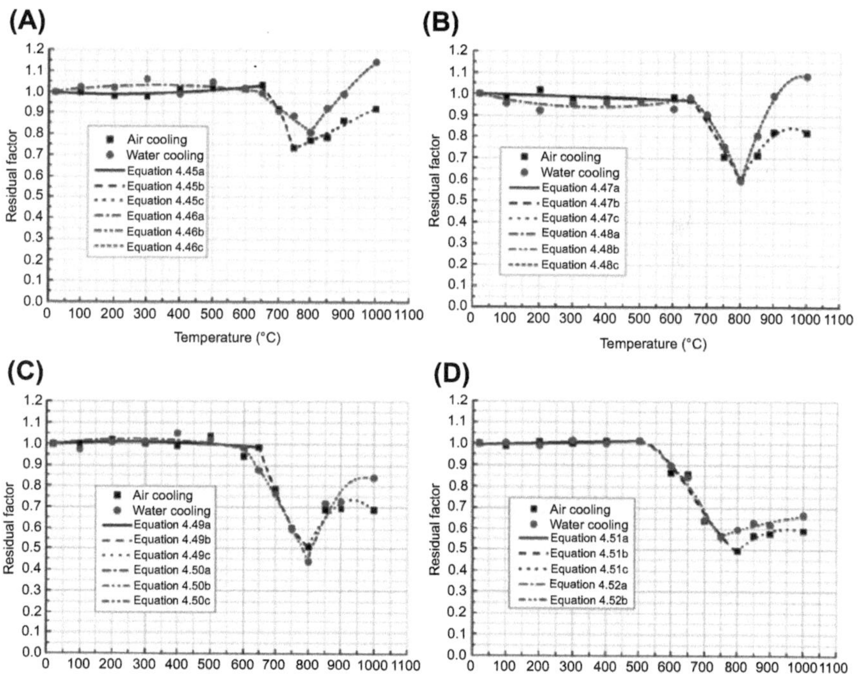

Figure 4.91 Comparison of predicted yield strength residual factors with the experimental results. (A) GLG460; (B) GLG550; (C) GLG650; and (D) GLG835.

GLG550:
For air cooling,

$$20^{\circ}\text{C} \leq T \leq 650^{\circ}\text{C}, \frac{f_{yPT}}{f_y} = 1.002 - 4.79 \times 10^{-5}T \quad (4.47a)$$

$$650^{\circ}\text{C} < T \leq 800^{\circ}\text{C}, \frac{f_{yPT}}{f_y} = 1.229 + 1.43 \times 10^{-3}T - 2.79 \times 10^{-6}T^2 \quad (4.47b)$$

$$800^{\circ}\text{C} < T \leq 1000^{\circ}\text{C}, \frac{f_{yPT}}{f_y} = -8.610 + 1.98 \times 10^{-2}T - 1.04 \times 10^{-5}T^2 \quad (4.47c)$$

For water cooling,

$$20^{\circ}\text{C} \leq T \leq 650^{\circ}\text{C}, \frac{f_{yPT}}{f_y} = 1.007 - 3.79 \times 10^{-4}T + 5.22 \times 10^{-7}T^2 \quad (4.48a)$$

$$650^{\circ}\text{C} < T \leq 800^{\circ}\text{C}, \frac{f_{yPT}}{f_y} = -1.528 + 9.13 \times 10^{-3}T - 8.10 \times 10^{-6}T^2 \quad (4.48b)$$

$$800^{\circ}\text{C} < T \leq 1000^{\circ}\text{C}, \frac{f_{yPT}}{f_y} = -12.869 + 2.83 \times 10^{-2}T - 1.43 \times 10^{-5}T^2 \quad (4.48c)$$

Fig. 4.91B shows a comparison of the postfire yield strength residual factors of GLG550 predicted by Eqs. (4.47a)–(4.47c) and (4.48a)–(4.48c) with the experimental results, and a good agreement can be observed between them.

GLG650:
For air cooling,

$$20^{\circ}\text{C} \leq T \leq 650^{\circ}\text{C}, \frac{f_{yPT}}{f_y} = 0.998 + 8.27 \times 10^{-5}T - 1.61 \times 10^{-7}T^2 \quad (4.49a)$$

$$650^{\circ}\text{C} < T \leq 800^{\circ}\text{C}, \frac{f_{yPT}}{f_y} = 8.623 - 1.87 \times 10^{-2}T - 1.06 \times 10^{-5}T^2 \quad (4.49b)$$

$$800^{\circ}\text{C} < T \leq 1000^{\circ}\text{C}, \frac{f_{yPT}}{f_y} = -9.787 + 2.25 \times 10^{-2}T - 1.21 \times 10^{-5}T^2 \tag{4.49c}$$

For air cooling,

$$20^{\circ}\text{C} \leq T \leq 600^{\circ}\text{C}, \frac{f_{yPT}}{f_y} = 0.996 + 2.00 \times 10^{-4}T - 3.75 \times 10^{-7}T^2 \tag{4.50a}$$

$$600^{\circ}\text{C} < T \leq 800^{\circ}\text{C}, \frac{f_{yPT}}{f_y} = 0.482 + 3.50 \times 10^{-3}T - 4.46 \times 10^{-6}T^2 \tag{4.50b}$$

$$800^{\circ}\text{C} < T \leq 1000^{\circ}\text{C}, \frac{f_{yPT}}{f_y} = -11.444 + 2.53 \times 10^{-2}T - 1.30 \times 10^{-5}T^2 \tag{4.50c}$$

Fig. 4.91C presents the comparison of the predicted postfire yield strength residual factors of GLG650 from Eqs. (4.49a)–(4.49c) and (4.50a)–(4.50c) with the experimental results, and the agreement between them is good.

GLG835:

For air cooling,

$$20^{\circ}\text{C} \leq T \leq 500^{\circ}\text{C}, \frac{f_{yPT}}{f_y} = 0.994 + 3.43 \times 10^{-5}T \tag{4.51a}$$

$$500^{\circ}\text{C} < T \leq 800^{\circ}\text{C}, \frac{f_{yPT}}{f_y} = -5.918 + 3.44 \times 10^{-2}T - 5.45 \times 10^{-5}T^2 + 2.70 \times 10^{-8}T^3 \tag{4.51b}$$

$$800^{\circ}\text{C} < T \leq 1000^{\circ}\text{C}, \frac{f_{yPT}}{f_y} = -3.187 + 7.93 \times 10^{-3}T - 4.15 \times 10^{-6}T^2 \tag{4.51c}$$

For water cooling, when the exposure temperature is below 500°C, the predictive equation proposed is equal to the corresponding equation of air cooling (Eq. 4.51a) because of the slight difference between them. When the exposed temperature exceeds 500°C, Eqs. (4.52a) and (4.52b) are suggested as follows:

$$500^\circ\mathrm{C} < T \le 750^\circ\mathrm{C}, \frac{f_{yPT}}{f_y} = 0.059 + 4.45 \times 10^{-3}T - 5.07 \times 10^{-6}T^2 \tag{4.52a}$$

$$800^\circ\mathrm{C} < T \le 1000^\circ\mathrm{C}, \frac{f_{yPT}}{f_y} = -0.348 + 1.83 \times 10^{-3}T - 8.17 \times 10^{-7}T^2 \tag{4.52b}$$

Fig. 4.91D compares the predictions of Eqs. (4.51a)—(4.51c) and (4.52a) —(4.52b) with the experimental results on the postfire yield strength residual factors of GLG835. The proposed equations agree well with the experimental results.

4.10.4.3 *Residual ultimate strength*

Similar to the postfire yield strength, piecewise equations were proposed to predict the residual ultimate strength residual factors of steel tie rods studied in this experiment, and the influence of different cooling methods was also considered.

GLG460:

For air cooling,

$$20^\circ\mathrm{C} \le T \le 650^\circ\mathrm{C}, \frac{f_{uPT}}{f_u} = 0.994 + 2.25 \times 10^{-5}T \tag{4.53a}$$

$$650^\circ\mathrm{C} < T \le 1000^\circ\mathrm{C}, \frac{f_{uPT}}{f_u} = 36.957 - 0.132T + 1.59 \times 10^{-4}T^2 - 6.26 \times 10^{-8}T^3 \tag{4.53b}$$

For water cooling,

$$20^\circ\mathrm{C} \le T \le 650^\circ\mathrm{C}, \frac{f_{uPT}}{f_u} = 0.997 - 1.74 \times 10^{-4}T - 2.86 \times 10^{-7}T^2 \tag{4.54a}$$

$$650^\circ\mathrm{C} < T \le 1000^\circ\mathrm{C}, \frac{f_{uPT}}{f_u} = 18.784 - 6.43 \times 10^{-2}T + 7.52 \times 10^{-5}T^2 - 2.84 \times 10^{-8}T^3 \tag{4.54b}$$

Fig. 4.92A shows a comparison of the postfire ultimate strength residual factors of GLG460 predicted by Eqs. (4.53a)–(4.53b) and (4.54a)–(4.54b) with the experimental results, and a good agreement can be observed between them.

GLG550:

For air cooling,

$$20^\circ\text{C} \le T \le 650^\circ\text{C}, \frac{f_{uPT}}{f_u} = 1.004 - 3.92 \times 10^{-5} T \tag{4.55a}$$

$$650^\circ\text{C} < T \le 800^\circ\text{C}, \frac{f_{uPT}}{f_u} = 4.794 - 9.51 \times 10^{-3} T + 5.63 \times 10^{-6} T^2 \tag{4.55b}$$

$$800^\circ\text{C} < T \le 1000^\circ\text{C}, \frac{f_{uPT}}{f_u} = -9.777 + 2.31 \times 10^{-2} T - 1.23 \times 10^{-5} T^2 \tag{4.55c}$$

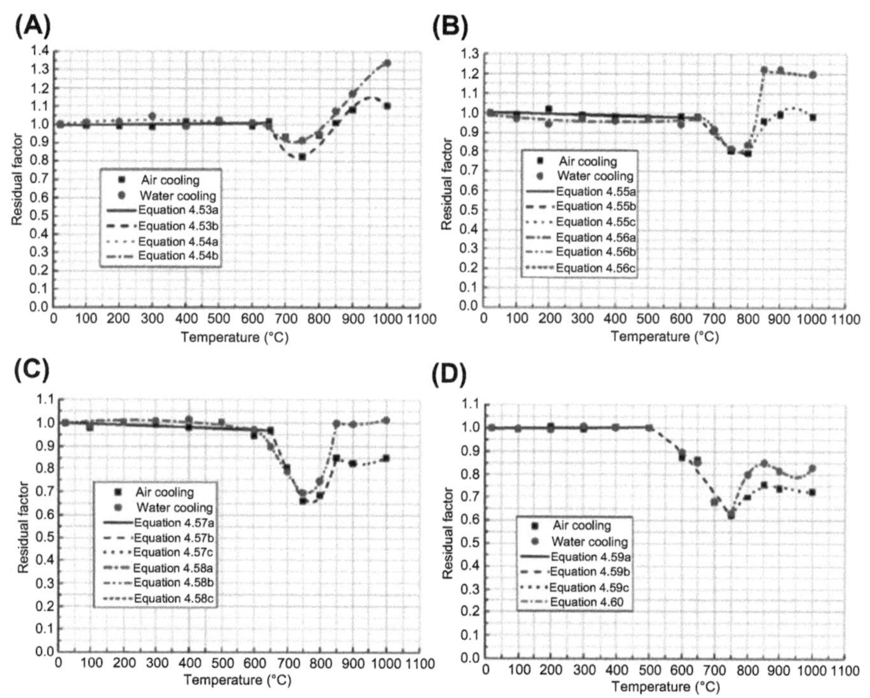

Figure 4.92 Comparison of predicted ultimate strength residual factors with the experimental results. (A) GLG460; (B) GLG550; (C) GLG650; and (D) GLG835.

For water cooling,

$$20^\circ\text{C} \leq T \leq 800^\circ\text{C}, \frac{f_{uPT}}{f_u} = 0.994 - 1.79 \times 10^{-4}T + 2.19 \times 10^{-7}T^2 \tag{4.56a}$$

$$650^\circ\text{C} < T \leq 850^\circ\text{C}, \frac{f_{uPT}}{f_u} = -93.748 + 0.40T - 5.67 \times 10^{-4}T^2 + 2.65 \times 10^{-7}T^3 \tag{4.56b}$$

$$850^\circ\text{C} < T \leq 1000^\circ\text{C}, \frac{f_{uPT}}{f_u} = 0.421 + 1.85 \times 10^{-3}T - 1.08 \times 10^{-6}T^2 \tag{4.56c}$$

Fig. 4.92B shows a comparison of the ultimate strength residual factors predicted by Eqs. (4.55a)–(4.55c) and (4.56a)–(4.56c) with the test results, and a good agreement can be observed between them.

GLG650:

For air cooling,

$$20^\circ\text{C} \leq T \leq 650^\circ\text{C}, \frac{f_{uPT}}{f_u} = 1.002 - 5.12 \times 10^{-5}T \tag{4.57a}$$

$$650^\circ\text{C} < T \leq 800^\circ\text{C}, \frac{f_{uPT}}{f_u} = -21.696 - 0.11T - 1.72 \times 10^{-4}T^2 + 8.70 \times 10^{-8}T^3 \tag{4.57b}$$

$$850^\circ\text{C} < T \leq 1000^\circ\text{C}, \frac{f_{uPT}}{f_u} = 4.948 - 8.92 \times 10^{-3}T + 4.82 \times 10^{-6}T^2 \tag{4.57c}$$

For water cooling,

$$20^\circ\text{C} \leq T \leq 600^\circ\text{C}, \frac{f_{uPT}}{f_u} = 0.997 + 1.34 \times 10^{-4}T - 2.97 \times 10^{-7}T^2 \tag{4.58a}$$

$$600^\circ\text{C} < T \leq 850^\circ\text{C}, \frac{f_{uPT}}{f_u} = -34.033 + 0.160T - 2.38 \times 10^{-4}T^2 + 1.17 \times 10^{-7}T^3 \tag{4.58b}$$

$$850^{\circ}\mathrm{C} < T \leq 1000^{\circ}\mathrm{C}, \frac{f_{uPT}}{f_u} = 2.450 - 3.24 \times 10^{-3}T + 1.81 \times 10^{-6}T^2 \tag{4.58c}$$

Fig. 4.92C presents the comparison of the ultimate strength residual factors predicted by Eqs. (4.57a)−(4.57c) and (4.58a)−(4.58c) with the experimental results, and a good agreement can be observed between them.

GLG835:

For air cooling,

$$20^{\circ}\mathrm{C} \leq T \leq 500^{\circ}\mathrm{C}, \frac{f_{uPT}}{f_u} = 0.998 + 1.12 \times 10^{-5}T \tag{4.59a}$$

$$500^{\circ}\mathrm{C} < T \leq 750^{\circ}\mathrm{C}, \frac{f_{uPT}}{f_u} = 0.556 + 2.53 \times 10^{-3}T - 3.28 \times 10^{-6}T^2 \tag{4.59b}$$

$$750^{\circ}\mathrm{C} < T \leq 1000^{\circ}\mathrm{C}, \frac{f_{uPT}}{f_u} = -20.429 + 6.83 \times 10^{-2}T - 7.31 \times 10^{-5}T^2 + 2.60 \times 10^{-8}T^3 \tag{4.59c}$$

For water cooling, when the exposed temperature is below 750°C, the predictive equations are equal to the equations of air cooling, i.e., Eqs. (4.59a) and (4.59b) because of the slight difference between them. When the exposed temperature exceeds 750°C, Eq. (4.60) is recommended as follows.

$$750^{\circ}\mathrm{C} < T \leq 1000^{\circ}\mathrm{C}, \frac{f_{uPT}}{f_u} = -65.095 - 0.22T - 2.46 \times 10^{-4}T^2 + 9.13 \times 10^{-8}T^3 \tag{4.60}$$

Fig. 4.92D compares the predicted postfire ultimate strength residual factors from Eqs. (4.59) and (4.60) with the experimental results, and a good agreement can be observed between them.

References

AISC, 2005. American Institution of Steel Construction. Specification for structural steel buildings, Chicago.

Arrayago, I., Real, E., Gardner, L., 2015. Description of stress−strain curves for stainless steel alloys. Mater. Des. 87, 540−552.

ASM, 2002. Atlas of Stress-Strain Curves, second ed. ASM International.
ASTM E8, 2005. Standard Test Methods for Tension Testing of Metallic Materials. West Conshohocken, PA.
ASTM, 2009. Standard Specification for High-Yield-Strength, Quenched and Tempered Alloy Steel Plate, Suitable for Welding. West Conshohocken, PA.
ASTM E21-09, 2009. Standard Test Methods for Elevated Temperature Tension Tests of Metallic Materials. West Conshohocken, PA.
Australia Standards AS4100, 1998. Steel Structures. Sydney, Australia.
Australian/New Zealand Standard AS/NZS 4600, 2005. Cold-Formed Steel Structures, Standards Australia. Sydney, Australia and Standards New Zealand, Wellington, New Zealand.
AWS, 2008. Structural Welding Code. Steel. American National Standards Institute, Miami.
Azhari, F., Heidarpour, A., Zhao, X.L., Hutchinson, C.R., 2015. Mechanical properties of ultra-high strength (Grade 1200) steel tubes under cooling phase of a fire: an experimental investigation. Constr. Build. Mater. 93, 841–850.
Bhadeshia, H.K.D.H., Honeycombe, R.W.K., 2006. Steels: Microstructure and Properties, 3th ed. Elsevier Science & Technology, Oxford, United Kingdom.
Brohez, S., Delvosalle, C., Marlair, G., 2004. A two-thermocouples probe for radiation corrections of measured temperatures in compartment fires. Fire Saf. J. 39, 399–411.
BS EN 10025-6, 2004. Hot Rolled Products of Structural Steels: Part 6. Technical Delivery Conditions for Flat Products of High Yield Strength Structural Steels in the Quenched and Tempered Condition. British Standards Institution, London.
CEN, 2009. Hot rolled products of structural steels. Part 6: technical delivery conditions for flat products of high yield strength structural steels in the quenched and tempered condition. Brussels.
Chen, J., Young, B., Uy, B., 2006. Behavior of high strength structural steel at elevated temperatures. J. Struct. Eng. 132 (12), 1948–1954.
Chiew, S.P., Zhao, M.S., Lee, C.K., 2014. Mechanical properties of heat-treated high strength steel under fire/post-fire conditions. J. Constr. Steel Res. 98, 12–19.
Chen, Z., Lu, J., Liu, H., Liao, X., 2016. Experimental study on the post-fire mechanical properties of high-strength steel tie rods, J. Constr. Steel Res. 121, 311–329.
EN 10002–5, 1992. Metallic Materials-Tensile Testing-Part5: Method of Testing at Elevated Temperature. European Committee for Standardization, Brussels, Belgium.
EN10025-3, 2004. Hot Rolled Products of Structural Steels-Part 3: Technical Delivery Conditions for Normalized/Normalized Rolled Weldable Fine Grain Structural Steels. European Committee for Standardization, Brussels, Belgium.
European Code 3, 2010. Eurocode 3: Design of Steel Structures. European Committee for Standardization, Brussels, Belgium.
Fonseca, R., Dario, I., 2014. Correction of the Post -necking True Stress - Strain Data Using Instrumented Nanoindentation. The University of North Carolina at Charlotte.
GB 50045-95, 2005. Code for Fire Protection Design of Tall Buildings. Standardization Administration of the China, Beijing.
GB/T 1591-2008, 2008. High Strength Low Alloy Structural Steels. Standardization Administration of the China, Beijing.
GB/T 20934-2007, 2007. Steel Tie Rod. Standardization Administration of the China, Beijing.
GB/T 22315-2008, 2008. Metallic Material-Determination of Modulus of Elasticity and Poisson's Ratio. Standardization Administration of the China, Beijing.
GB/T 228.1-2010, 2010. Metallic Materials-Tensile Testing-Part 1: Method of Test at Room Temperature. Standardization Administration of the China, Beijing.
GB/T 3077-1999, 1999. Alloy Structure Steels. Standardization Administration of the China, Beijing.

GB/T 4338-2010, 2010. Metallic Material-Tensile Testing at Elevated Temperature. Standardization Administration of the China, Beijing.
GB/T 700-2006, 2006. Carbon Structural Steels. Standardization Administration of the China, Beijing.
Gunalan, S., Mahendran, M., 2014. Experimental investigation of post-fire mechanical properties of cold-formed steels. Thin-Walled Struct. 84, 241−254.
Harada, K., Yabuki, Y., Terai, T., 1997. A theoretical consideration on heat transfer in fire resistance furnaces for furnace harmonization. Fire Saf. Sci. 5, 1033−1044.
Lee, J., Engelhardt, M.D., Taleff, E.M., 2012. Mechanical properties of ASTM A992 steel after fire. Eng. J. 49, 33−44.
Li, G.Q., Lyu, H., Zhang, C., 2017. Post-fire mechanical properties of high strength q690 structural steel. J. Constr. Steel Res. 132, 108−116.
Ling, Y., 1996. Uniaxial True stress-strain after necking. AMP J. Technol. 5.
Lu, J., Liu, H., Chen, Z., Liao, X., 2016. Experimental investigation into the post-fire mechanical properties of hot-rolled and cold-formed steels, J. Constr. Steel Res. 121, 291−310.
Makelainen, P., Outinen, J., Kesti, J., 1998. Fire designmodel for structural steel S420Mbased upon transient-state tensile test results. J. Constr. Steel Res. 48 (1), 47−57.
Mirambell, E., Real, E., 2000. On the calculation of deflections in structural stainless steel beams: an experimental and numerical investigation. J. Constr. Steel Res. 54, 109−133.
Outinen, J., 2007. Mechanical Properties of Structural Steels at High Temperatures and after Cooling Down [Ph.D. Thesis]. Helsinki University of Technology, Helsinki, Finland.
Outinen, J., Makelainen, P., 2004. Mechanical properties of structural steel at elevated temperatures and after cooling down. J. Fire Mater 28, 237−251.
Outinen, J., Kesti, J., Makelainen, P., 1997. Fire design model for structural steel S355 based upon transient state tensile test results. J. Constr. Steel Res. 42 (3), 161−169.
Qiang, X., Bijlaard, F.S.K., Kolstein, H., 2012. Post-fire mechanical properties of high strength structural steels S460 and S690. Eng. Struct. 35, 1−10.
Qiang, X., Bijlaard, F.S.K., Kolstein, H., 2013. Post-fire performance of very high strength steel S960. J. Constr. Steel Res. 80, 235−242.
Rasmussen, K.J.R., 2003. Full-range stress−strain curves for stainless steel alloys. J. Constr. Steel Res. 59, 47−61.
Smith, C.I., Kirby, B.R., Lapwood, D.G., et al., 1981. The reinstatement of fire damaged steel framed structures. Fire Saf. J. 4 (1), 21−62.
Tao, Z., Wang, X.Q., Uy, B., 2012. Stress-strain curves of structural and reinforcing steels after exposure to elevated temperatures. J. Mater. Civ. Eng. 25 (9), 1306−1316.
Wang, L., Tong, W., 2015. Identification of post-necking strain hardening behavior of thin sheet metals from image-based surface strain data in uniaxial tension tests. Int. J. Solids Struct. 75 (76), 12−31.
Zhao, K., Wang, L., Chang, Y., Yan, J., 2016. Identification of post-necking stress−strain curve for sheet metals by inverse method. Mech. Mater. 92, 107−118.

CHAPTER 5

Creep behavior of steels at elevated temperatures

Contents

Material Properties of Steel in Fire Conditions
ISBN 978-0-12-813302-6
https://doi.org/10.1016/B978-0-12-813302-6.00005-9

5.1 Introduction

When exposed to elevated temperatures, steel structures undergo increasing permanent deformations with time even when the applied stress level is below that of yield stress, and this time-dependent deformation is referred to as creep. Thus, creep can be defined as an increase in strain in a solid material under constant stress over a period.

Significance of creep deformations under fire conditions has been highlighted by various researchers. Morovat et al. (2014) examined the effect of creep on the buckling strength of steel columns using Harmathy creep model (Harmathy, 1967) and Fields and Fields creep model (Fields and Fields, 1989). It was shown that the buckling load capacity of columns not only depends on load level and temperature range but also on time duration for which the load was applied. In a series of recent studies, Kodur and Dwaikat, (2010); Kodur et al., (2010) studied the influence of high-temperature creep on midspan deflections and axial forces in steel beams by utilizing ANSYS software package. It was concluded that the exclusion of creep leads to lower midspan deflections, an overestimation of fire induced compressive axial forces, and an underestimation of fire induced tensile axial forces in restrained steel beams. Furthermore, accounting for high-temperature creep in the numerical model leads to good predictions of deflections and restraint forces and compares well with measured data from tests on the beams, again highlighting the importance of creep. In addition, recent studies have clearly shown that the extent of creep is influenced by the chemical composition of steel, and this varies for different grades of steel (Kodur and Aziz, 2014), that is, different types of steel exhibit different creep response.

Currently, creep, in most structural fire engineering applications, is modeled using Harmathy creep model (Harmathy, 1967), AYSYS creep models (Kodur et al., 2010; Zienkiewicz and Cormeau, 1974), or Fields and Fields creep model (Fields and Fields, 1989). Although a few other models have been proposed in literatures (Brnic et al., 2010; Toric et al., 2013; Cowan and Khandelwal, 2014), they are rarely used. Because of lack of creep test data, the commonly employed creep models were developed based on creep tests on Australian A149 steel or ASTM A36 structural steel. In addition, the creep models, used in structural fire engineering applications, have been proposed on an ad hoc basis and are not usually presented in a consistent mechanics-based framework. As the creep models are highly phenomenological, without a consistent description, it is difficult to evaluate and extend the application of existing creep models to different structural steels. In addition, much of the available creep data in literature are for carbon steels. Therefore, for better understanding of creep effects in new type steels, which are widely used in structural engineering applications, relevant experimental creep data and the corresponding constitutive creep models are needed.

5.2 Typical creep–time curve

Creep is the inelastic response in a material when subjected to elevated temperatures. Temperature has an important effect on deformation phenomena. When a material is subjected to stress due to loading, it undergoes some level of strain. In contrast, creep strain is induced in the material as time-dependent mechanical response. The standard tensile strength test on materials, with appropriate modifications, can be applied to investigate the creep response of materials at high temperatures. There are different approaches to undertake stress–strain tests at elevated temperatures (Kodur et al., 2010). The less complex approach to measure creep is through tension tests at constant load and temperature. The resulting strain is measured as a function of time. This is often referred to as creep test, and the resulting strain–time curves include the effects of creep. If the creep test extends over a sufficiently long period of time, a stage at which the specimen weakens rapidly and eventually ruptures may be reached. Tests aimed at investigating this final stage of the creep phenomenon are referred to as creep rupture tests.

Typical strain–time curves obtained from creep tests at constant loading, applied over sufficiently long periods of time, often exhibit three characteristic stages as shown in Fig. 5.1. In stage 1, referred to as primary or transient creep, following setting in of instantaneous elastic strain, ε_0, the material deforms rapidly but at a decreasing rate. The duration of this stage is typically relatively short in relation to the total creep curve. In stage 2, referred to as secondary or steady-state creep, the creep strain rate reaches a

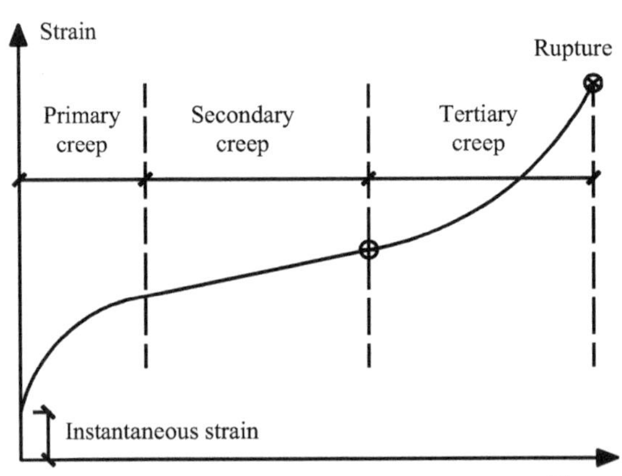

Figure 5.1 Creep response of steel at elevated temperature.

minimum value and remains approximately constant over a relatively longer period. In stage 3, referred to as tertiary creep, the creep strain rate accelerates rapidly and attains ruptures when the material is unable to withstand the applied loading.

5.3 Creep behavior of high strength Q460 steel

A comprehensive test program was designed to undertake creep tests on Q460 steel. The creep tests were carried out at nine temperatures in the range of 300−900°C, namely, 300, 400, 450, 500, 550, 600, 700, 800, and 900°C and at various stress levels ranging from 13 to 509 MPa.

5.3.1 Test setup

For undertaking high-temperature creep tests, high-temperature creep test equipment was specially designed and fabricated. A schematic of test setup for high-temperature creep test, shown in Fig. 5.2A, comprises of a tensile testing equipment, an electric furnace, a strain measurement device, and a data acquisition system. The tensile testing equipment is MTS loading system with a load capacity of 50 kN. In this equipment, the axial deformation in the test coupon at elevated temperature can be measured through two linear variable differential transducers (LVDTs) (±10 mm), with 0.001 mm sensitivity, that are mounted outside the furnace through two connecting rods placed inside the furnace (Fig. 5.2B). The electric furnace comprises of cylindrical chamber with a maximum heating length of 150 mm. In creep tests, the specimen is mounted between two cylinder

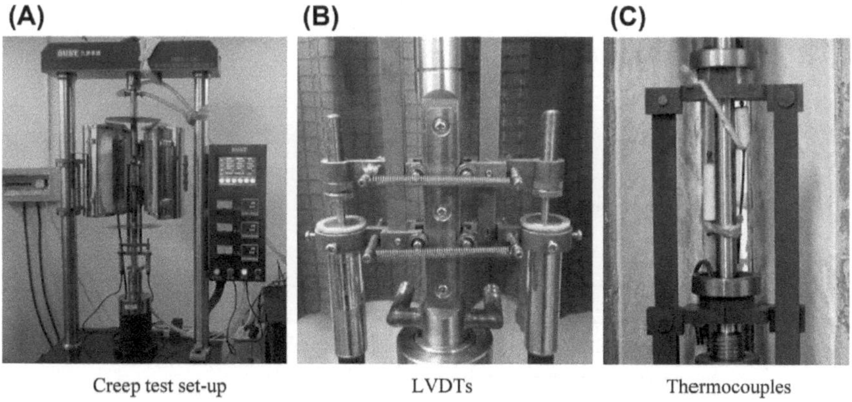

Figure 5.2 Creep test setup and instrumentations on test specimen. A) Creep test setup, (B) LVDTs, and (C) thermocouples.

grips with threads inside. The programmable furnace enables generation of target temperatures up to 1100°C at a specified rate of temperature rise and keeps the target temperature inside the furnace constant for the duration of creep test. Three built-in thermocouples are mounted inside the furnace to measure temperatures at upper, middle, and lower zones, and the average readings of these three thermocouples are taken as the furnace temperature (FT) (as shown in Fig. 5.2C). The furnace and LVDTs are connected to a data acquisition system and a computer wherein temperature, tension load and displacements during the creep test are recorded.

5.3.2 Test specimens

For high-temperature creep tests, some round steel coupons were cut from Q460 steel sheet of 20 mm in thickness. A total of 45 creep test specimens were fabricated. The test coupons measured 187 mm in total length, and the gage length for creep measurement is 100 mm. The diameter of cross section of the tapered section is 10 mm. The configuration of a typical test coupon used in creep test is shown in Fig. 5.3. The stress levels (load ratio) and target temperatures at which these tests were carried out are tabulated in Table 5.1, in which the yield strength (1.0% strain) of Q460 at elevated temperatures was determined by GB/T 228.1-2010 (2010). The load ratio is defined as the ratio of applied stress on the specimen to the yield strength at a given temperature. When stress level in steel is too high, the coupon experiences rupture in a very short time. Conversely, a coupon exhibits inconspicuous creep response under low stress level. The stress levels during creep tests were varied to observe the creep effect on mechanical response at various temperatures.

5.3.3 Test procedure

Mechanical properties of steel, including the creep, at a specific temperature can be evaluated through transient- or steady-state tests. Although transient-state tests are more realistic in simulating real fire exposure conditions, steady-state tests are more commonly employed in evaluating

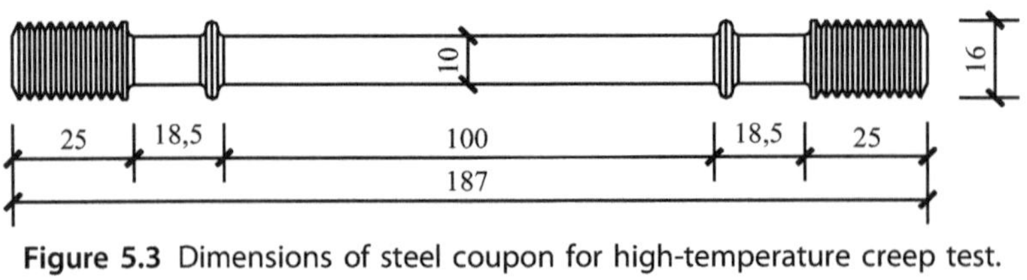

Figure 5.3 Dimensions of steel coupon for high-temperature creep test.

Table 5.1 Selected temperature and stress level for creep tests.

Temperature/ °C	Yield strength (MPa)	Stress levels/MPa	Load levels
300	575	457, 483, 495, 509	0.79–0.89
400	518	406, 432, 457, 476	0.78–0.92
450	532	369, 382, 394, 400, 406	0.69–0.76
500	430	242, 254, 267, 280	0.56–0.65
550	374	165, 178, 204, 210	0.44–0.56
600	367	89, 102, 114, 127, 178	0.24–0.49
700	182	25.5, 38,51, 64	0.14–0.35
800	89	13, 19, 25.5, 32, 38	0.15–0.43
900	-	13, 19, 25.5, 32	–

mechanical properties due to ease of undertaking these tests. In the current study, to enable comparison with previously developed test data, steady-state test method was utilized for undertaking high-temperature creep tests. Creep tests were conducted at different temperatures, and stress levels using the test setup are described above.

For these creep tests, steel coupons were installed into the grips of the MTS machine and then three thermocouples were mounted at top, bottom, and midlength of the gage length of specimen as shown in Fig. 5.2C. These thermocouples are to monitor actual temperature in the furnace during the creep test. Then, an electric furnace was set up around the specimen with special care being taken to ensure that the coupon is placed at the center of the furnace. Following this, the furnace is turned on to be heated to a specified target temperature at a heating rate of 10°C/min. Once the target temperature is attained, the steel coupon was maintained at this target temperature for 30 min to attain a steady-state condition, and then the coupon was maintained at the target temperature for the entire test duration. Then a predefined load was applied at a rate of 10 kN/min. The maximum load applied in the creep tests is 40 kN; hence, the loading phase lasts no more than 5 min. After the applied loading reaches the target load, the creep displacements were measured using two highly sensitive LVDTs. The total length of the coupon that is exposed to high temperature during a creep test is 150 mm, and of this, only 100 mm in the middle portion is taken to be the gage length to calculate corresponding creep strain ($\Delta L/L$).

Using this procedure, creep tests were conducted at different stress levels and at different temperatures. When the creep deformation reaches the maximum allowable displacement of LVDTs, namely 10 mm, the LVDTs were adjusted by hand quickly, and the data were continuously recorded. In this way, the creep deformation was obtained within a wide displacement range, till the maximum movement of hydraulic jack.

5.3.4 Creep–time curves at various stress levels

The measured creep strains at various stress levels and at different temperatures, from 300 to 900°C, are plotted in Fig. 5.4. Note that the creep strain versus time curves presented here does not include the instantaneous strain and only contains the resulting creep strain. Zero time on the X-axis corresponds to the point at which full loading was applied on the coupon. In Fig. 5.4, symbol "⊕" represents the initiation (boundary point) of secondary creep and tertiary creep, and symbol "⊗" represents the occurrence of fracture at the termination of tertiary creep. Generally, at a given temperature, the creep strain at a higher stress level is larger than that at a lower stress level. Increased creep strain at higher stress level can be attributed to dislocation movement due to diffusion of atoms that occurs at higher temperatures and under high stress levels.

The general trend of creep response in lower temperatures (300–400°C), shown in Fig. 5.4A,B, can be grouped into two stages, namely primary (stage I) and secondary (stage II). The increase in secondary creep strain with time during stage II is dependent both on temperature and stress levels. At low stress levels, there is little secondary creep generated at 300 and 400°C. However, at temperatures of 400°C and higher stress level of 476 MPa, the tertiary creep occurs.

During the moderate temperature range of 450–550°C (Fig. 5.4C–E), secondary creep and tertiary creep increased at a rapid rate, and most specimens experienced rupture. The maximum creep strain is relatively low before rupture when the stress levels are very high. However, at a lower stress level, the maximum creep strain becomes much bigger before rupture. The increasing of maximum creep strain with lower stress level can be attributed to better development of plasticity in steel with longer time duration, and the time lasting of creep in steel is longer with the lower stress levels. For example, at temperature of 500°C, the maximum creep strain at stress of 267 MPa is much higher than that at stress of 280 MPa, even though both specimens experienced rupture. The same trend can be found at temperatures of 450 and 550°C.

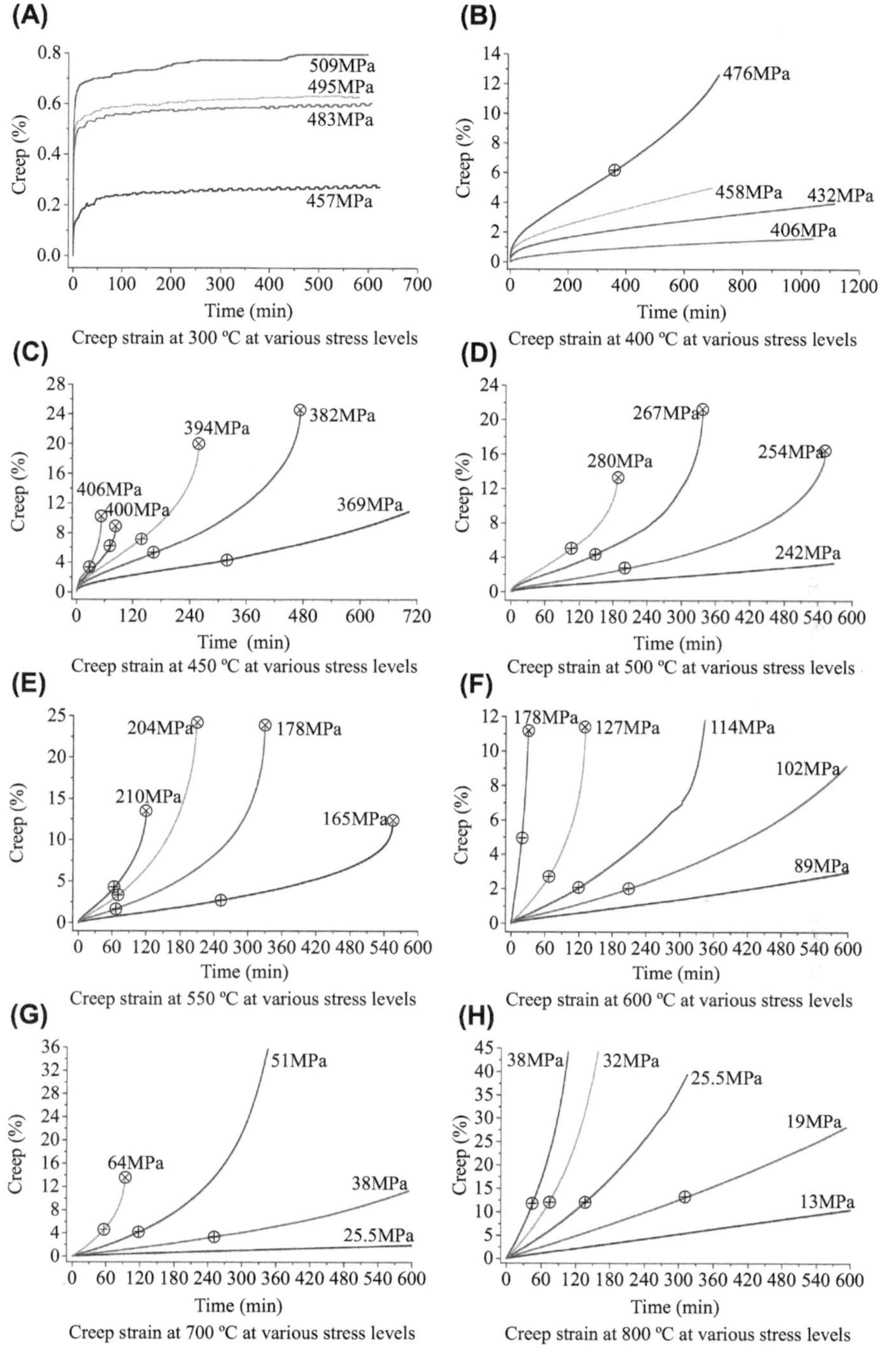

Figure 5.4 Creep strain response in Q460 steel at various stress levels. (A) Creep strain at 300°C at various stress levels; (B) creep strain at 400°C at various stress levels; (C) creep strain at 450°C at various stress levels; (D) creep strain at 500°C at various stress levels; (E) creep strain at 550°C at various stress levels; (F) creep strain at 600°C at various stress levels; (G) creep strain at 700°C at various stress levels; (H) creep strain at 800°C at various stress levels; and (I) creep strain at 900°C at various stress levels.

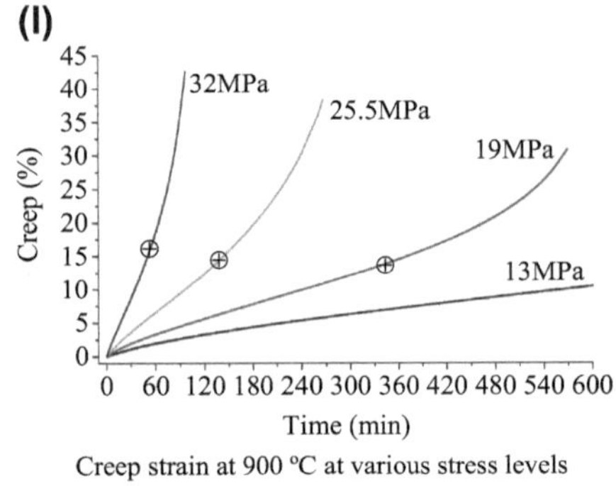

Figure 5.4 Cont'd

When the temperature is very high (600–900°C), the trend of creep response can be also generally grouped into two stages (Fig. 5.4F–I), namely secondary (stage II) and tertiary (stage III). The primary creep becomes very short, and the deformation fully developed in very few minutes. The quick rise in creep strain with time in stage II is due to thermal softening. Secondary creep is highly important under fire exposure conditions because it dominates the creep response and occurs in a constant rate and over short period of time. In stage III, the tertiary creep increases at a faster rate due to reduced cross section of the specimen resulting from necking phenomenon that results in higher stresses for the same level of applied load. With the increase of temperature, the ductility of steel increases, and maximum creep values become higher. The same phenomenon can also be seen in higher temperature tension test (Wang et al., 2013). Because of the limitation of movement of hydraulic jack, the rupture of some specimens during the temperature range of 600–900°C was not obtained. However, the limited data have little effect on the creep response of steel because the tertiary creep is not important as the secondary creep in structural analysis.

5.3.5 Creep–time curves at various temperatures

The effect of temperature on creep response of Q460 steel is illustrated in Fig. 5.5. It can be seen from Fig. 5.5A–C, for temperatures range of 300–700°C, creep at a certain stress level increased significantly with temperature, even the elevation of temperature is only 50°C. For example, at stress level of 406 MPa, the creep at temperature of 450°C is much bigger than that at temperature of 400°C. Therefore, at a given stress level, the creep deformation increases with temperature. This is due to diffusion of

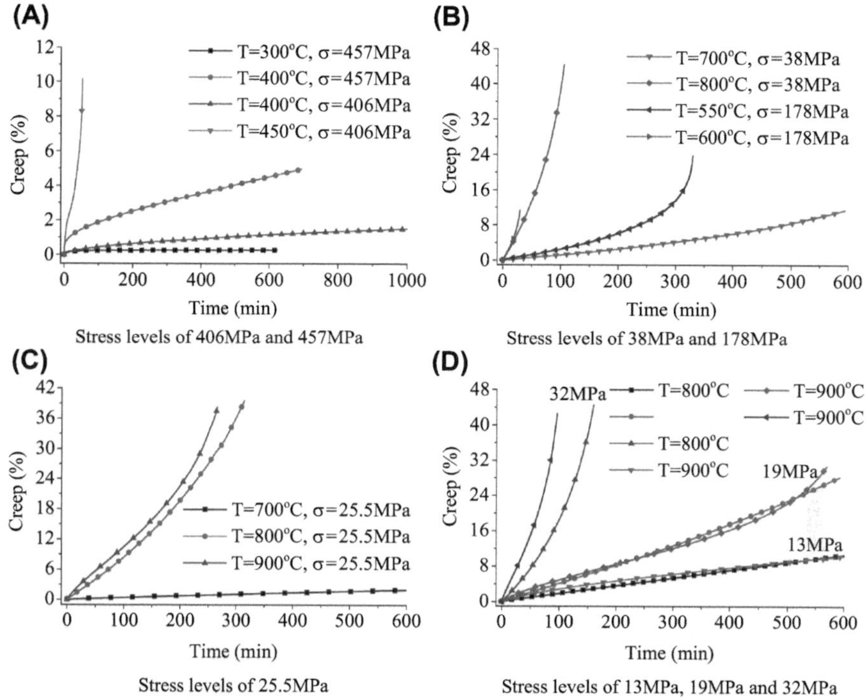

Figure 5.5 Creep strain response in Q460 steel at various temperatures. (A) Stress levels of 406 and 457 MPa; (B) stress levels of 38 and 178 MPa; (C) stress levels of 25.5 MPa; (D) stress levels of 13, 19, and 32 MPa.

atoms within material grains resulting from temperature rise. When temperatures are in the range of up to 800°C, which represents 50% of steel melting temperature, grain boundary diffusion tends to be the dominant mechanism that results in accelerated creep strain (Ashby and Jones, 2005). However, at temperature range of 800–900°C, the general trend observed above is not applicable, as is shown in Fig. 5.5D. The creep strain at 900°C is similar with that at 800°C under the same stress level, no matter what stress levels are. The trends in the figure indicate that creep deformation increases substantially with increasing temperature up to 800°C and then keep constant with the increase of temperature.

5.3.6 Rupture of specimens

The failure modes in test specimens at various temperatures and stress levels are shown in Fig. 5.6. These illustrations, at each temperature, correspond to the test coupons with stress levels ranging from low to high, as tabulated in Table 5.1. As can be seen from Fig. 5.6, at temperature of 900°C, the failure of test specimens exhibited a nonsignificant necking phenomenon, but a very large ductile response was observed. However, during the temperatures range of 450–550°C, brittle fracture was observed for most specimens and the ductility varies greatly with different stress level. To compare the creep strain before rupture, all the specimens with occurrence of rupture are listed in Table 5.2. From the table, the creep strains at rupture are within the range of 9%–25% in different stress levels and different temperatures.

5.3.7 Proposed creep model

Some empirical models (Harmathy, 1967; Fields and Fields, 1989) to predict creep strain of carbon steel at elevated temperature are available in the literature. However, because of the facts that creep response in steel is highly material dependent, current models are not applicable to Q460 steel. Test data generated from the above creep test can be utilized to develop

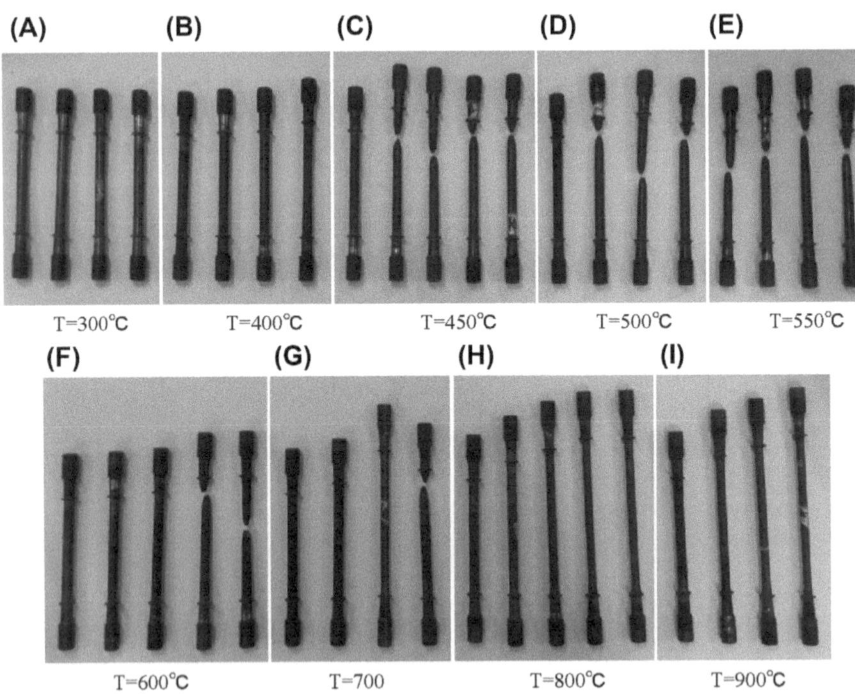

Figure 5.6 Failure modes in creep test coupons at various temperatures. (A) $T = 300°C$; (B) $T = 400°C$; (C) $T = 450°C$; (D) $T = 500°C$; (E) $T = 550°C$; (F) $T = 600°C$; (G) $T = 700$; (H) $T = 800°C$; and (I) $T = 900°C$.

Table 5.2 Ultimate creep in specimens with fracture.

Temperature (°C)	Stress (MPa)	Ultimate creep (%)	Temperature (°C)	Stress (MPa)	Ultimate creep (%)
450	382	24.4	700	64	13.2
550	204	24.2	500	280	13.1
550	178	23.7	550	165	12.2
500	267	21.1	600	127	11.4
450	394	19.8	600	178	11.2
550	254	16.4	450	406	10.1
550	210	13.4	450	400	8.9

creep models as a function of temperature and stress levels for Q460 steel. As can be seen from the test data, little creep strain was observed at the temperature of 300°C. Therefore, only the test data for temperature above 400°C were selected in the creep models. Such models can be useful to evaluate fire response of Q460 steel structures.

5.3.7.1 Fields and Fields model

One of the widely used creep models in structural fire engineering applications is the one proposed by Fields and Fields (Fields and Fields, 1989). This model incorporates a power law and represents general expression for creep strain, ε_{cr}, in the form of a Norton–Bailey equation, which is given as

$$\varepsilon_{cr} = at^b\sigma^c \tag{5.1}$$

where ε_{cr} is the creep strain in steel; t is time (in min); σ is stress (in ksi); and a, b, and c are temperature-dependent material properties.

As Fields and Fields creep model is very simple in its form, this model can easily be incorporated in numerical analysis of fire resistance of steel structures. Utilizing Eq. (5.1) as the basis for temperature induced creep, Fields and Fields (1989) derived expressions for "a," "b," and "c" based on test data on ASTM A36 steel.

$$a = \begin{cases} 10^{-6.10-0.0053T} & 350°\text{C} \le T < 500°\text{C} \\ 10^{-13.25+0.00851T} & 500°\text{C} \le T \le 650°\text{C} \end{cases} \tag{5.2}$$

$$b = -1.1 + 0.0035T \quad 350°\text{C} \le T \le 650°\text{C} \tag{5.3}$$

$$c = 2.1 + 0.0064T \quad 350°\text{C} \le T \le 650°\text{C} \tag{5.4}$$

Cowan and Khandelwal (2014) calibrated the above equations for creep of Australian A149 structural steel and Japanese SS41 steel (both like ASTM

A36 steel), where the creep data were obtained for temperatures between 350 and 650°C and for a time period of 4 h.

If stress and time are independent, Eq. (5.1) can be rewritten in rate form as

$$\varepsilon'_{cr} = At^{B}\sigma^{C} \tag{5.5}$$

where $A = a \times b$, $B = b - 1$, and $C = c$.

In the rate form, it can be observed that this formulation (Eq. 5.5) is a time-hardening model. In addition, tertiary creep is not considered in this model.

To simulate high-temperature creep response in Q460 steel using the form of Fields and Fields creep model, a nonlinear regression fitting tool was employed and the values of coefficients a, b, and c were evaluated at different temperatures, using the creep test data in Q460 steel. For example, when fitting the test data at 300°C, four creep—time cures are selected, and predefined function of Fields and Fields creep model is employed. Then, value ranges of a, b, and c are set up. After fitting is finished, the fitting values are accepted if the adjusted R-square is higher than 0.95. Otherwise, other ranges are set up again until the adjusted R-square reaches 0.95. These fitting values of a, b, c and adjust R-square are tabulated in Table 5.3.

Like the expressions in Eq. (5.2—5.4), a nonlinear regression fitting tool was employed, and the values of coefficients a, b, and c were fitted as a function of temperature, using the creep test data in Q460 steel. The expressions of a, b, and c for Q460 steel are given in Eq. (5.6—5.8).

$$\mathrm{Log}a = \begin{cases} -56.797 + 0.2265T - 3.562 \times 10^{-4}T^{2} & 300n \le T \le 450n \\ 2286.225 - 13.65927T + 2.65644 \times 10^{-2}T^{2} - 1.6964 \times 10^{-5}T^{3} & 450n < T \le 600n \\ -69.01 + 0.155T - 1.040 \times 10^{-4}T^{2} + 1.420 \times 10^{-8}T^{3} & 600n < T \le 900n \end{cases} \tag{5.6}$$

$$b = \begin{cases} -2.71 + 0.0128T - 1.20 \times 10^{-5}T^{2} & 300n \le T \le 450n \\ -14.96 + 0.08T - 1.36 \times 10^{-4}T^{2} + 7.81 \times 10^{-8}T^{3} & 450n < T \le 600n \\ 1.18 - 7.07 \times 10^{-3}T + 1.9 \times 10^{-5}T^{2} - 1.31 \times 10^{-8}T^{3} & 600n < T \le 900n \end{cases} \tag{5.7}$$

Table 5.3 Coefficients of *a*, *b*, and *c* in Fields and Fields creep model for Q460 steel.

Temperature	*a*	*b*	*c*	Adjust R-square
400	6.48E-22	0.49	11.25	0.986
450	1.00E-25	0.62	13.94	0.992
500	1.55E-21	0.80	12.37	0.960
550	9.38E-12	0.89	6.75	0.971
600	4.20E-09	0.95	5.53	0.970
700	2.52E-05	1.05	3.50	0.982
800	5.00E-03	0.98	2.25	0.991
900	4.00E-02	0.66	1.98	0.994

$$c = \begin{cases} 39.25 - 0.18T + 2.75 \times 10^{-4}T^2 & 300n \leq T \leq 450n \\ -1557.61 + 9.21T - 1.78 \times 10^{-2}T^2 + 1.132 \times 10^{-5}T^3 & 450n < T \leq 600n \\ 22.52 - 2.07 \times 10^{-2}T - 3.34 \times 10^{-5}T^2 + 3.45 \times 10^{-8}T^3 & 600n < T \leq 900n \end{cases} \tag{5.8}$$

Fig. 5.7 shows a comparison of high-temperature creep strains computed using Eqs. (5.1, 5.6—5.8) with those of measured test data for different stress levels at various temperatures. As can be seen from the figure, the proposed creep model based on fitting to creep data shows good agreement with the test data for Q460 steel.

To further study the influence of steel types on resulting coefficients, the parameters *a*, *b*, and *c* obtained in this paper for Q460 steel are compared with Eq. (5.2—5.4) proposed by Fields and Fields (1989) for conventional carbon A36 steel, and this is shown in Fig. 5.8. As can be seen in Fig. 5.8, the coefficients obtained from fitting Q460 steel creep test data are quite different with those for A36 steel, and this further infers that different types of steel exhibit varying creep response. This phenomenon can be attributed to the different content of chemical compositions in two steels and different condition of heat treatment. The main chemical compositions of the two steels are tabulated in Table 5.4. As can be seen from the table, the value of manganese in Q460 steel is much higher than that in A36 steel. In addition, the heat treatment for Q460 steel includes hot rolling and normalizing; however, for A36 steel, only hot rolling is conducted.

Therefore, coefficients generated for A149 steel or A36 steel cannot be directly utilized to predict the creep behavior of other steels, such as Q460 steel. It can also be seen from Fig. 5.8 that the parameter b for both A36 steel and Q460 steel is quite close. However, for parameters *a* and *c*, the

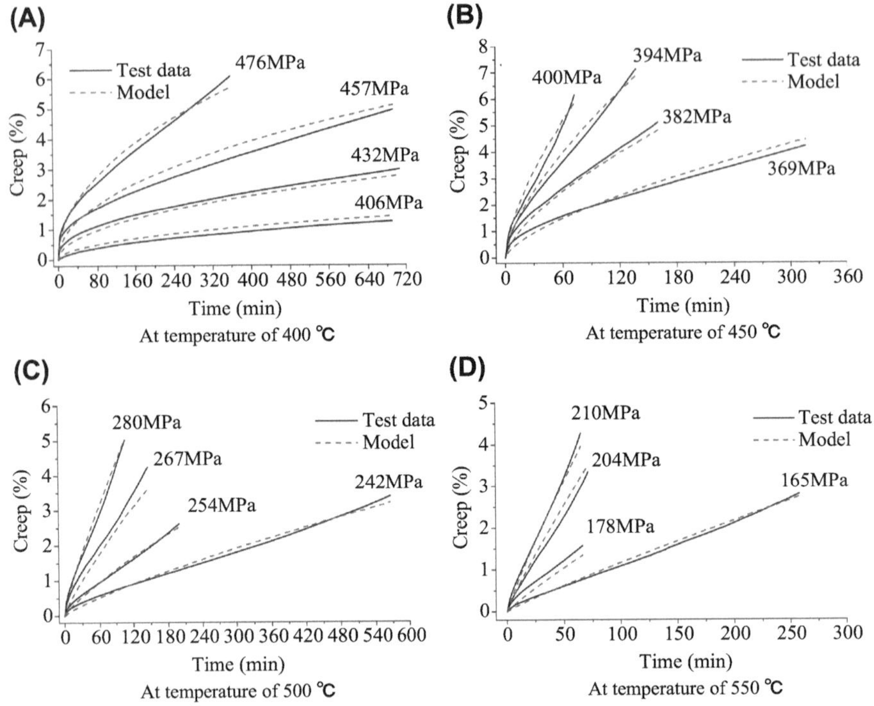

Figure 5.7 Calibration of fields and fields creep model for Q460 steel at various stress levels. (A) At temperature of 400°C; (B) at temperature of 450°C; (C) at temperature of 500°C; (D) at temperature of 550°C; (E) at temperature of 600°C; (F) at temperature of 700°C; (G) at temperature of 800°C; (H) at temperature of 900°C.

trends are quite different. For Q460 steel, at 450°C, parameter a reaches the minimum value and parameter c reaches the maximum value.

5.3.7.2 ANSYS model

ANSYS is a versatile finite element analysis program that provides 13 creep models for evaluating response of steel structures in fire. ANSYS also provides a nonlinear regression tool option for fitting a creep model based on specific data generated from creep experiment. The creep models in ANSYS account for the primary and secondary stages of creep that occurs in steel structures under fire conditions. Tertiary creep is usually not accounted for as it implies impending failure due to the gross element distortion at that stage as well as the complexity involved in capturing this phenomenon (Kodur and Dwaikat, 2009). ANSYS analyzes creep using two time integration schemes, namely explicit and implicit methods. These two integration schemes are relatively equivalent to time-hardening and

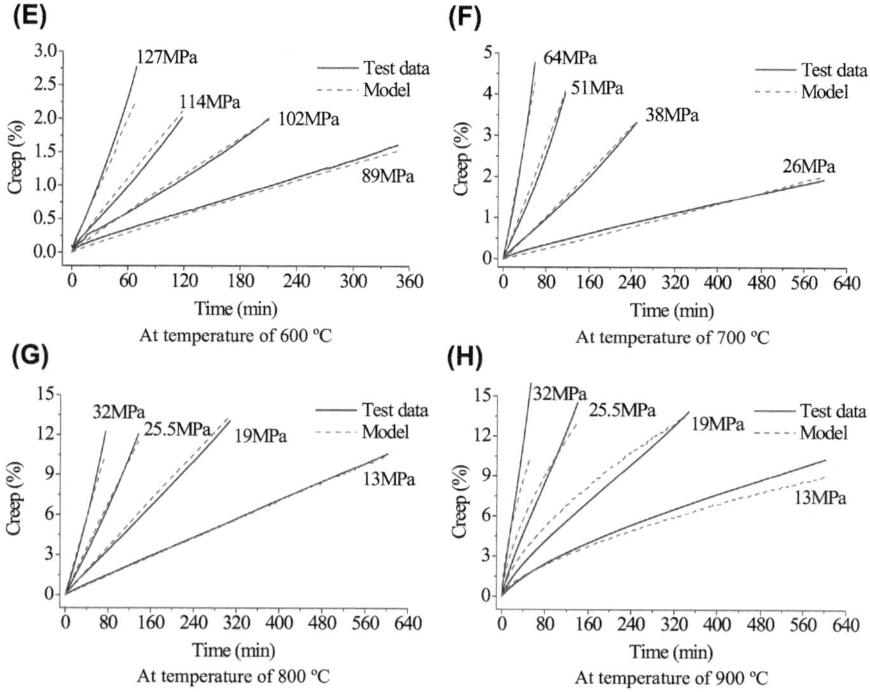

Figure 5.7 Cont'd

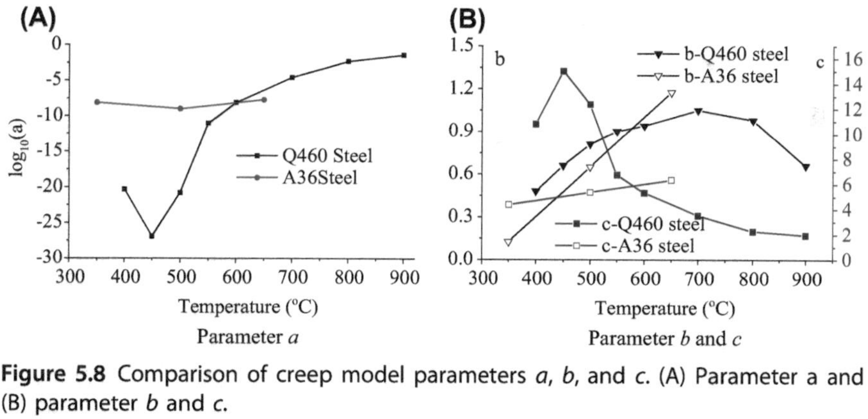

Figure 5.8 Comparison of creep model parameters *a*, *b*, and *c*. (A) Parameter a and (B) parameter *b* and *c*.

strain-hardening rules, respectively. Both are applicable to static or transient analysis.

The implicit method is robust, fast, accurate, and recommended for general use, especially with problems involving large creep strains and large deformations. It has provisions for including temperature-dependent

Table 5.4 Comparative chemical composition limit in Q460 steel and A36 steel (in %).

Steel	Carbon	Silicon	Manganese	Phosphorous	Sulfur	Vanadium
Q460	≤0.2	≤0.55	≤1.70	≤0.035	≤0.035	≤0.20
A36	≤0.26	≤0.4	≤0.40	≤0.05	≤0.04	–

constants. ANSYS can model pure creep, creep with isotropic hardening plasticity, and creep with kinematic hardening plasticity, using both von Mises and Hill potentials. As the creep and plasticity are modeled simultaneously (no superposition), the implicit method is more accurate and efficient than the explicit method. The explicit method is useful for cases involving very small-time steps, such as in transient analyses. There are neither provisions for temperature-dependent constants nor simultaneous modeling of creep with any other material models such as plasticity. However, there is temperature dependency option in Arrhenius function, and this option can be utilized combining explicit creep with other plasticity options through nonsimultaneous modeling (superposition). In these cases, ANSYS first performs the plastic analysis and then undertakes the creep calculations.

For evaluating material coefficients, implicit creep model "11" available in ANSYS was employed. Implicit creep model number "11" is selected because it can capture both the primary (utilizing ANSYS creep model "2," as expressed in Eq. (5.9)) and secondary creep (utilizing ANSYS creep model "10," as expressed in Eq. (5.10)) and is capable of predicting creep strain regardless of any coupling between time and either stress or temperature of steel. This is because ANSYS employs a variation of "strain-hardening rule" for integrating creep strains.

Time hardening:

$$\varepsilon_{cr}^{'} = c_1 \sigma^{c_2} t^{c_3} e^{-c_4/t} \tag{5.9}$$

Norton:

$$\varepsilon_{cr}^{'} = c_5 \sigma^{c_6} e^{-c_7/T} \tag{5.10}$$

The ANSYS creep model "11," in simplified format, is obtained through a combination of integration of Eq. (5.9) and Eq. (5.10) and is expressed as

$$\varepsilon_{cr}(\sigma, T, t) = c_1 \sigma^{c_2} t^{c_3+1} e^{-c_4/t} / (c_3 + 1) + c_5 \mathrm{t} \sigma^{c_6} e^{-c_7/T} \tag{5.11}$$

where c_1 through c_7 are coefficients that are dependent on type of steel and are obtained through nonlinear regression analysis to high-temperature creep test data.

To use this creep model to analysis the creep behavior of Q460 steel in ANSYS directly, the coefficients c_1 through c_7 are fitting and tabulated in Table 5.5. The results of the calibration for fitting parameters c_1 to c_7 are plotted in Fig. 5.9, which shows a comparison of high-temperature creep strain computed using Eq. (5.11) and coefficients c_1 through c_7 in Table 5.5 with that of test data for various stress levels at various temperatures. As is shown in the figure, the model also shows good agreement with the above test data generated for Q460 steel.

5.4 Creep behavior of high strength Q690 steel

A test program was carefully designed to carry out creep tests on Q690 steel to generate the creep strains at elevated temperatures. The creep tests were undertaken at various temperatures in 450—900°C range, as creep strains dominate at temperature above 450°C.

Table 5.5 Parameters of c_1 to c_7 in ANSYS creep model "11" for Q460 steel.

Temperature (°C)	c_1	c_2	c_3	c_4	c_5	c_6	c_7
400	3.30E-10	10.85	−0.5	18,800	1.30E-10	4.75E-04	40,000
450	6.33E-11	14.05	−0.45	28,000	1.03E-10	18	40,000
500	3.45E+00	6.5	−0.8	20,300	6.95E-10	14.35	31,500
550	3.50E-03	3.35	−0.5	8,600	1.75E-09	7.5	12,600
600	2.47E-09	4.35	−0.05	2,900	1.70E-08	16.55	40,000
700	1.46E-02	0.02	−0.27	1,200	4.50E-07	3.65	2,200
800	2.85E-02	4.75	−0.05	12,500	6.75E-04	1.98	1,450
900	7.60E+01	3.8	−0.2	16,400	1.10E-01	1.00E-10	1,900

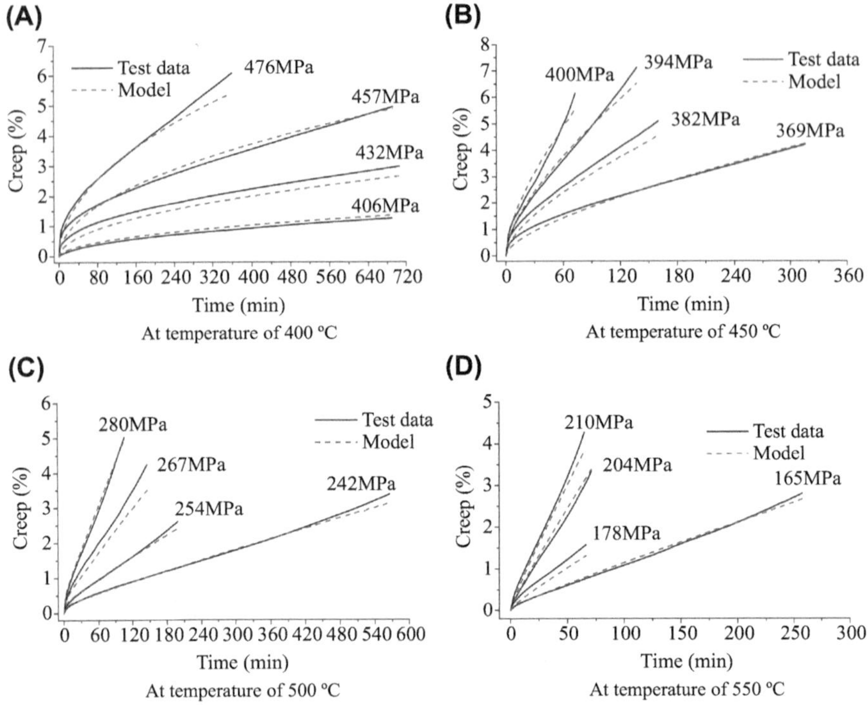

Figure 5.9 Calibration of ANSYS creep model "11" for Q460 steel at various stress levels. (A) At temperature of 400°C; (B) at temperature of 450°C; (C) at temperature of 500°C; (D) at temperature of 550°C; (E) at temperature of 600°C; (F) at temperature of 700°C; (G) at temperature of 800°C; and (H) at temperature of 900°C.

5.4.1 Test setup

The creep tests were performed by using electronic high-temperature creep testing machine, which consists of an electric furnace, tensile testing setup, a strain measurement device, and a data processing system, as shown in Fig.5.10. The tensile testing setup is MTS loading system with 50 kN capacity. The axial extension in the specimen at elevated temperature can be evaluated by two linear variable differential transformers (LVDTs) (range of ±10 mm), with 0.001 mm sensitivity, that are installed outside the heating zone of furnace by using two connecting rods arranged inside the furnace. The electric furnace can reach the maximum temperature of 1100°C. Three armored thermocouples are installed inside the furnace to obtain temperatures at lower, middle, and upper zones, to ensure the temperature raise in the furnace is uniform. The furnace and displacement transducers are connected to a data processing system and a computer through which tension load, displacements, and temperature during the creep test are recorded.

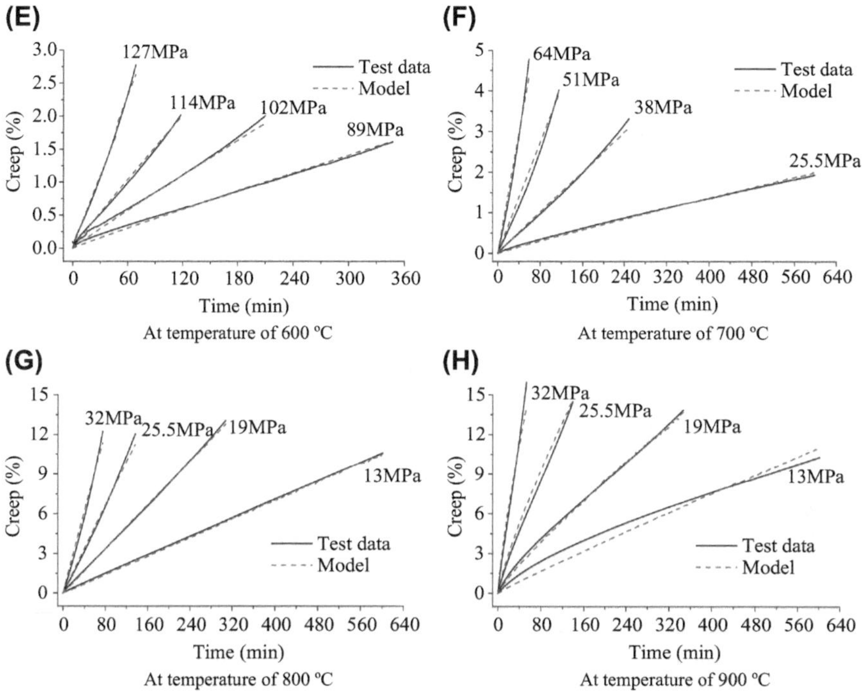

Figure 5.9 Cont'd

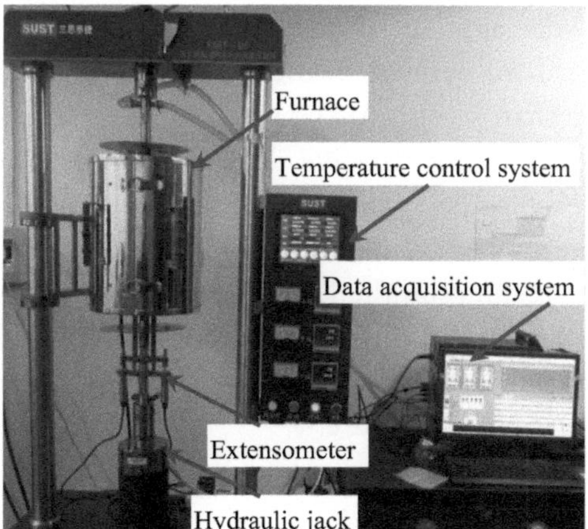

Figure 5.10 Creep test setup.

5.4.2 Test specimens

The creep specimens were fabricated from Q690 steel sheet of 20 mm in thickness. The steel plate used for creep specimen is same as that used for preparing tension specimen. The dimensions of creep specimens are shown in Fig. 5.11. The load ratio (representing stress level) and target temperatures at which these creep tests were conducted are given in Table 5.6. The load (stress) ratio represents the ratio of applied stress on the specimen to yield strength at the corresponding target temperature. Both yield strength $f_{T,0.2}$ and $f_{T,2.0}$ are calculated at various temperature. These loads allow the total creep time of specimens evenly distributed within 1−10 h at the same temperature.

5.4.3 Test procedures

To enable comparison with available test results, steady-state test method was utilized for carrying out high-temperature creep tests. Creep tests were carried out at various temperatures and stress levels.

In the creep tests, steel coupons were mounted into the grips of the tensile setup and then three build-in thermocouples were connected at bottom, middle, and top length of gage length of specimen. Then the electric furnace is turned on to be heated to a predefined temperature with a predefined heating rate of 10°C/min. When the target temperature is reached, the specimen is held at this temperature for 20 min to ensure that the full length of specimen achieves uniform temperature. Then a predefined load was applied at a rate of 10 kN per minute. As the maximum value of load defined in the creep tests is 44.5 kN, the loading procedure lasts less than 5 min. At the same time, creep deformations were recorded by the two highly sensitive displacement transducers. When the creep extension reaches the maximum allowable displacement of LVDTs, namely 10 mm, the support system of LVDTs was quickly adjusted by hand to confirm that the readings in the LVDTs return to zero, and then the data were continuously recorded. In this method, the creep displacement was recorded within an extended range, till the ultimate displacement of hydraulic jack or specimen fracture.

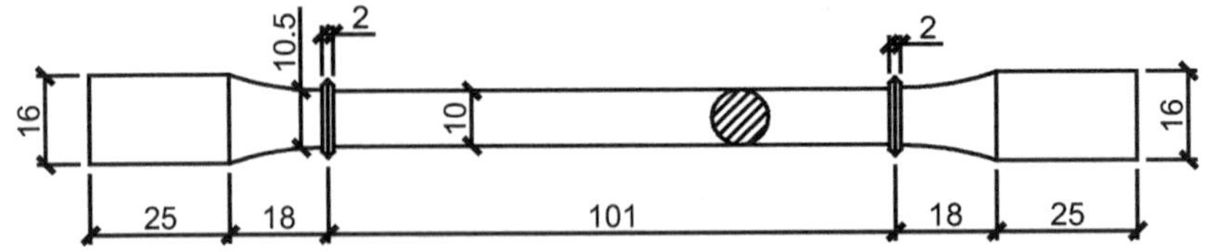

Figure 5.11 Schematics of elevated temperature creep specimen (unit: mm).

Table 5.6 Elevated temperature creep test conditions for Q690 steel.

Temperature(°C)	σ (MPa)	$f_{0.2}$ (MPa)	$\frac{\sigma}{f_{0.2}}$	$f_{2.0}$ (MPa)	$\frac{\sigma}{f_{2.0}}$
450	500, 515, 539, 567	567.1	0.88–1.0	625.8	0.80–0.91
500	400, 410, 429	506.4	0.78–0.85	556.5	0.72–0.77
550	242, 252, 278, 297	318.9	0.76–0.93	429.5	0.56–0.69
600	139, 157, 175	179.9	0.77–0.97	279.5	0.49–0.63
700	39, 43, 50	47.9	0.81–1.04	97.5	0.4–0.51
800	25, 32, 40	36.81	0.67–1.08	66.1	0.38–0.61
900	26, 29, 32	31.7	0.82–1.0	43.35	0.6–0.74

5.4.4 Test results

In 450–600°C temperature range (Fig.5.12A–D), the creep strain increases with temperature. For example, the ultimate creep strain at 450°C is about 10%. The ultimate creep strain at 600°C increases to more than 30%. When the temperature reaches higher range (700–900°C, Fig.5.12E–G), the plasticity of Q690 steel increases and creep deformations values become much higher. Because of the limitation of movement of hydraulic jack, the rupture of some specimens during the temperature range of 700–900°C could not be captured. The ultimate creep strain at the temperatures range of 700–900°C is more than 45%. However, the limited data have little effect on the creep response prediction of steel structure as tertiary creep effects is not included in fire resistance analysis as these data are not reliable. Only the primary and the secondary creep are considered in fire resistance analysis. At the same temperature, the stress has large influence on the creep rate of specimen. For example, the total creep time at 600°C increases from 620 min at $\sigma = 139$ MPa to 150 min at 175 MPa. When the temperature is higher, the stress has greater impact on creep.

In general, at the same stress level, the creep of steel increases with temperature. However, the creep strain evolution at temperature of 800°C is almost identical for the equivalent stress level on coupons subjected to 900°C. This means that the creep mechanism is not sensitive to stress after 800°C for Q690 steel alloy because of austenitizing in steel after the temperature reaching 800°C. The yield strength reduction factor of Q690 steel also infers that the strength almost keeps constant between 800 and 900°C.

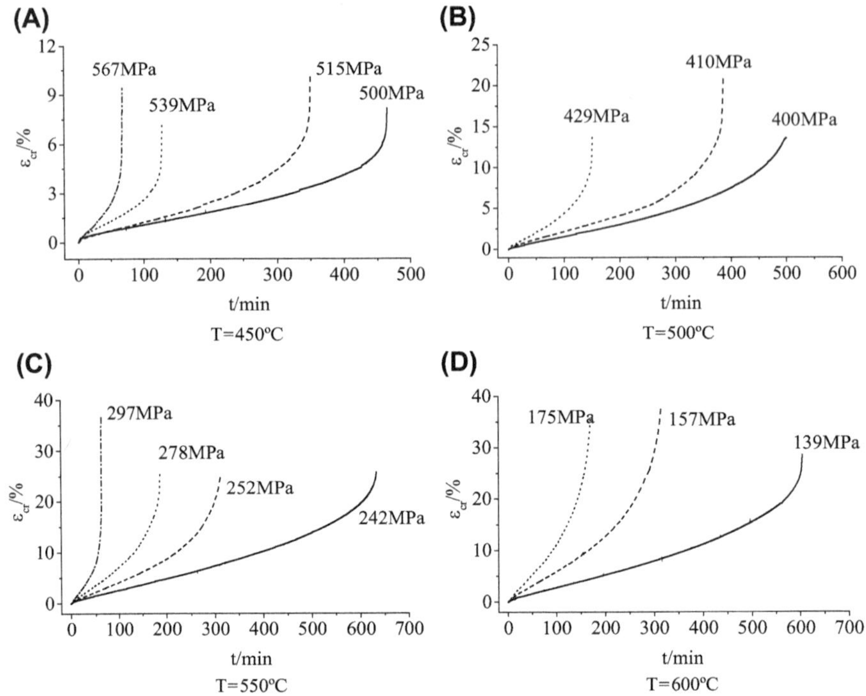

Figure 5.12 Creep strain curves of Q690 steel at various stress and temperature levels. (A) $T = 450°C$; (B) $T = 500°C$; (C) $T = 550°C$; (D) $T = 600°C$; (E) $T = 700°C$; (F) $T = 800°C$; and (G) $T = 900°C$.

The photographs of test coupons at various stress levels and temperatures after testing are shown in Fig. 5.13. These photos correspond to the test specimens with load ratios ranging from low to high, as illustrated in Table 5.6. It is indicated from Fig. 5.13, at temperature range of 450–600°C, obvious necking phenomenon was observed when the specimens fracture. At temperature range of 700–900°C, the test specimens exhibited a very large ductile response before the onset of significant necking phenomenon. As the loading of specimens at 700–900°C has not reached to failure yet, the necking effect is not seen at this stage.

5.4.5 Comparison with other steels

To study the differences in high-temperature creep between Q690 steel and other steels, creep data from Q690 steel are compared with that of Q460 steel (Wang et al., 2017) and low alloy Q345 steel (Wang et al., 2016). In general, at the same temperature and stress levels, the creep strain

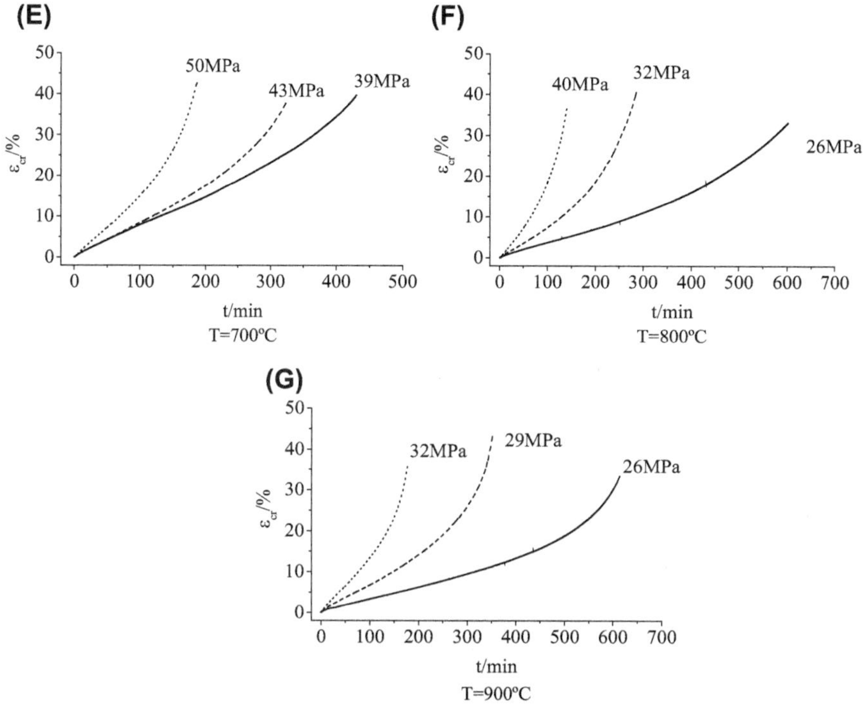

Figure 5.12 Cont'd

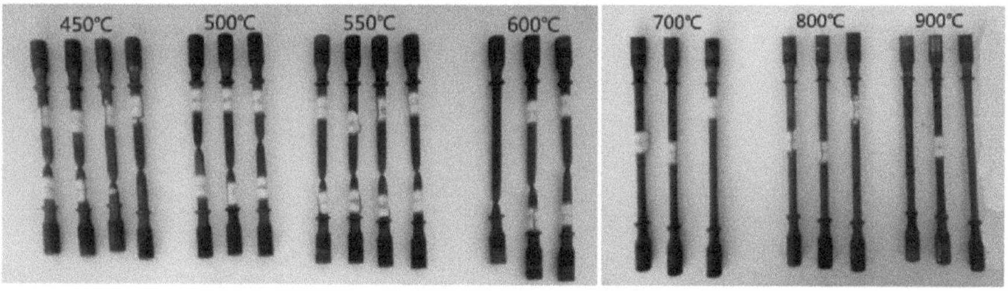

Figure 5.13 Photos of specimens after elevated temperature creep test for Q690 steel.

in higher strength steel (Q690) develops at a slower rate. To compare with other types of steels, stress ratio (*R*), defined as the ratio of applied stress to that of the yield strength at ambient conditions, is used.

Comparison of creep for Q690 steel, Q460 steel, and low alloy Q345 steel at elevated temperature range of 450–900°C is shown in Fig. 5.14. As shown in Fig. 5.14A–D, the ultimate creep strain of Q690 steel is lower than that of Q460 steel and low alloy Q345 steel at the temperature range of 450–500°C. At 550 and 600°C, the ultimate creep strain of Q690 steel

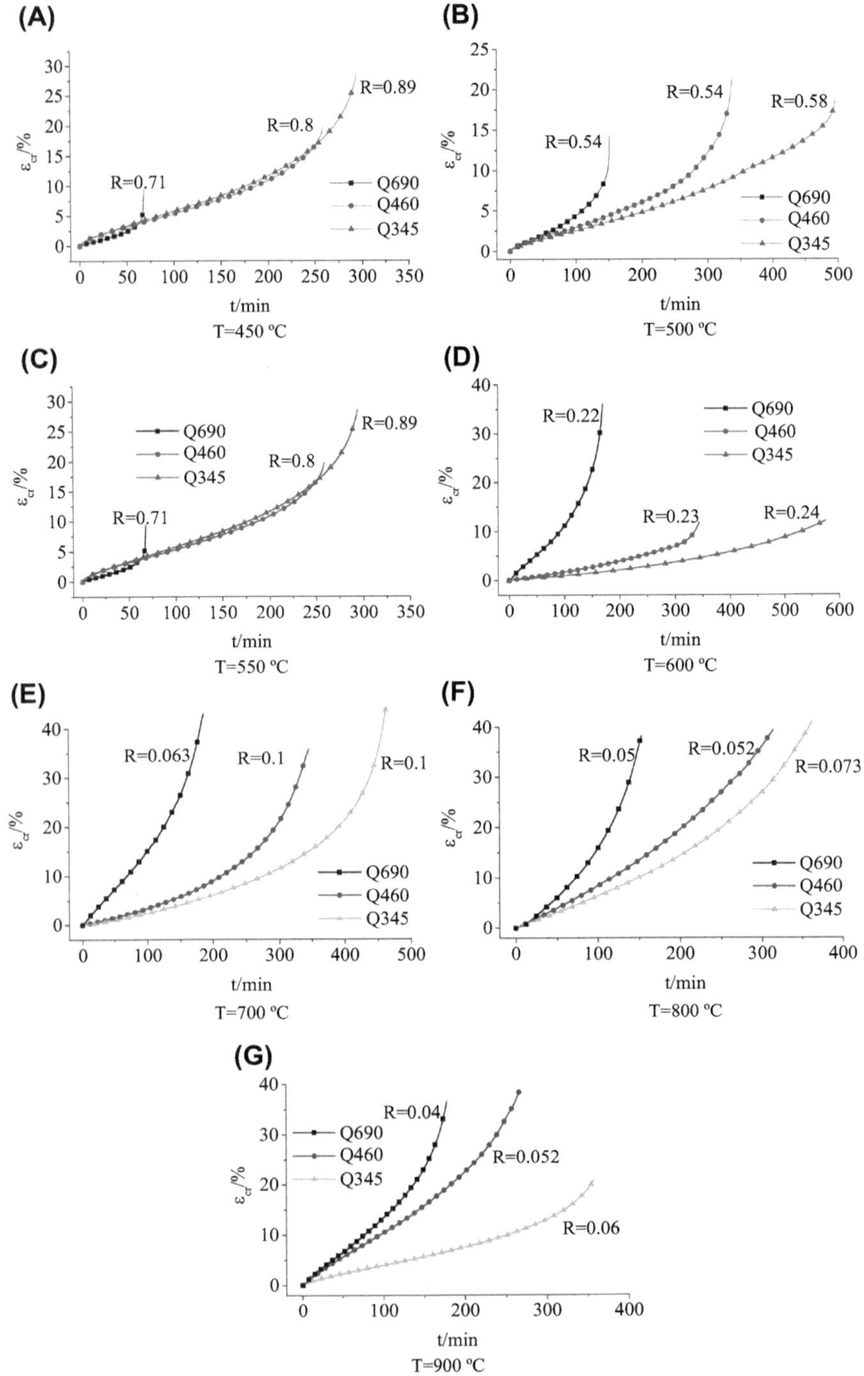

Figure 5.14 Comparison of creep strain between Q690 steel and Q345 and Q460 steels. (A)$T = 450°C$; (B) $T = 500°C$; (C) $T = 550°C$; (D) $T = 600°C$; (E) $T = 700°C$; (F) $T = 800°C$; and (G) $T = 900°C$.

is higher than that of Q460 steel and low alloy Q345 steel. Because of the limitation of movement of hydraulic jack, the rupture of some specimens during the temperature range of 700–800°C could not be obtained. The ultimate creep strain of three types of steel at the temperatures range of 700–800°C is more than 35%. At 900°C, the ultimate creep strain of Q690 steel and Q460 steel could not be measured and is bigger than that of Q345 steel.

The stress ratio of Q690 steel is not higher than that of Q460 steel and low alloy Q345 steel at the same temperature. The total creep time is lower than that of Q460 steel and low alloy Q345 steel. It is significant to note that the stress ratio of Q690 steel is half of that of other two steels, and the total creep time is only half of that of Q460 steel and one-third of that of low alloy Q345 steel. This means that at the same stress ratio and temperature, creep effects Q690 steel develop to full extent. Therefore, high-temperature creep effects should be considered in evaluating fire response of steel structures made of Q690 steel.

5.4.6 Creep model

To develop similar relations for Q690 steel as Q460 steel, test data recorded from the above creep test are applied to generate creep relations as a function of stress and temperature.

Fields and Fields (1989) Eq. 5.13a–d to of A36 steel.

To predict high-temperature creep response in Q690 steel using Fields and Fields creep model, a nonlinear regression fitting tool was employed. The new expression is shown in Eq. (5.12):

$$\varepsilon_{cr} = at^b\left(\frac{\sigma}{d}\right)^c \tag{5.12}$$

The values of coefficients a, b, and c were fitted as a function of temperature. In fitting the parameters, temperature range of 450–900°C was taken as the primary consideration. These fitting expressions of a, b, c, and d were shown in Eq. (5.13).

$$a = \begin{cases} 10^{-1.4868+0.00439T} & 450n \leq T < 500n \\ 10^{1.347+0.000378T} & 550n \leq T \leq 900n \end{cases} \tag{5.13a}$$

$$b = 0.8937 + 9.68 \times 10^{-5} \tag{5.13b}$$

$$c = 84.76 - 0.2238T + 0.0001524T^2 \tag{5.13c}$$

$$d = 1399.6 - 1.47119T \tag{5.13d}$$

A comparison of high-temperature creep strains computed using Eq. (5.13) with those of measured test data for different stress levels at various temperatures are shown in Fig. 5.15. As can be seen from the figure, the proposed creep relations based on fitting to test data show good agreement with the test data for Q690 steel.

5.5 Creep behavior of ASTM A572 steel

A comprehensive test program was designed to undertake creep tests on ASTM A572 steel. The creep tests were carried out at various temperatures in 400–800°C range and at various stress levels in the range 11%–55% of room temperature yield stress.

5.5.1 Test specimens

For high-temperature creep tests, rectangular steel coupons were cut from A572 Gr. 50 steel sheets. The chemical composition of A572 steel is different from that of conventional A36 steels. The chemical composition of both types of steel, together with actual composition of A572 steel used in current test program (as specified by the manufacturer), is tabulated in Table 5.7. The test coupons measured 609.6 mm in total length, and the length of tapered section is 304.8 mm. The cross section of the reduced section is 9.53 × 12.7 mm. The configuration of a typical test coupon used in creep tests is shown in Fig. 5.16.

5.5.2 Test setup

For undertaking high-temperature creep tests, a custom-built, integrated, load-heating furnace equipment was used. This test setup comprises of a steel frame for applying tensile loading, a loading lever arm, an electric furnace, and a data acquisition system. The axial tensile load is applied using suspended concrete blocks via the lever arm. The concrete blocks are weighed to determine the magnitude of load. This loading system allows the load to be maintained constant during the entire duration of strength test. The axial deformation of the specimen is measured by a displacement transducer (±38 mm LVDT), with 0.0254 mm sensitivity, that is placed outside the furnace.

The electric furnace comprises of cylindrical chamber with an inner diameter of 203 mm and a height of 305 mm. Therefore, the heating length of the test specimen is to be 305 mm. The test specimen is mounted

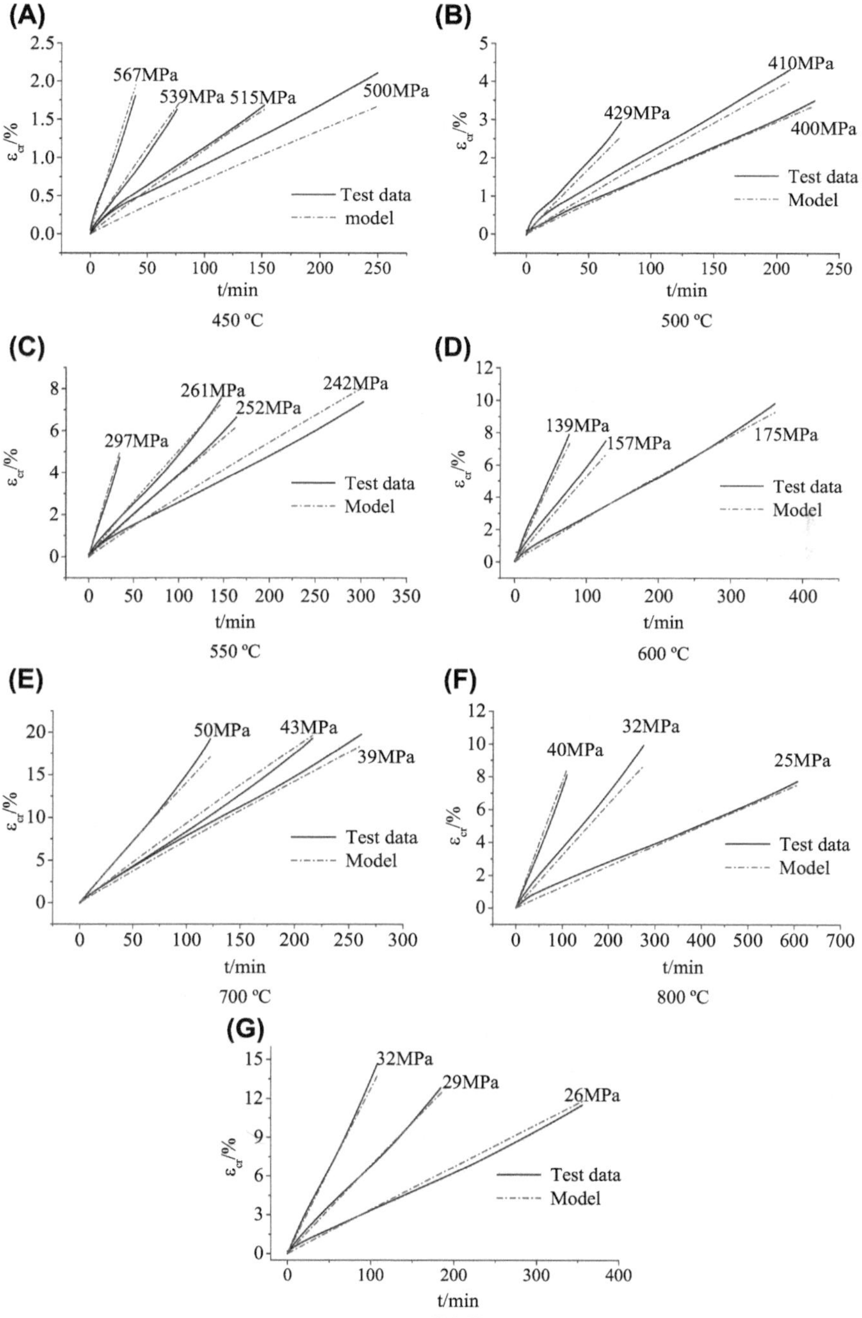

Figure 5.15 Test results and fitted values of creep strain in Q690 steel. (A) 450°C; (B) 500°C; (C) 550°C; (D) 600°C; (E) 700°C; (F) 800°C; and (G) 900°C.

Table 5.7 Chemical composition of A572 Gr 50 steel as compared with A36 steel.

Chemical composition	A36 steel (ASTM limits)	A572 grade 50 steel (ASTM limits)	A572 grade 50 steel (actual)
Carbon(C)%	0.25–0.290	0.23 max	0.06
Manganese(Mn)%	1.03	1.35 max	0.91
Phosphorous(P)%	0.04	0.04 max	0.009
Silicon(Si)%	0.28	0.40 max	0.02
Sulfur(S)%	0.05	0.05 max	0.009
Copper(Cu)%	0.2	–	0.12
Vanadium(V)%	–	0.06 max	0.0818
Cobalt(Co)%	–	0.05 max	–
Iron(Fe)%	The rest	The rest	The rest

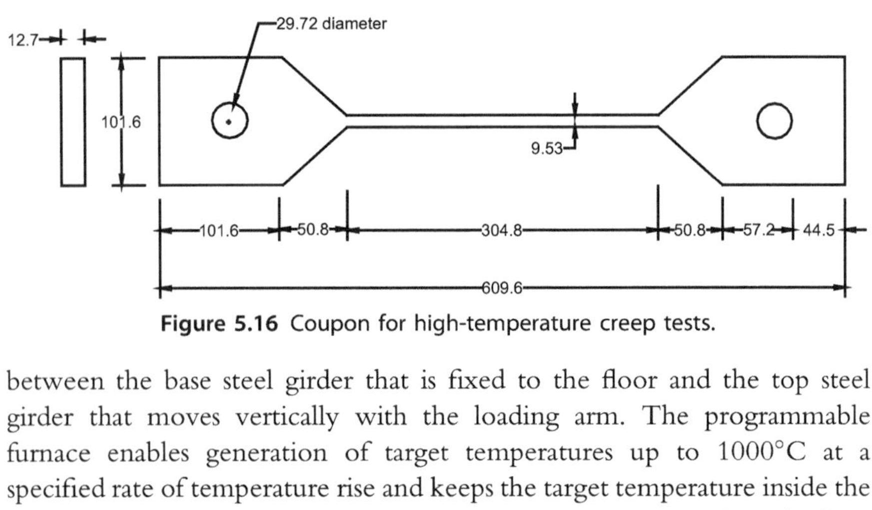

Figure 5.16 Coupon for high-temperature creep tests.

between the base steel girder that is fixed to the floor and the top steel girder that moves vertically with the loading arm. The programmable furnace enables generation of target temperatures up to 1000°C at a specified rate of temperature rise and keeps the target temperature inside the furnace constant for the duration of strength (creep) test. Three built-in thermocouples are mounted inside the furnace to measure temperatures at upper, middle, and lower zones, and the average readings of these three thermocouples is taken as the FT. The furnace and LVDT are connected to a data acquisition system and a computer wherein temperature and displacements during the test are recorded. A schematic of test setup for high-temperature creep test is shown in Figs. 5.17 and 5.18.

5.5.3 Test procedure

Strength and creep properties of steel at a specific temperature can be evaluated through transient- and/or steady-state tests. Although transient-state tests are more realistic in simulating real fire conditions, steady-state

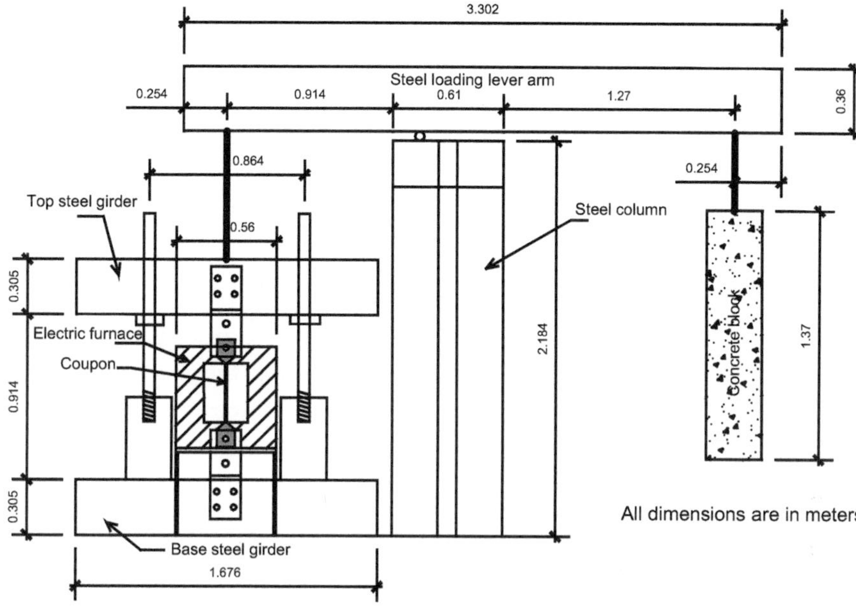

Figure 5.17 Schematics of test setup for high-temperature creep tests.

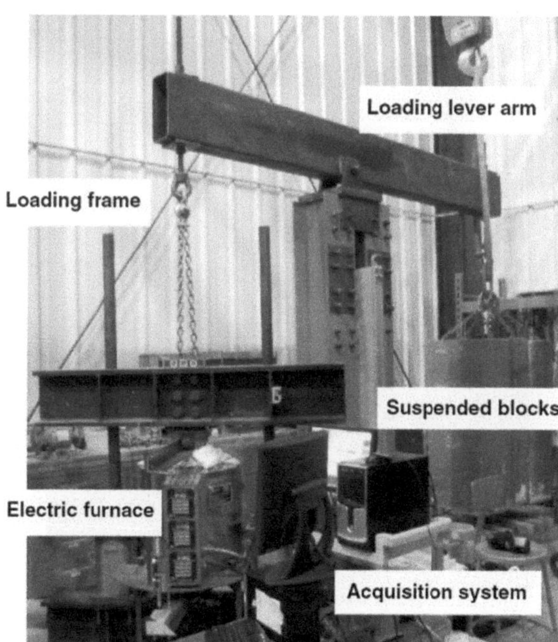

Figure 5.18 Custom-built creep test equipment for high-temperature creep tests.

tests are more commonly applied due to ease of undertaking these tests. In the current study, steady-state test method was utilized for undertaking high-temperature creep tests.

Before undertaking creep tests, tensile strength tests were carried out to evaluate stress—strain response of A572 steel at ambient conditions. These tests were conducted using MTS-810 test machine (depicted in Fig. 5.19) with 250 kN loading capacity.

Creep tests were conducted at different temperatures, and stress levels in the custom-built test setup were described above. For these creep tests, steel coupons were prepared by mounting a thermocouple at midlength of the tapered cross section as shown in Fig. 5.20. This thermocouple is to monitor actual temperature progression in the specimen during the creep test. Then, the steel coupon was placed inside the electric furnace, and special care was taken to ensure the coupon to be at the center of the furnace and the loading frame. Following this, a predefined load was gradually applied through a suspended concrete block and this tensile force gets transferred to the coupon through the steel beam (acting as a lever arm). Special care was taken in the design of loading frame to ensure the loading to be perfectly axial. The total load, which is required to develop a

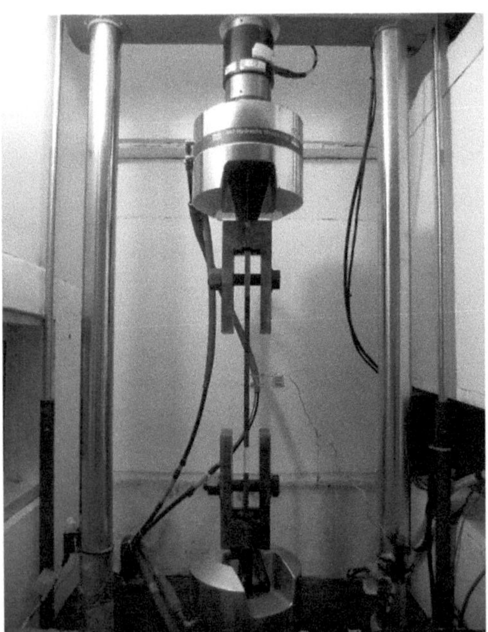

Figure 5.19 Test setup for tensile strength tests.

Figure 5.20 Coupon placement in the furnace for creep test.

target stress level at the tapered cross section, was applied before heating the specimen.

After few minutes of application of loading, the furnace is turned on to be heated to a target temperature at a heating rate of 10°C/min. Once the target temperature is attained, the steel coupon was maintained at this temperature for 30 min to attain a steady-state condition and then the specimen was maintained at the target temperature for the entire test duration. The creep displacement during the test is measured using highly sensitive LVDT that is placed outside the furnace. The total length of tapered section of the coupons that is exposed to high temperature during creep tests is 304.8 mm, and this is taken to be the gage length to calculate corresponding creep strain (DL/L). Using this procedure, creep tests were conducted at different stress levels and at different temperatures.

5.5.4 Room temperature stress—strain response

The stress—strain response of A572 steel at room temperature is evaluated using data from uniaxial tensile strength test. The displacements recorded at various load levels, during tension test, were used to generate stress—strain

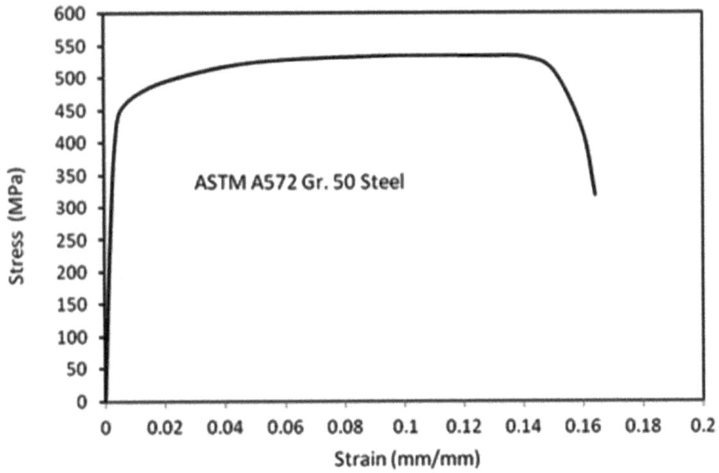

Figure 5.21 Stress–strain response of A572 Gr50 steel at room temperature.

response of A572 steel, and this is plotted in Fig. 5.21. It can be seen in the figure that the general trend of stress–strain curve at room temperature is linear elastic up to yielding of steel at a stress of 453 MPa, followed by nonlinear response. No well-defined yield plateau is recorded in the current test on A572 steel unlike that reported in literature (ASM, 2012). This can be attributed to very low amount of carbon content and also presence of alloy elements such as vanadium in the chemical composition of steel as shown in Table 5.7. Once the ultimate stress is reached at a stress of 535 MPa, steel undergoes plastic deformation, through unloading phase up to rapture, which occurs at an ultimate strain of 0.165 mm/mm.

5.5.5 Creep response at elevated temperatures

Data from high-temperature creep tests are utilized to derive creep response of A572 steel at various stress levels and temperatures. The stress levels and target temperatures at which these tests were carried out are tabulated in Table 5.8. The stress level represents ratio of stress due to applied loading to room temperature yield stress.

Table 5.8 Selected temperatures and stress levels for creep tests.

Stress (MPa)/stress level (%)	Temperatures (°C)			
50/11	600	700	750	800
115/25	500	600	650	700
182/40	500	600	650	—
250/55	400	500	550	—

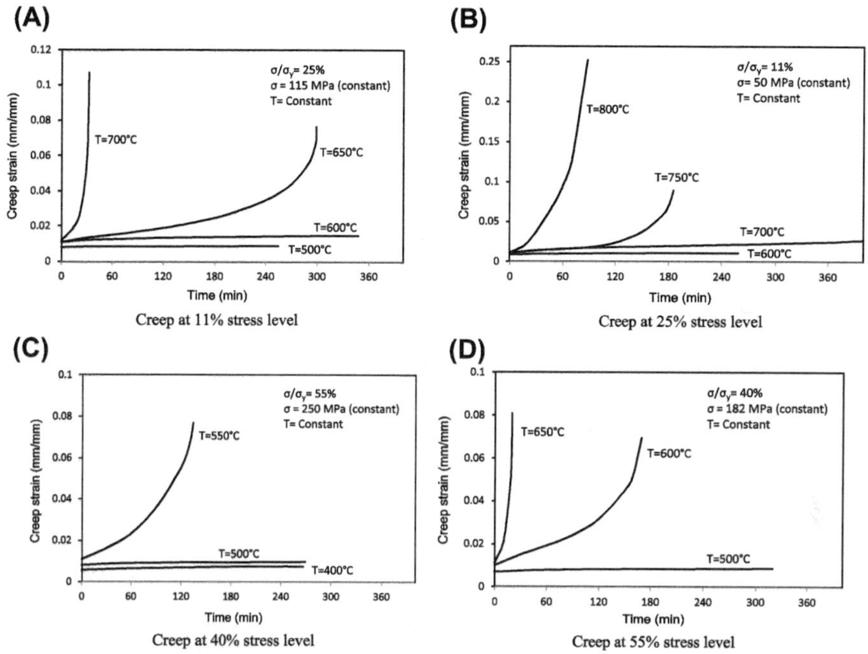

Figure 5.22 Creep deformations at various temperatures and stress levels. (A) Creep at 11%stress level, (B) creep at 25%stress level, (C) creep at 40% stress level, and (D) creep at 55% stress level.

The measured creep strain at four stress levels is plotted as a function of time in Fig. 5.22. The general trend of creep response, shown in Fig. 5.22A–D, can be grouped into two stages, namely secondary (stage II) and tertiary creep (stage III). Any primary creep (stage I) that resulted in these specimens has occurred before heating of specimens. Therefore, the initial total strain at the start of heating (time = 0) in creep response curves shown in Fig. 5.22 represents the sum of mechanical strain from the applied load, thermal strain from heating to target temperature, and strain due to primary creep.

The increase in secondary creep strain with time during stage II is dependent both on temperature and stress levels. At low stress level of 11%, there is little secondary creep generated at 500, 600, and 700°C. However, at higher temperatures of 750 and 800°C, secondary creep increased at a high rate at this stress level. For moderate stress level of 25%, the secondary creep starts to dominate at slightly lower temperature of 700°C. For high stress level of 40% and 55%, secondary creep increased substantially at temperatures of 600 and 550°C, respectively. The slow rise in creep strain with time in stage II is due to movement of dislocations that counteract the strain hardening resulting in a balance between strain hardening and thermal softening. Secondary creep is highly important under fire exposure conditions because it dominates the creep response, and occurs in a constant rate and over short period of time.

In stage III, the tertiary creep increases at a faster rate due to reduced cross section of the specimen resulting from necking phenomenon that results in higher stresses for the same level of applied load. Finally, the material flows, and this leads to rupture of specimen under the combined effect of mechanical loading and temperature. At high temperature and stress levels, it becomes difficult to distinguish between stage II and stage III of creep because of high creep rate and accelerated flow of material at high temperatures.

A review of results in Fig. 5.22 indicates that creep deformations are appreciable at temperatures above 500°C. At a given stress level, the creep deformation increases with temperature. This is due to diffusion of atoms within material grains resulting from temperature rise. When temperatures are in the range of 600–800°C, which represents 30%–50% of steel melting temperature, grain boundary diffusion tends to be the dominant mechanism, which results in accelerated creep strain, while bulk diffusion tends to dominate creep deformation in temperature range of 800–1400°C (50%–99% of steel melting temperature) (Ashby and Jones, 2005). Hence, for the case of 11% stress level, when steel temperatures rise from 600 to 750°C, creep strains increase significantly, and this leads to rupture of the specimen. Fig. 5.22A–D clearly show that for any stress level, there is a critical temperature at which creep deformations tends to accelerate at a rapid pace and produces failure. This critical temperature typically corresponds to initiation of tertiary creep. Higher the stress level, lower will be the critical temperature.

Different failure modes that resulted in test specimens at various temperature and stress levels are shown in Fig. 5.23. At higher temperature and under low stress levels, the failure of test specimens was through a well-defined necking phenomenon, indicating ductile response. For example, failure at 800°C and stress level of 11% is in ductile mode as the specimen experienced significant elongation, mainly due to very high creep deformations, before failure. However, at lower temperatures and higher stress levels, the failure was through brittle fracture, and this can be seen from flat surface at the midlength of specimens (see Fig. 5.23). For example,

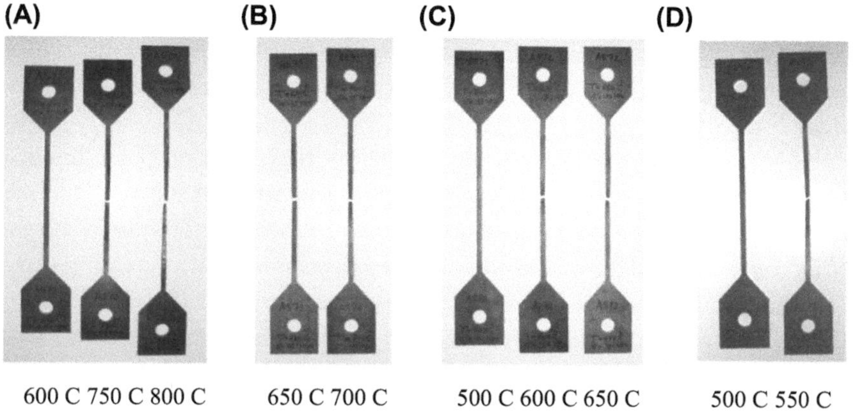

Figure 5.23 Creep failure modes at various temperatures and stress levels. (A) Stress level = 11%, (B) stress level = 25%, (C) stress level = 40%, and (D) stress level = 55%.

at 550°C and stress level of 55%, the failure is through brittle fracture and mainly due to relatively lower creep deformations.

5.5.6 Creep response at various stress levels

The effect of stress level on creep response of A572 steel is illustrated in Fig. 5.24. The creep strain is plotted at temperature of 650°C for stress levels of 25% and 40%. It can be seen, for moderate stress level of 25%, secondary creep increased with time at a constant rate up to 240 min, then the creep strain increased rapidly in the tertiary creep stage up to rupture. However, at 40% stress level, it becomes difficult to distinguish between secondary and tertiary creep resulting from high creep rate generated because of higher stress level. The trends in the figure indicate that creep deformation increases substantially with increasing stress level. In addition, data plotted in Fig. 5.22D show that creep deformation can be critical for

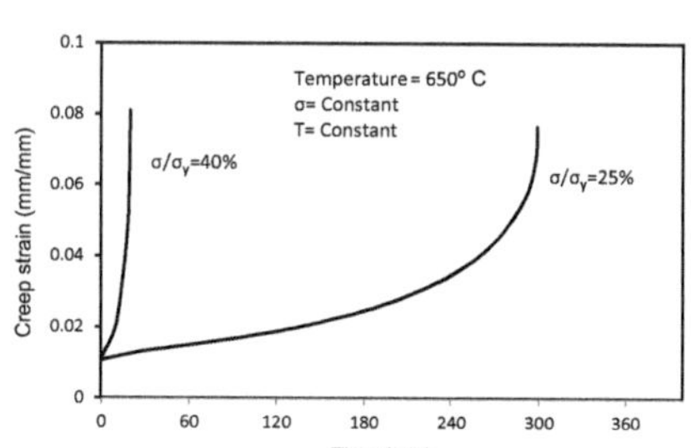

Figure 5.24 Creep strain at 650°C at various.

stress levels above 50% and will produce failure of specimens. Increased creep strain at stress level of 40% or more can be attributed to dislocation movement due to diffusion of atoms that occur at higher temperatures and under high stress levels.

Fig. 5.25 shows total creep strain just before rupture in steel as a function of temperature at four stress levels. The total creep strain at rupture increases with temperature at a given stress level. Furthermore, ductility, which is the ability of steel to deform before rupture, decreases with increasing stress level. As it can be seen in Fig. 5.25, the specimens under high stress levels of 40% and 55% ruptured at lower creep strain than those subjected to lower stress level of 11% and 25%.

5.5.7 Design recommendations

Creep deformations can dominate structural response of steel structures under fire conditions, especially when steel temperatures exceed 500°C. For a specific type of steel, creep strain is mainly dependent on temperature and stress level. Therefore, defining a critical temperature limit for creep at different stress levels can be highly useful for evaluating fire response of steel structural members. Critical temperature for creep can be defined as the temperature that corresponds to onset of tertiary creep, and this point is an indication of imminent failure of a structural member exposed to fire. Based on test data generated in this study, the critical temperature for four stress levels are derived as the temperature corresponding to onset of tertiary creep, and these values are tabulated in Table 5.9. Under fire conditions, structural members are typically assumed to be subjected to a loading of about 50% of room temperature capacity, which corresponds to a stress level of 50% (Franssen et al., 2009). Hence, stress level beyond 50% is not of too much interest from the point of fire design of structures.

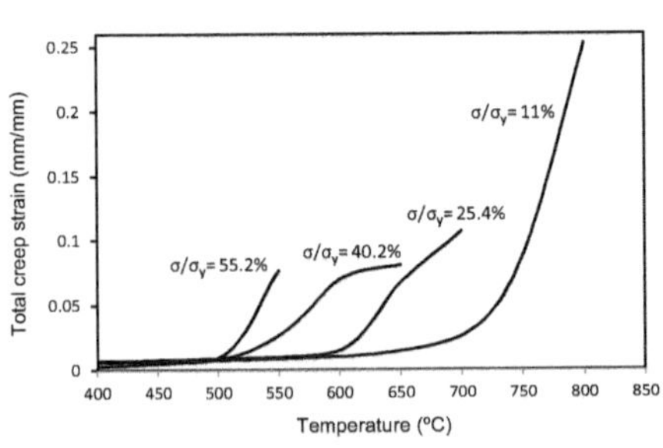

Figure 5.25 Total creep strain as a function stress levels of temperature.

Table 5.9 Critical temperatures for A572 steel at different stress levels.

Stress level (%)	Critical temperature (°C)
11	750
25	650
40	600
55	550

5.6 Creep buckling of high strength Q460 steel columns

The creep buckling is a phenomenon in which the buckling load for a column depends not only on slenderness and temperature but also on the duration of the temperature and applied load ratio. When a building is subjected to a fire, creep strain developed in a steel column can significantly influence the behavior of the column. As the result, the column deformation will be increased and the strength will be decreased. However, the effect of creep in steel structures has not been fully considered in current fire resistance design codes. Study on creep bucking of steel columns in fire is a critical issue for the fire safety design of steel structures as steel columns may experience buckling under a very small applied load ratio if the column is subjected to elevated temperature for a relatively long period of time. To address this issue, this section presents a comprehensive experimental and numerical investigation on creep buckling of high strength Q460 steel columns subjected to constant temperature and compressive load.

5.6.1 Specimen preparation

To evaluate the effect of load ratio, elevated temperature, and slenderness ratio on creep buckling time of high strength Q460 steel column, six welded H-shape specimens with a length of 2700 mm were fabricated with 14 mm thick high strength Q460 steel plates. Q460 steel is a typical high strength low-alloy steel, and its chemical compositionsspecified by GB/T 1591-2008 (2008) are listed in Table 5.10. At each end of the specimen, a 15 mm thickness steel end plate was welded to the specimen, and the plate was connected to the test setup. Two elevated temperatures, 600 and 800°C, were selected. Two column cross sections, H200 × 150 × 14 × 14 and H200 × 200 × 14 × 14 (unit: mm), were tested. Two applied load ratios, 0.09 and 0.12, were chosen for the tests. The selected applied load level has to be relatively low as the primary objectives of the study are to investigate the effect of creep strain on the column buckling and provide test data for the validation of the numerical model. In the case that the load

Table 5.10 Chemical composition of Q460 steel.

Chemical compositions	C	Mn	Si	S	P	V	Nb	Ti	Al	Cr	Ni
Content (%)	≤0.2	1.00 −1.70	≤0.55	≤0.035	≤0.035	0.02 −0.20	0.015 −0.060	0.02 −0.20	≤0.015	≤0.74	≤0.7

level is high, the columns may buckle prematurely within a short period of time at the elevated temperature without the occurrence of creep. The detailed information of the specimens is tabulated in Table 5.11. The load ratio R was determined as the ratio of the applied load to the load-bearing capacity of the steel columns at elevated temperature, and the capacity is obtained from a calculation procedure specified in GB51249 (2017). The mechanical properties of high strength Q460 steel at temperatures 600 and 800°C were obtained from the tests conducted by Wang et al. (2013). Carbon dioxide gas shielded welding was used in fabrication of the specimens with weld size of 8 mm. The yield strength and tension strength of electrode is 550 and 640 MPa, respectively. The yield strength, tension strength, and elastic modulus of Q460 steel plate are 587 , 628 and 199,440 MPa, respectively. These properties are obtained by the standard tensile coupon test according to ASTM A370 test protocol (ASTM, 2009).

The effect of initial crookedness of the specimens on the load capacity of steel columns is accounted for in the investigation. The geometric imperfection of the specimens was defined as the maximum deviation of the column longitudinal axis from the straight line connecting centroids of both ends of the column, and the values at a quarter, midheight, and three-quarter points were measured, labeled as v_1, v_2, and v_3, respectively. The measurement procedure was repeated twice through rotating the specimen by 180 degrees along the column axis in the buckling plane, and the measured results were averaged to reduce possible errors in measurements (Ban, 2012). The maximum measured initial crookedness of the specimens, v_0, which is located at the midheight of the specimen, is shown in Table 5.12. It can be seen that the maximum initial crookedness of the specimens is 2.83 mm, and the minimum one is 1.38 mm, which corresponds to 1.03% and 0.5% of the length of the specimens, respectively.

5.6.2 Test setup

As shown in Fig. 5.26, a test setup was designed for the experimental investigation of the creep response of high strength Q460 steel columns under constant load at elevated temperature. The setup consists of a reaction frame (constructed by two columns and two reaction beams) with an overall dimension of 5.85 m (height) × 2 m (width), a transfer girder, an electric furnace, and a hydraulic jack. The specimen was installed between the lower reaction beam and transfer girder with hinge supports. The hydraulic jack with a maximum load capacity 500 kN was installed

Table 5.11 Specimen details.

Specimen Number	Temperature/ °C	Cross-sectional dimensions	Length/ mm	Slenderness ratio λ	Load (kN)	Load ratio R
S-1	800	H200 × 150 × 14 × 14	2730	86	45	0.12
S-2	800	H200 × 150 × 14 × 14	2730	86	35	0.09
S-3	800	H200 × 200 × 14 × 14	2730	62	62	0.12
S-4	800	H200 × 200 × 14 × 14	2730	62	45	0.09
S-5	600	H200 × 150 × 14 × 14	2730	86	113	0.12
S-6	600	H200 × 200 × 14 × 14	2730	62	192	0.12

Table 5.12 Measurements of specimens' initial crookedness.

Specimen Number	S-1	S-2	S-3	S-4	S-5	S-6
v_o/mm	2.2	2.83	1.74	1.38	2.18	1.39

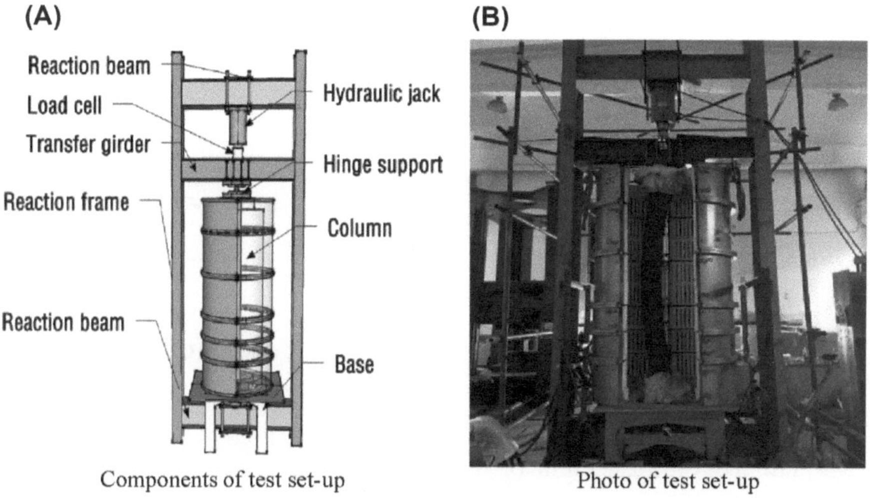

Figure 5.26 Test setup. (A) Components of test setup and (B) photo of test setup.

underneath of the top reaction beam to push the transfer girder downward when the load was applied. The column hinge supports and the ends of transfer girder were lubricated to minimize friction between contact surfaces. The both ends of the specimen and the hinge supports were protected by fireproof blanket. The electric furnace was lifted using a height adjustable steel table to ensure the entire specimen being heated. The maximum temperature that the furnace can produce is 1200°C, and the temperature elevation can be set up in according to any predefined temperature—time curve.

5.6.3 Instrumentation

In the test, FT, specimen temperature, actual load applied on the specimen, axial displacement, and lateral deflection at the midheight of the specimen were recorded. The location and arrangement of the instrumentation are shown in Fig. 5.27. Three thermocouples, labeled as FT1, FT2, and FT3, were placed at different height of the furnace to measure the FT during heating. Nine thermocouples, labeled as CT1-CT9, were mounted on the specimen to monitor the temperature variations of section C-1, C-2, and

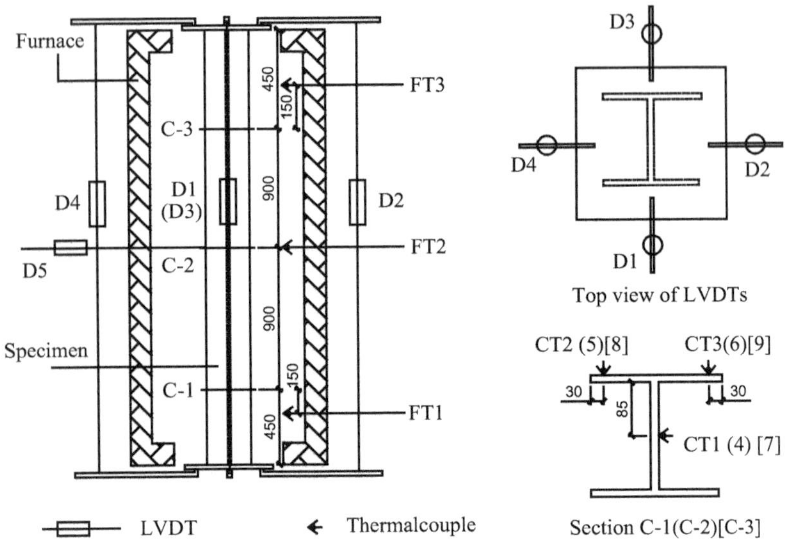

Figure 5.27 Measurement arrangement.

C-3. The axial displacement of the specimen was recorded by using four linear variable differential transducers (LVDTs) installed vertically around four sides of the specimen shown as D1to D4. Another LVDT, labeled as D5, was installed horizontally to measure lateral deflection at midheight of the specimen.

5.6.4 Test procedures

The test was conducted based on the following procedure. (1) The specimen was first installed into the reaction frame by connecting two end plates of the specimen to the hinge supports. (2) The furnace was moved to encase the specimen with the instrumentation such as LVDTs and thermocouples being installed before the furnace was closed. (3) The specimen was gradually loaded up to 10% of its ultimate load capacity, and the load was maintained for 2 minutes before the load was released. This process is important to ensure the specimen and instrumentation were installed and worked appropriately. (4) A target temperature, either 600 or 800°C, and the temperature increase rate 30°C/min were set up before the furnace was turned on to heat the specimen. (5) When the temperature of the specimen reached the target temperature, the hydraulic jack was turned on with a loading rate of 4–25 kN/min until the apply load reached to the specified level. The loading rate was adjusted according to the magnitude of the specified load to ensure the loading phase was completed within 10 min (6) Once the applied load reached to the specified level, both of the applied load and specimen's temperature were maintained as constant, and the axial and lateral deflection were recorded until the creep buckling occurred. (7) If the lateral displacement at the midheight of the specimen reached $L/20$, the test was stopped and the data were saved.

5.6.5 Experimental results

5.6.5.1 Furnace temperature

The measured FT−time curves for the six tests are plotted in Fig. 5.28, where FT1, FT2, and FT3 are the temperatures of the furnace at the three locations shown in Fig. 5.27. It is observed that the temperature of the

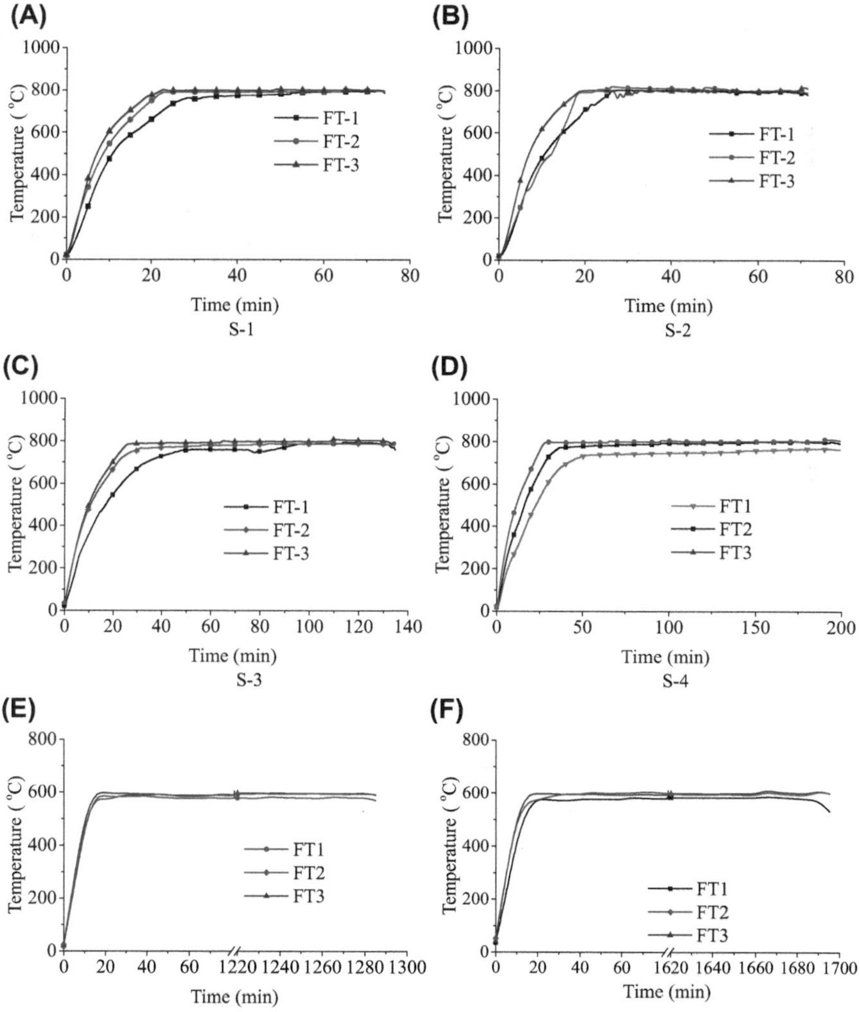

Figure 5.28 Furnace temperature. (A) S-1, (B) S-2, (C) S-3, (D) S-4, (E) S-5, and (F) S-6.

furnace increased linearly at the beginning of the heating phase and then the heating rate decreased gradually until the temperature approaching the target temperature. The temperature of top segment of the furnace was slightly higher than that of bottom segment as the heated air always moves upward. The maximum temperature difference along the height of the furnace is about 150°C during the heating phase. The temperature was kept as constant after approximately 30 and 20 min of heating for the target temperature 800 and 600°C, respectively.

5.6.5.2 Column temperature and compression load

The specimen temperature and the compressive load applied on the columns are presented in Fig. 5.29. The temperature evolution trend of the specimen is similar to that of furnace. The temperature distributions of the cross section along the column length are almost uniform, and the corresponding temperature differences are not significant for the majority of specimens. However, it was observed the temperature distribution of cross section of specimen S-2 was not quite uniform. A temperature fluctuation of 70°C was observed during the loading phase. After testing of specimen S-2, the furnace was carefully inspected, the sealing of fire protection at the top of the furnace was not very good, and more attention was paid in the rest of tests.

Once the column temperature (CT) was kept as constant, the load was applied almost linearly until it reached the specified level within about 10 min. The scale of the load is shown by the left vertical axis of each figure. The interior vertical lines in the figure defines the three phases of the loading process, namely, loading, load maintaining, and creeping buckling, in which the vertical line on the right indicates when creep buckling was occurred. As demonstrated in Fig. 5.29, all specimens experienced creep buckling with both compressive load and temperature being kept as constant for a certain period. Once the creep buckling occurred, the load dropped quickly because of the lost load-bearing capacity of the specimen.

5.6.5.3 Axial and lateral displacement

Fig. 5.30 presents the measured axial and lateral displacements of the six specimens. In general, the creep deformation was significant at a temperature of 800°C even then the applied load ratio was as low as 10%. In each figure, the curves can be divided into three stages, namely heating, loading, and creeping. In the heating stage, the axial displacement increased as the elongation of the specimen because of thermal expansion. Once the

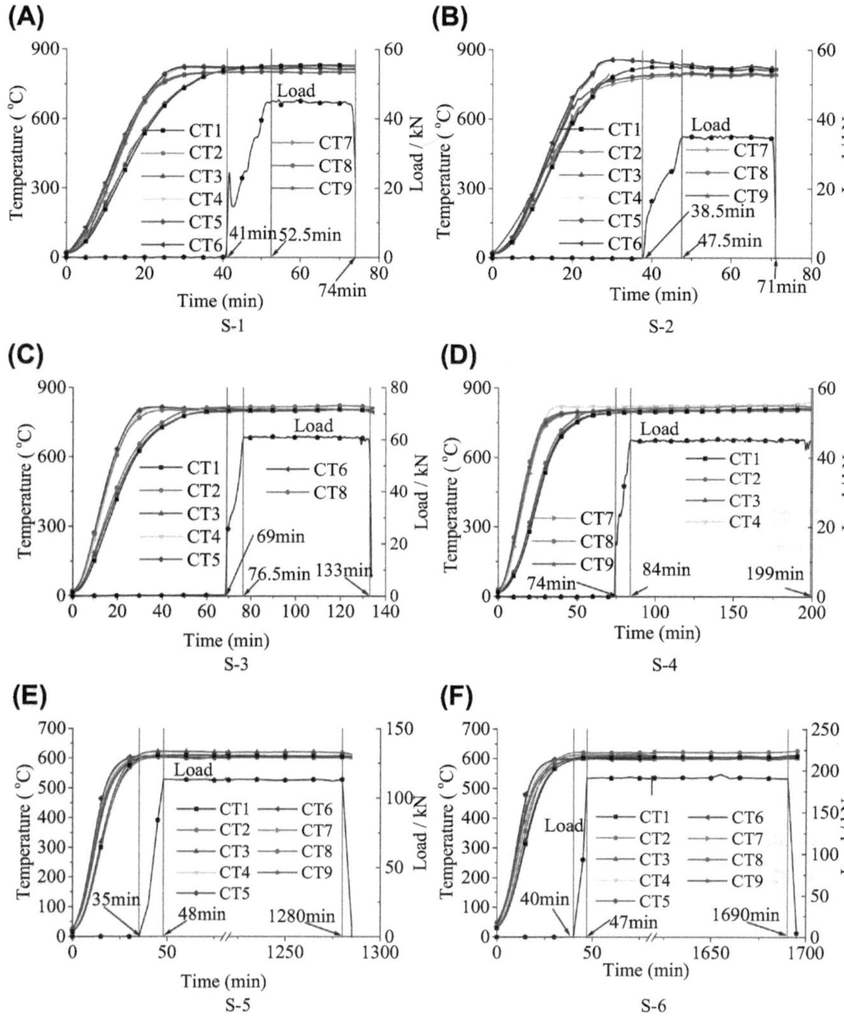

Figure 5.29 Temperature and compression load of specimens. (A) S-1, (B) S-2, (C) S-3, (D) S-4. (E) S-5, and (F) S-6.

temperature was kept as constant, the axial displacement of the specimen became almost as constant since the thermal expansion was completed. The lateral displacement for most specimens in this stage was negligible, and this could be attributed to the fact that the temperature distribution along the section was uniform and little thermal bending occurred. But for specimen S-2, the lateral displacement reached to 10 mm due to nonuniform temperature distribution in the column cross section as previously discussed. In the loading stage, the axial displacement decreased gradually because of

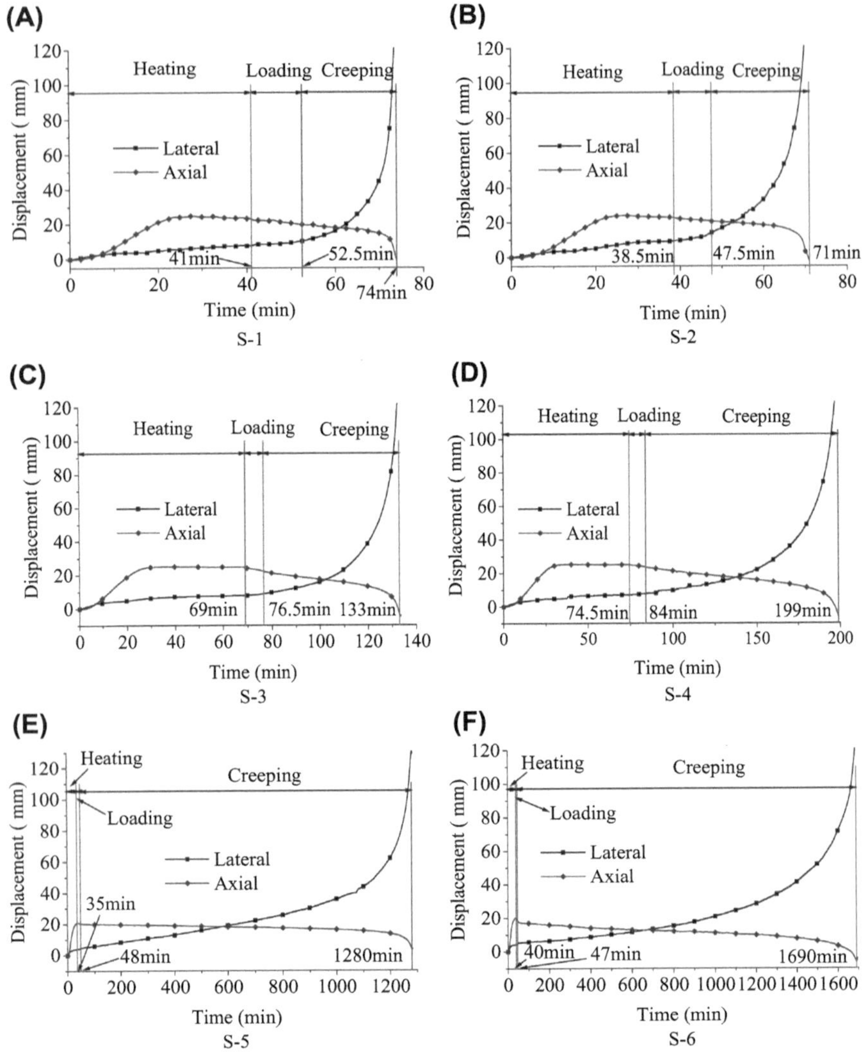

Figure 5.30 Displacement–time curves of steel columns. (A) S-1, (B) S-2, (C) S-3, (D) S-4, (E) S-5, and (F) S-6.

compression of the column, and the lateral displacement increased gradually because of the second-order effect. The displacements in this stage were not obvious because the load ratio was very low. In the creeping stage, the lateral displacement increased gradually in beginning and became significant later. In this stage, the deformation of the specimen was mainly resulted from the creep strain developed in the steel column.

5.6.5.4 Creep bucking time

In this investigation, the creep buckling time is defined as the period between the time when the applied load reaches to the specified level and the time when the specimen column failed by creep buckling. To compare the evolutions of both axial and lateral displacements among the specimens, the time–displacement relationships of the specimens are presented in Fig. 5.31. As the creep buckling time of individual specimen varies with a large range between 21.5 and 1643 min, the natural logarithm of time is adopted to present the scale of X-coordinate axis. The creep buckling time of each column was also shown in Fig. 5.31. First, it can be observed that the creep buckling time increases as the decrease of the load ratio. For example, the cross section, length, and temperature of specimen S-3 and S-4 were similar, but the load ratio was different, respectively, 0.12 and 0.09. The corresponding creep buckling time of specimen S-3 and S-4 was 56.5 and 115 min, respectively. Second, the creep buckling time increases as the decrease of the slenderness ratio of the specimen. For example, the cross section, load ratio, and temperature of specimen S-1 and S-3 were similar, but the slenderness ratio was different, respectively, 86 and 62. The creep buckling time of S-1 and S-3 was 21.5 and 56.5 min, respectively. The comparison on the creep buckling time between specimen S-5 and S-6 at a temperature of 600°C resulted in a similar conclusion. Third, the buckling time decreases as the elevation of temperature. This trend can be illustrated by a comparison between specimen S-3 and S-6 as all the parameters such as the cross section, length, and load ratio are similar for both specimens except the target temperature is 800 and 600°C, respectively. The creep buckling time of S-3 and S-6 is 56.5 and 1643 min, respectively.

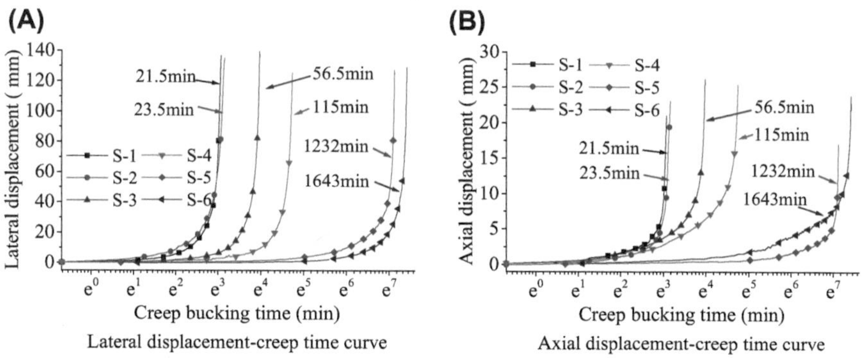

Figure 5.31 Comparison of creep buckling time of steel columns. (A) Lateral displacement–creep time curve and (B) axial displacement–creep time curve.

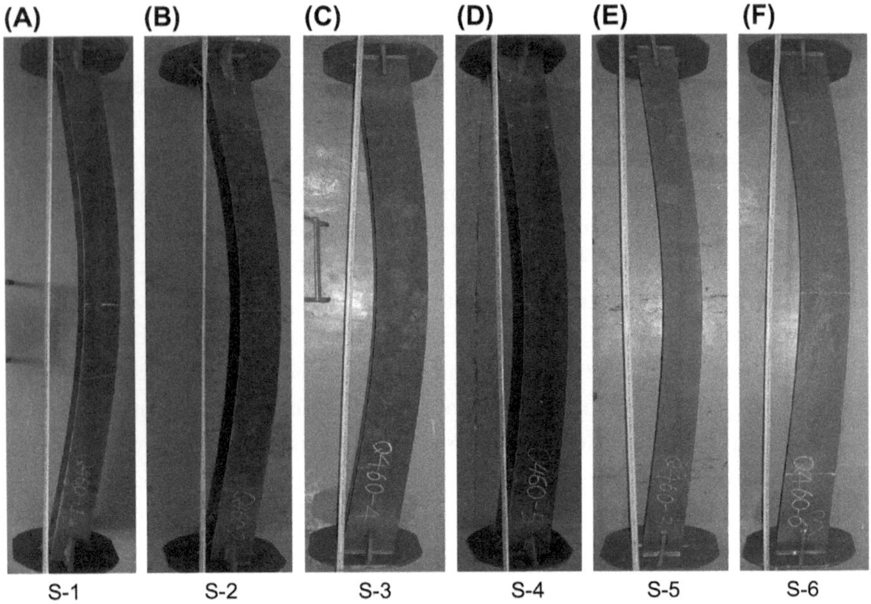

Figure 5.32 Failure pattern of the specimens. (A) S-1, (B) S-2, (C) S-3, (D) S-4, (E) S-5, and (F) S-6.

5.6.5.5 Failure patterns and visual observation

The failure patterns of the specimens are illustrated in Fig. 5.32. All steel columns buckled globally and eventually failed about the weak axis, and no local buckling was observed. The maximum bending appeared in the midheight of the column as where the initial imperfection is the largest. The lateral displacement of all specimens associated with creep bucking exceeded 135 mm, which was 1/20 of the column length.

5.6.6 Finite element model

To predict creep response of high strength Q460 steel columns at elevated temperature, a finite element model was established by employing software package ANSYS. The dimensions of the model were based on the nominal dimensions of the specimens shown in Table 5.2, and the boundary conditions at ends of the model were hinge supports. The values of yield strength, tension strength, and modulus of elasticity adopted in the model were obtained from results of the coupon tests. The test data, i.e., temperature distribution of steel columns and applied compressive load shown in Fig. 5.29, were directly input to simulate the actual elevated temperatures and load scenarios in the test. The stress—strain relationship of steel at elevated temperatures derived by Poh (2001), which did not account for the effect of the creep strain, was employed, whereas the effect of creep strain of Q460 steel at elevated temperature was mimicked by Norton creep model proposed by Wang et al. (2017). The Norton creep model is embedded in ANSYS, which is a widely used creep model, and is capable of simulating the secondary creep stage. The comparison on creep strains obtained by Norton creep model and the test is shown in Fig. 5.33. It can be seen that creep strains obtained from Norton model are in good agreement with the experimental results. The thermal expansion coefficient and Poisson's ratio of the steel were input in the model as $1.2 \times 10^{-5}/°C$ and 0.3, respectively.

The effect of welding residual stress was also considered in the finite element simulation by adopting the residual stress pattern shown in Fig. 5.34, which was proposed by Ban (2012). Considering the influence of elevated temperature on the variation of the residual stress, a reduction factor on the residual stress at elevated temperature derived by Wang and Qin (2016) was adopted in the simulation.

SHELL181 element in ANSYS, which has 6 degrees of freedom (three translations and three rotations) per node, was used to simulate the columns and end plates. The finite element model of the steel columns is shown in Fig. 5.35. The initial geometric imperfections adopted in the finite element analysis were based on the measured values shown in Table 5.4.

5.6.7 Validation of finite element modeling

To validate the finite element model that accounts for the effect of creep, axial and lateral displacements of the specimen obtained from the finite element analysis were compared with the test results as that shown in Fig. 5.36. It is observed that the axial and lateral displacements of specimens

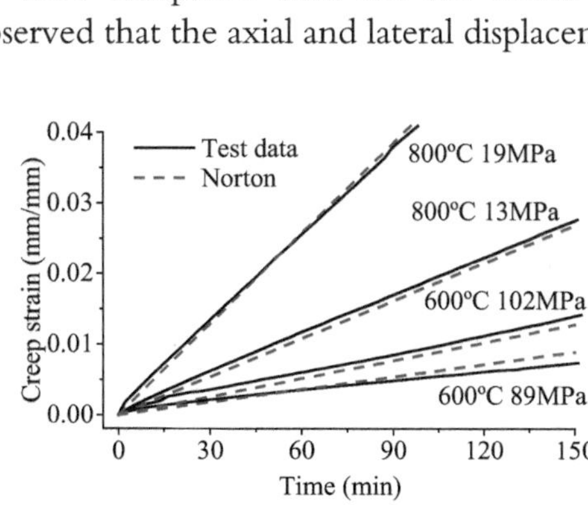

Figure 5.33 Comparison of creep model with test.

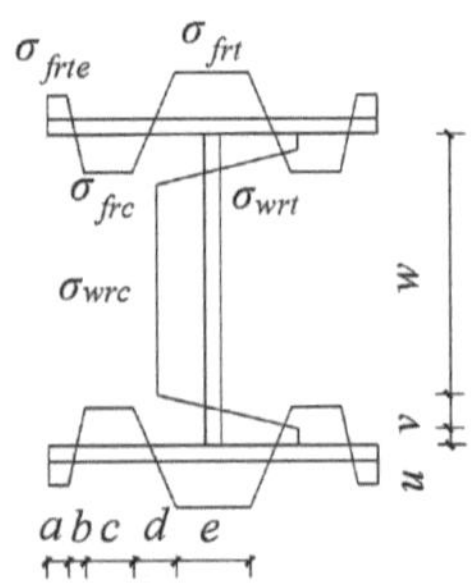

Figure 5.34 Residual stress pattern of Q460 steel H-section.

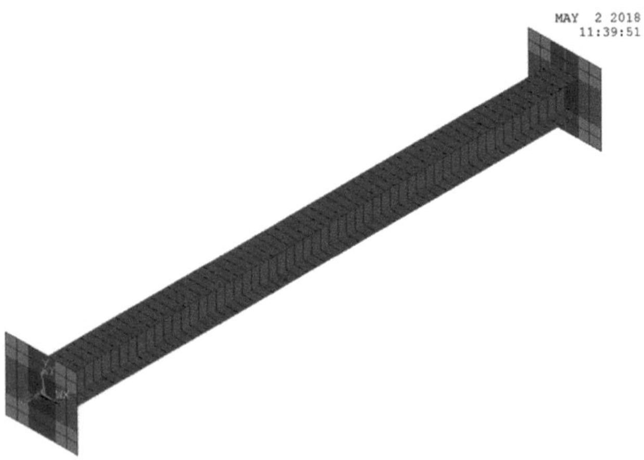

Figure 5.35 Finite element model of the specimen.

S-1 and S-3—S-6 are agreed well between that of tested and predicted. However, considerable discrepancies on the displacements of specimen S-2 were observed. As previously explained in Section 3.2, such occurrence that can be attributed to the temperature distribution in specimen S-2 was nonuniform.

The creep buckling time of specimens obtained from the tests and the finite element simulations are tabulated in Table 5.13. The percentage of the maximum difference is within 12% except for specimen S-2. The percentage was defined as the ratio of difference between the results of predicted and tested to the average value of these two values. Therefore, the established finite element model is reliable to simulate the creep response of high strength Q460 steel columns in fire conditions.

Comparison on the typical failure mode obtained from the test and the simulation is illustrated in Fig. 5.37. It is clearly seen that the failure pattern

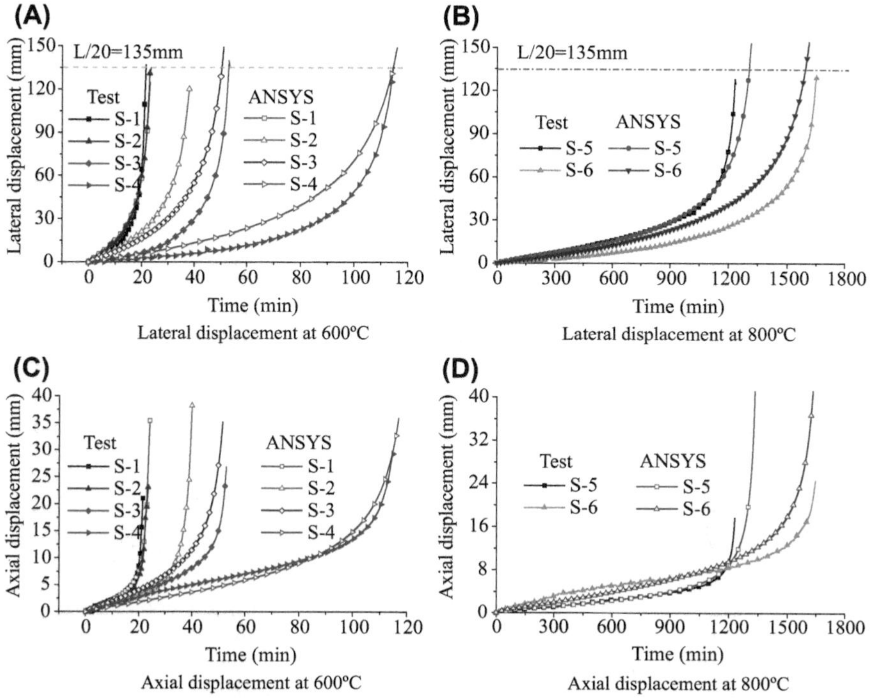

Figure 5.36 Comparison on axial displacements—creep time curves of steel columns. (A) Lateral displacement at 600°C; (B) lateral displacement at 800°C; (C) axial displacement at 600°C; and (D) axial displacement at 800°C.

Table 5.13 Comparison of creep buckling time between predictions and tests.

Column number	Test (min)	ANSYS (min)	Difference %
S-1	21.5	23	6.7
S-2	23.5	38	47.2
S-3	56.5	50	12.2
S-4	115	114.5	0.4
S-5	1232	1305	5.8
S-6	1643	1592	3.2

of the specimens in the test and the simulation were global buckling about the weak axis of the specimens, and no local buckling was occurred. Therefore, the failure mode obtained by the finite element analysis agrees well with that of the tests.

Fig. 5.38 shows the Mises stress distribution obtained from the simulation of specimen S-3 at different times. It is clearly indicated that the

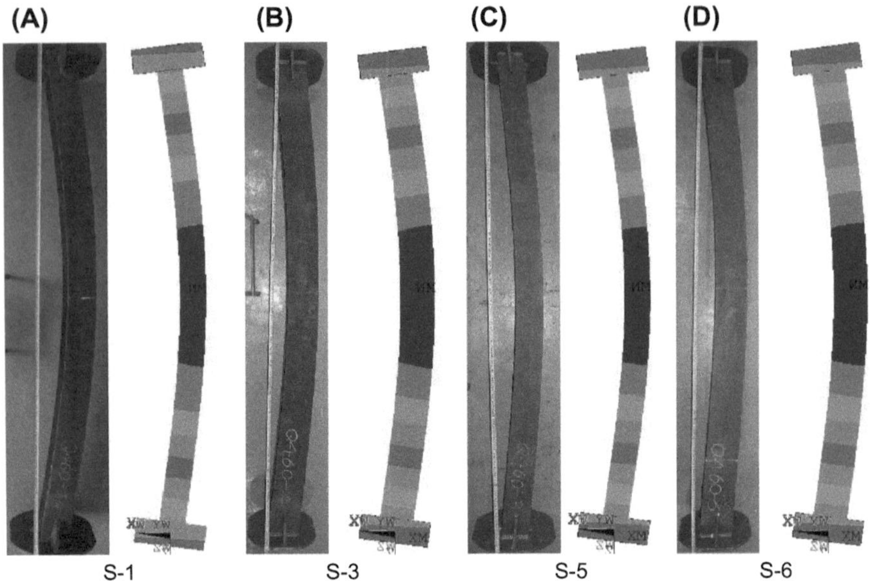

Figure 5.37 Failure mode comparison. (A) S-1, (B) S-3, (C) S-5, and (D) S-6.

stresses in the flanges gradually increased as increase of time, and the increasing rate of the stresses was gradually accelerated because of the increase of the second-order effect. This observation is consistent with the trend of deformation of the steel column. As the maximum stress in the steel column did not exceed the tension stress (70.6 MPa) of the steel, the resulting failure model of the steel columns is a typical global buckling instead of strength failure.

5.6.8 Parametric studies

The validated finite element model was utilized to study the effects of slenderness ratio, load ratio, temperature, and initial crookedness on the creep bucking time of Q460 steel columns. In parametric studies, the investigated parameters for high strength Q460 steel columns are listed in Table 5.14.

5.6.8.1 Slenderness ratio

Fig. 5.39A shows the comparison of the creep deformation of Q460 steel columns with different slenderness ratios obtained from the parametric studies. It is clearly seen that the higher slenderness ratio will lead to accelerate creep deformation and reduce corresponding creep buckling

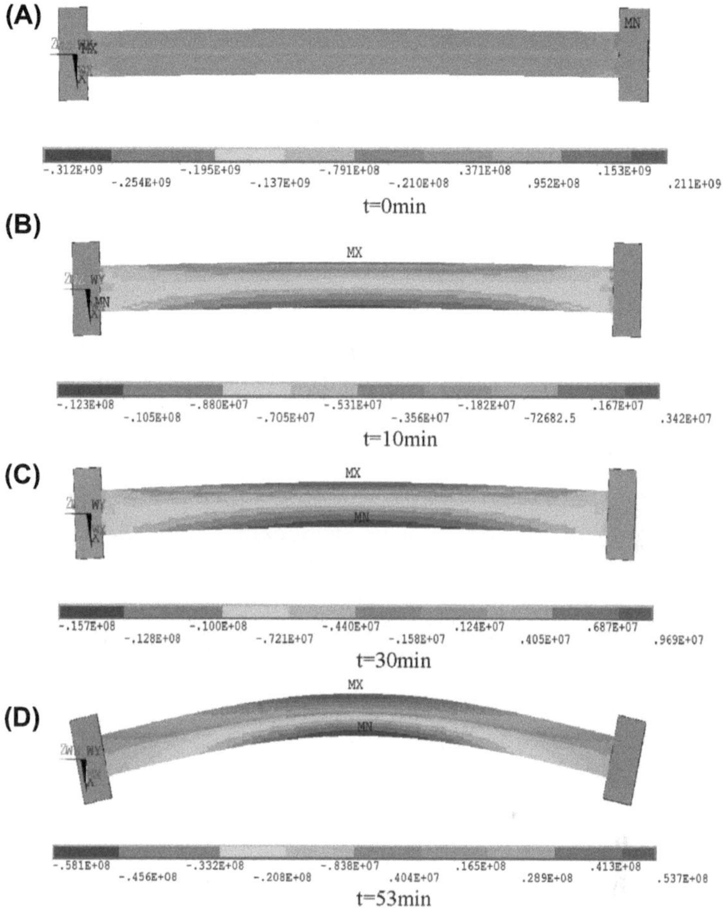

Figure 5.38 Stress distribution of S-3. (A) $t = 0$ min, (B) $t = 10$ min, (C) $t = 30$ min, and (D) $t = 53$ min.

Table 5.14 Values adopted from creep buckling parametric analysis of Q460 steel columns.

Parameters	Value range	Default value
Slenderness ratio	50, 60, 70, 80	60
Load ratio	0.1, 0.2, 0.3, 0.4	0.1
Temperature/°C	600, 700, 800	800
Initial crookedness	0.5‰, 1‰, 1.5‰, 2‰	1‰

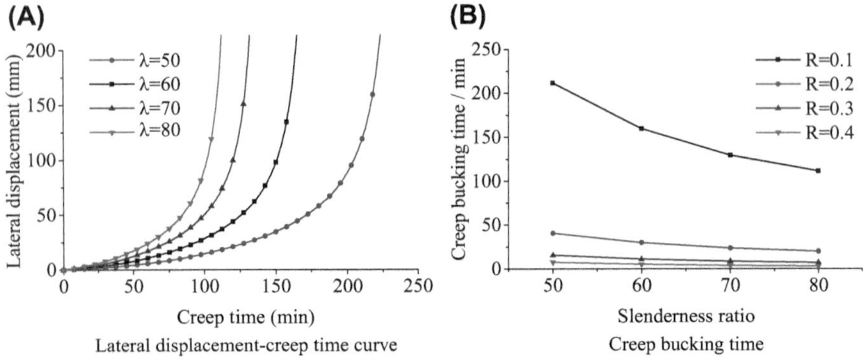

Figure 5.39 Comparison of creep buckling of Q460 steel columns with different slenderness ratios. (A) Lateral displacement—creep time curve and (B) creep bucking time.

time. This conclusion agrees well with the observations from the tests. Fig. 5.39B compares the trend of creep buckling time to the slenderness ratio under different load ratios. It is also found that with the increase of the slenderness ratio, the creep buckling time corresponding to a column under a smaller load ratio decreases more rapidly than that of a column subjected to a lager load ratio. Once the load ratio exceeds 0.3, the creep buckling time of steel columns is almost independent to the slenderness ratio.

5.6.8.2 Load ratios

Fig. 5.40 shows comparisons of the creep deformation and creep buckling time of Q460 steel columns with different load ratios. It is illustrated that the creep deformation increases more quickly for the column with higher load ratio. The creep buckling time declined with the increase of the load ratio. This trend is also consistent in the observations from the tests.

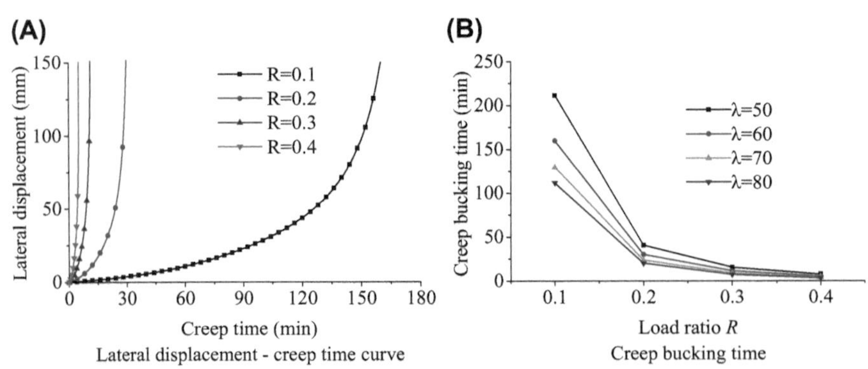

Figure 5.40 Comparison of creep buckling of steel columns with different load ratios. (A) Lateral displacement—creep time curve and (B) creep bucking time.

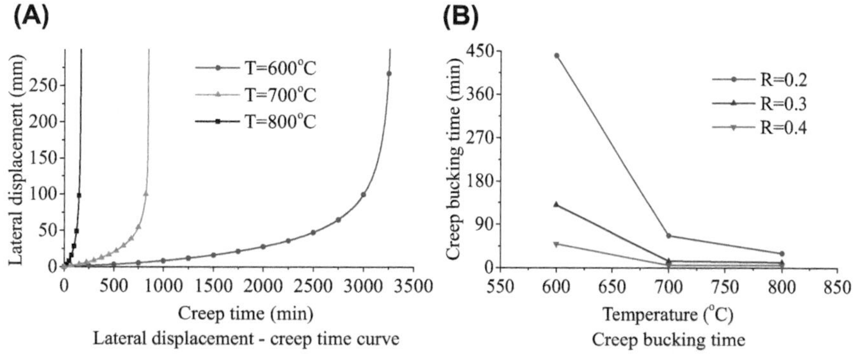

Figure 5.41 Comparison of creep buckling of steel columns with different temperatures. (A) Lateral displacement—creep time curve and (B) creep bucking time.

5.6.8.3 Temperature

Fig. 5.41 shows comparisons of creep deformation and creep buckling time of Q460 steel columns with different elevated temperatures. It is clearly observed that there is a significant difference in creep buckling time between the columns subjected to the target temperature of 600 and 800°C. The creep buckling time at 600°C is 3305 min, whereas the creep buckling time at 800°C is only 164.5 min. It is evident that the temperature has a significant influence on the creep bucking of the steel column, and it is a key factor for investigation of creep buckling of steel columns in fire conditions. Based on the test data on creep strain reported by Wang et al. (2017), when the stress was 20 MPa, the creep strain rate of Q460 steel at 800°C was 475 times greater than that at 600°C. Hence, the temperature elevation could accelerate the creep strain rate. This explains why the creep buckling time of steel columns reduce more rapidly at higher temperatures. It is also shown in Fig. 5.41B that the creep buckling time decreases along with temperature increase at different load ratios. At lower load ratio, the creep buckling time decreases more significantly.

5.6.8.4 Initial crookedness

The comparisons of the creep deformation and creep buckling time of Q460 steel columns with different initial crookedness are plotted in Fig. 5.42. It is shown that with the increase of the initial crookedness, the creep deformation rate increases, which leads to almost a proportional decrease of the creep buckling time of the steel column. Taking an example of the column with slenderness ratio being 60, the creep buckling time of the column corresponding to initial crookedness 0.5% and 1.0% is 188 and 160 min, respectively. Therefore, initial crookedness will amplify the effect of creep buckling.

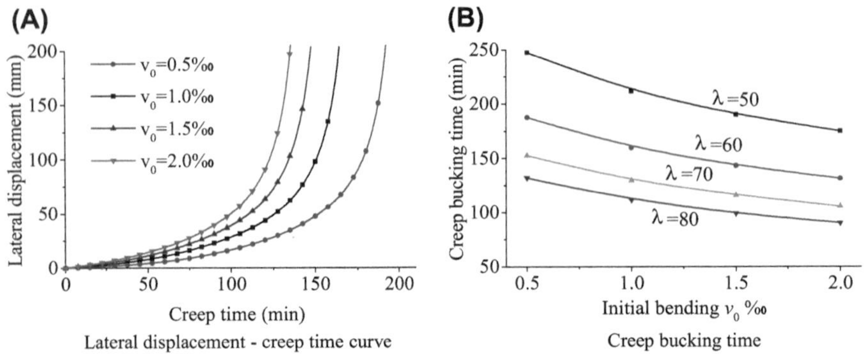

Figure 5.42 Comparison of creep buckling of steel columns with different initial crookedness. (A) Lateral displacement–creep time curve and (B) creep bucking time.

5.7 Creep buckling of high strength Q690 steel columns

A comprehensive test program was designed to undertake creep buckling tests on Q690 steel columns. Three factors that influence the creep buckling behavior of steel columns were considered: load level, temperature, and slenderness.

Steel coupons were extracted from Q690 steel plate respectively, and tensile test was carried out to obtain the actual mechanical properties. The shape and size of tensile strength coupons are prepared in accordance with GB/T 228.1-2010 (2010), and the test results of material properties are presented in Table 5.15. It can be seen that high strength steels (HSSs) have higher measured yield strength than nominal yield strength than that of mild steel.

5.7.1 Test specimens

To evaluate the influence of three factors, load ratio, temperature, and slenderness, on creep buckling of steel column, six welding H-shaped columns with the same length of 2700 mm were fabricated with Q690 HSS plate. The measured dimensions of specimens were tabulated in Table 5.16. Taking W690-800-0.3, for example, to illustrate the specimen

Table 5.15 Mechanical properties of high strength Q690 steel.

	Yield strength	Tensile strength	Elastic modulus	Elongation
Steel type	**f_y/MPa**	**f_u/MPa**	**E/GPa**	**Φ**
Q690	821	840	187.2	16.0

Table 5.16 Measured dimensions of test specimens.

	N	L	B	t_f	t_w	H	I	β
Specimen number	**kN**				**mm**			
N800-0.3	30	2731.50	151.8	13.81	13.90	200.5	0.47	1.58
N800-0.6	50	2734.75	150.3	14.07	14.09	199.9	2.61	1.63
W800-0.3	60	2732.50	199.7	13.84	13.95	200.2	2.55	1.56
W800-0.6	100	2732.00	199.9	13.91	14.15	200.4	2.23	1.07
N700-0.3	92	2731.70	200.4	14.01	14.10	199.6	3.61	1.44
W700-0.6	171	2731.15	200.2	14.22	14.17	200.8	3.00	1.92

labeling, where W denotes the wide section of column, which means small slenderness ratio (N denotes a narrow section); 800 is the temperature level (°C); and 0.3 is the load ratio (LR), LR was calculated by using Eq. (5.14a).

$$LR = \frac{N}{N_{cr,T}} = \frac{N}{\sigma_{cr,T}A} \tag{5.14a}$$

where N is the applied load and $N_{cr,T}$ is the bearing capacity of steel column at elevated temperature. $N_{cr,T}$ is obtained by using the calculation method specified in design code GB51249 (2017), as shown in Eq. (5.14b).

$$\sigma_{cr,T} = \frac{1}{2}\left\{(1+e_0)\sigma_{ET} + f_{yT} - \sqrt{[(1+e_0)\sigma_{ET} + f_{yT}]^2 - 4f_{yT}\sigma_{ET}}\right\} \tag{5.14b}$$

where A is cross-sectional area of column and f_{yt} is yield strength of steel at high temperature. It should be noted that $f_{1.0}$ (nominal yield stress when strain is 1%) was taken as f_{yt}, as no obvious yield platform existing in steel at high temperature. The measured mechanical properties of HSS at high temperature obtained by Wang et al. (2018) were utilized to get reduction factors of yield strength and elastic modulus of high strength Q690 steel in this paper. e_0 is initial eccentricity of steel column and σ_{ET} is Euler's critical stress of column at high temperature.

$$e_0 = 0.300\bar{\lambda} - 0.035 \tag{5.14c}$$

$$\bar{\lambda} = \lambda/\pi \cdot \sqrt{f_y/E} \tag{5.14d}$$

$$\sigma_{ET} = \pi^2 E_T/\lambda^2 \tag{5.14e}$$

where λ is slenderness; f_y is yield stress at ambient temperature; and E and E_T are elastic modulus at ambient and high temperature, respectively.

The simple method proposed by Shi (2014) was used to measure the imperfections of steel columns. The i shows the global geometric imperfections of column at midheight of the column length, and β is local geometric imperfection of the flange due to welding deformation, as shown in Fig. 5.43. It should be noted that the height of hinge supports (274 mm in total) was accounted for when using effective length of column to calculate its slenderness ratio.

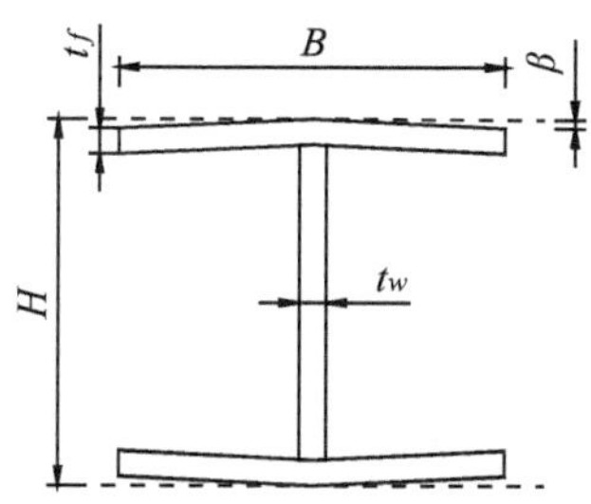

Figure 5.43 Notation of dimensions for steel column.

5.7.2 Test setup

The test setup to investigate the creep bucking of Q690 steel column under constant load and high temperature is similar to the test setup for carrying out creep buckling of Q460 steel columns, as shown in Fig. 5.26.

5.7.3 Instruments arrangement

Nine thermal couples were installed in each column to monitor the temperature variations in three sections (Fig. 5.44A,B). The temperatures of furnace were also measured by five thermocouples. The axial displacement was recorded in four locations using four LVDTs, which were fitted to the braces between two support hinges. As to lateral deflection, a nickel–chromium alloy wire was employed to link the lateral LVDT and the center of web through a hole on the furnace. This alloy wire is not sensitive to fire because of its low thermal expansion coefficient of $1.8 \times 10^{-5}/°C$.

5.7.4 Test procedures

The test procedure of creep buckling of Q690 steel columns is similar to Q460 steel columns as presented in Section 5.6.4.

5.7.5 Test results and discussion

5.7.5.1 Furnace temperatures

The FT curves recorded by thermal couples are shown in Fig. 5.45 and the corresponding height of each temperature measuring point are plotted in Fig. 5.44. Due to mechanical problem of the electronic furnace, the temperature of specimen W800-0.6 shows an unusual heating phase. It can be seen that the middle FT is the highest among measured results, followed by top and bottom FTs. The reason is that hot air would flow upward to

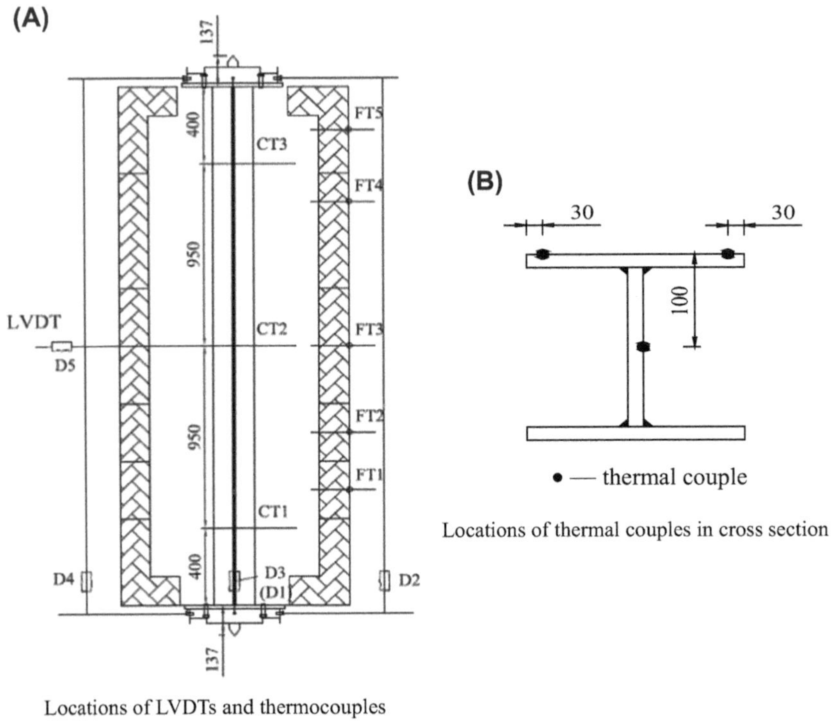

Figure 5. 44 Measurement arrangement of experiment. (A) Locations of LVDTs and thermocouples and (B) locations of thermal couples in cross section.

transfer heat, and meanwhile the top and bottom furnace cannot be completely sealed, resulting in an amount of heat loss. The malfunction of some thermal couples leads to abnormal measured results, like FT5 in Fig. 5.45C and FT1 in Fig. 5.45D. Generally, the FTs could reach to the target level within 30 min and then maintained constantly with discrepancies less than 20°C, which could provide ideal environment for creep buckling tests.

5.7.5.2 Column temperatures and load

Usually, the temperatures of steel column were regarded as uniformly distributed in previous researches. Therefore, the average temperatures of steel column in each section were used to reflect the CT evolution in tests, as shown in Fig. 5.46. The CTs at different height vary in accordance with the FTs, that is, CT2 is the highest, followed by CT3 and CT1. To directly illustrate the whole test process, the load evolutions are also plotted in

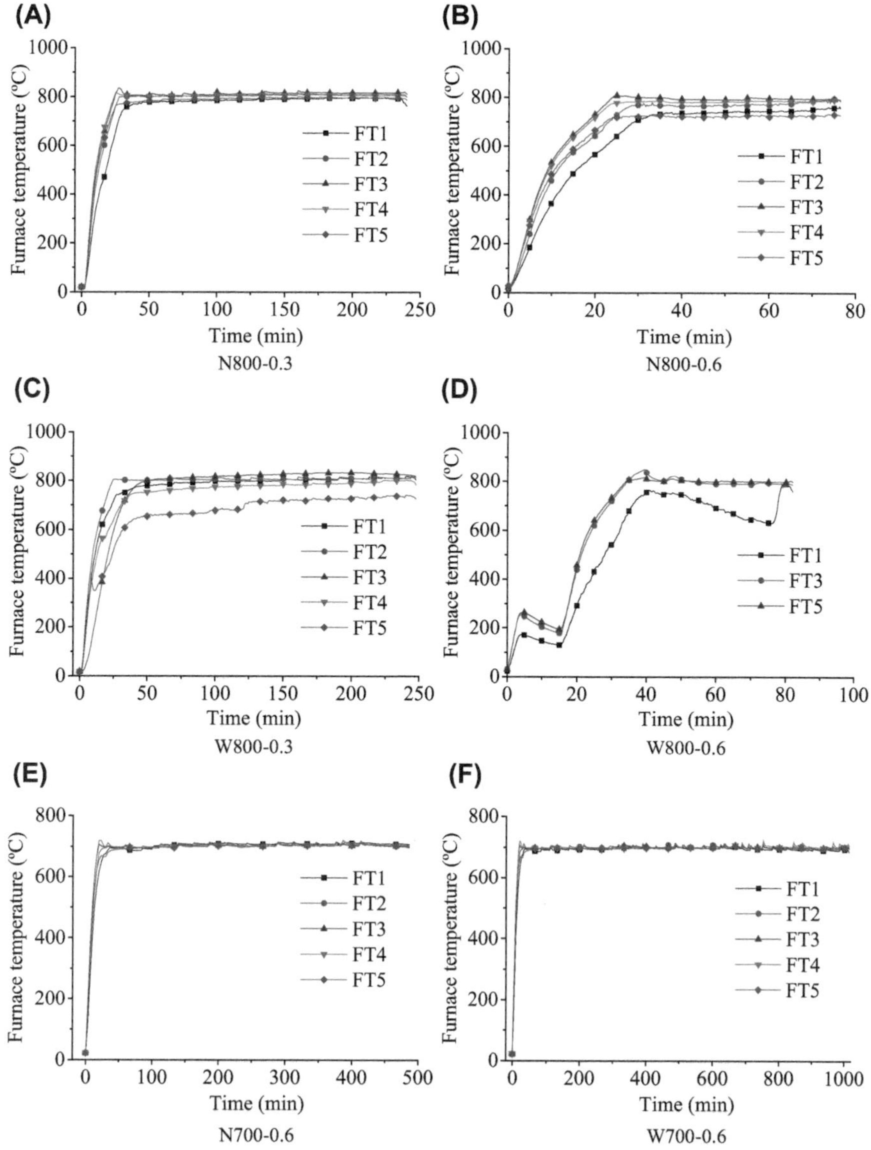

Figure 5.45 Measured furnace temperature curves at different locations versus time. (A) N800-0.3, (B) N800-0.6, (C) W800-0.3, (D) W800-0.6, (E) N700-0.6, and (F) W700-0.6.

Fig. 5.46. After all the CTs reached the target level, axial load was increased until being constant. Some load curves present slight fluctuation because the hydraulic jack is manually controlled and the axial displacement of column increases throughout the loading phase. It is difficult to maintain the load absolutely constant.

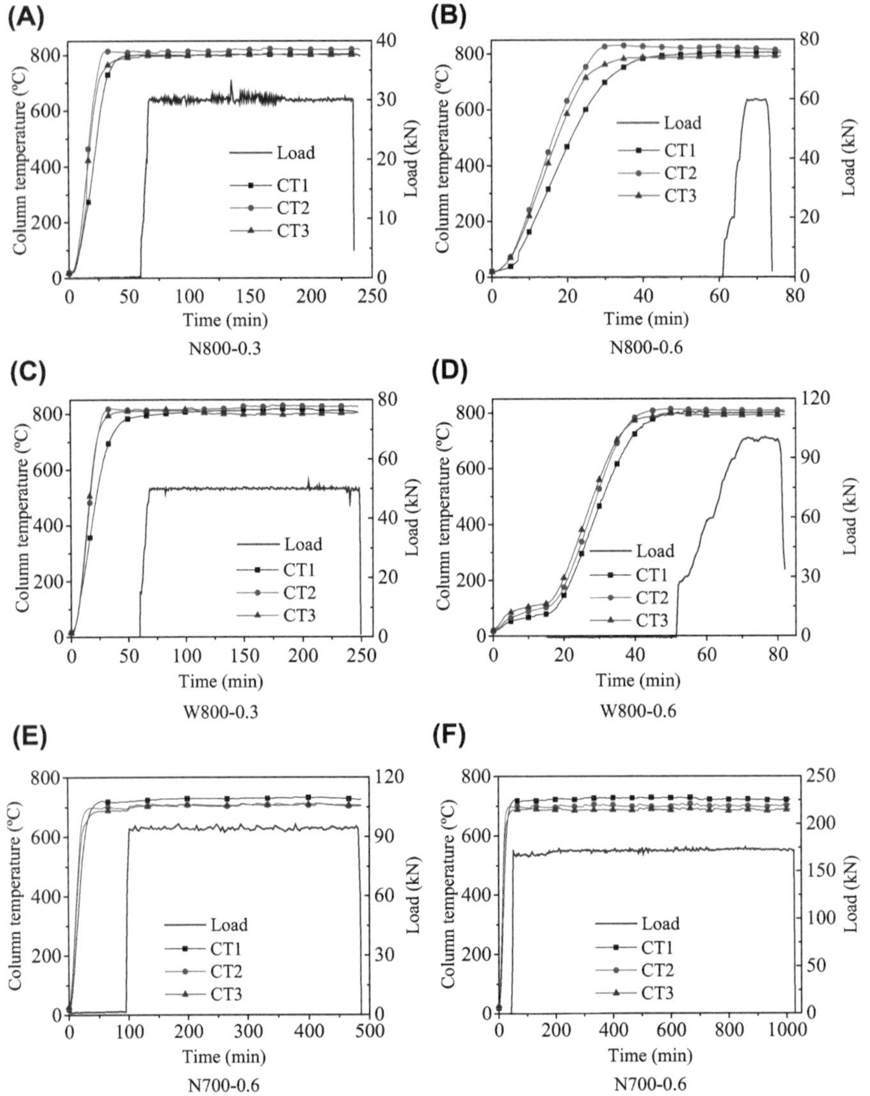

Figure 5.46 Measure column temperature at different locations and axial load curves versus time. (A)N800-0.3, (B) N800-0.6, (C) W800-0.3, (D) W800-0.6, (E) N700-0.6, and (F) N700-0.6.

5.7.5.3 Lateral deflection

In the test, creep effect brought about continual deformation of steel column at constant load and temperature level. The lateral deflection curves of each specimen were plotted in Fig. 5.47, wherein the boundary (red line)

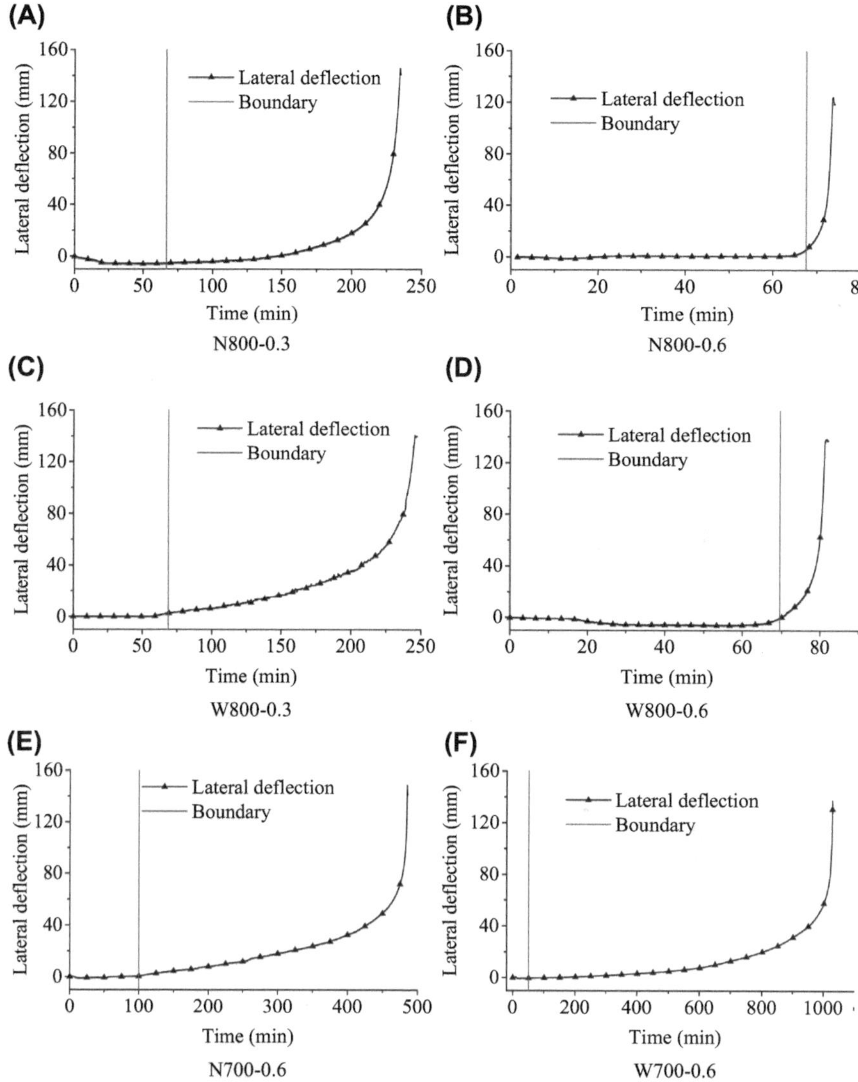

Figure 5.47 Lateral deflection versus time curve of Q690 high strength steel columns. (A)N800-0.3, (B) N800-0.6, (C) W800-0.3, (D) W800-0.6, (E) N700-0.6, and (F) W700-0.6.

indicates beginning of creep buckling phase, that is, the axial load reaches to a constant level. The change of lateral deflection is not obvious before the boundary (heating phase and load increasing phase), then lateral deflection increases gradually and accelerates sharply when increasing to $L/20$, which can be seen as a failure criterion of creep buckling test.

5.7.5.4 Axial displacements

The evolution of axial displacements in four directions (D1–D4 as plotted in Fig. 5.48) and average value are presented in Fig. 5.48. In the test, slight rotation would occur to transfer girder, which leads to larger discrepancy between D1 and D3 than that between D3 and D4. As to N700-0.6, some reversed changes may result from the loosening of bolts attached on the

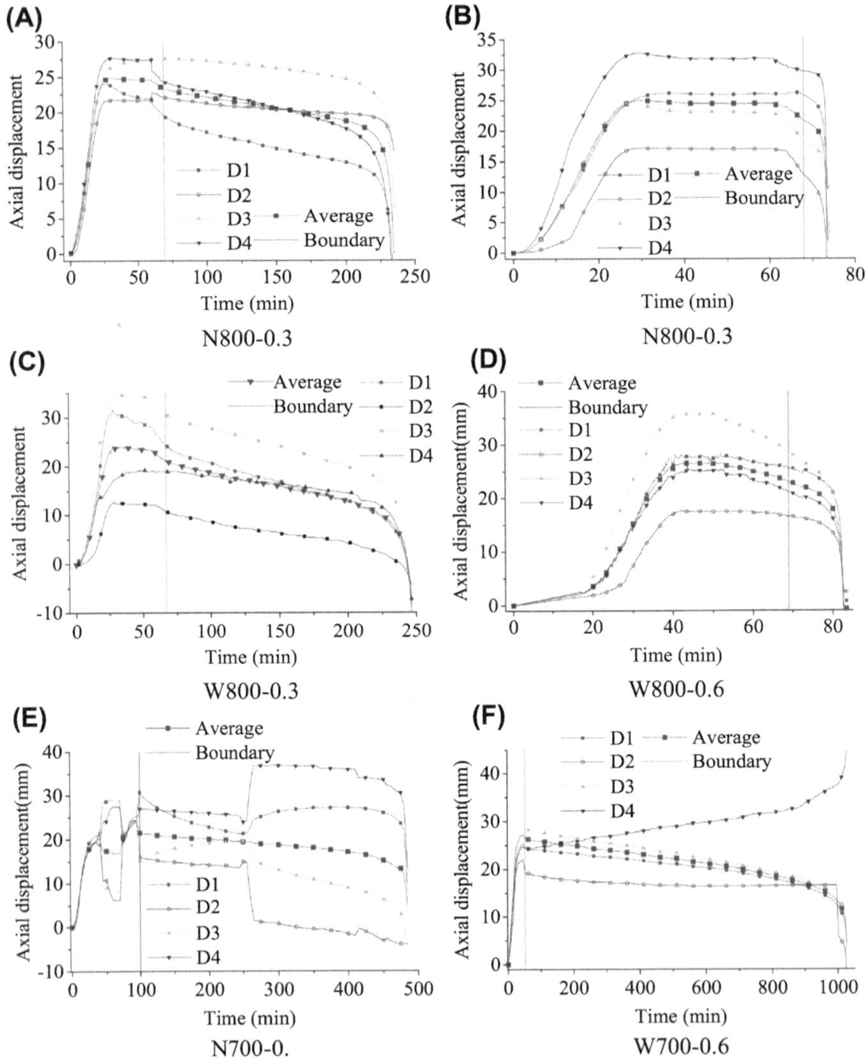

Figure 5.58 Axial displacement versus time curves of Q690 HSS columns. (A) N800-0.3, (B) N800-0.3, (C) W800-0.3, (D) W800-0.6, (E) N700-0, and (F) W700-0.6.

plate, which was used to fix the hinge support. Because of the axial displacement in D3 beyond the range of LVDT, the mean value of D2 and D4 was taken as average in Fig. 5.48E. Therefore, average of axial displacements is regarded as analysis subject in later content, more consistent with realistic axial displacement of steel column. It can be seen that, in the early stage, the evolution of axial displacement undergoes expansion phase because of heating temperature and slight decrease caused by the increasing applied load. When reaching to creep buckling phase, axial displacements decrease with a slow and steady rate and then speed up until failure.

5.7.5.5 Creep buckling failure time

The term creep buckling failure time, used in this chapter, refers to the duration from the time when applied load reaches to the target level to the time when column fails. The evolutions of axial and lateral displacements of columns in constant loading phase are plotted in Fig. 5.49. As the buckling time of each column varies from 6 to 973 min, the natural logarithm of time is adopted as X-axis scale, and the creep buckling failure times of each column are also presented in Fig. 5.49. With same load ratio and temperature level, the buckling time of steel column with larger slenderness is relevantly short. The difference induced by slenderness ratio is not obvious between W800-0.3 and N800-0.3 (168 and 177 min), which can be attributed that the smaller crookedness exists in N800-0.3 (0.47 mm) compared with that in W800-0.3 (2.55 mm). Besides, the evolution curves of axial displacement indicate that the steel columns with smaller slenderness ratio can maintain stability when experiencing larger deformation compared with that of bigger slenderness ratio. The load ratio has considerable influence on fire durance of column that buckling time

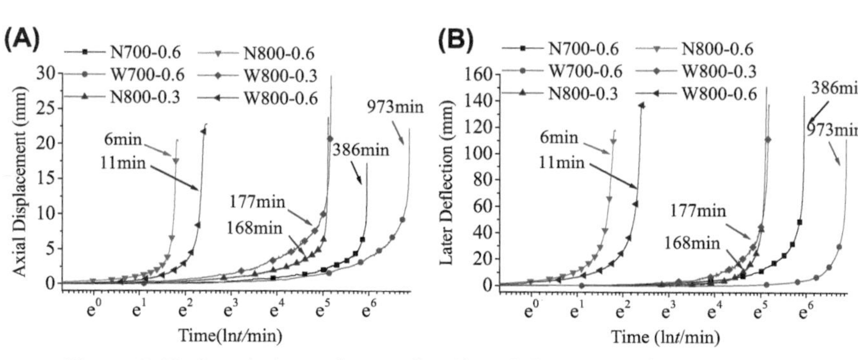

Figure 5.49 Comparison of creep buckling failure time of steel columns.

decreases with increasing load level under similar conditions. It can be seen that the buckling times at 800°C are much shorter than that at lower temperature, that is, the creep buckling failure times of Q690 steel columns vary greatly from different temperatures.

5.7.5.6 *Failure patterns*

All the high strength Q690 steel columns exhibit global flexural buckling about the weak axis, and no local buckling occurs, as shown in and Fig. 5.50. The maximum deflection of the columns located at the mid-height and the failure mode has no relation with load ratio and temperature.

5.7.6 Finite element modeling

To study the creep buckling behavior of steel columns, a numerical model was established using finite element software ABAQUS. The coupled thermal–mechanical analysis procedure was utilized in finite element mode, which has built-in creep power law to account for creep effect. The steel column was meshed with four node S4RT general-purposed shell

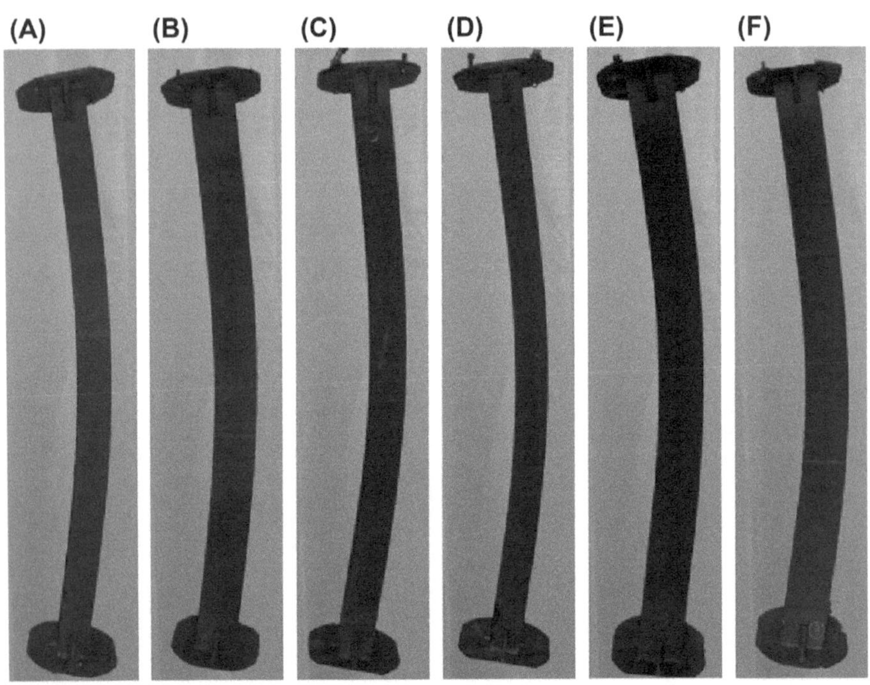

Figure 5.50 Specimens of Q690 columns after failure. (A) N690-700-0.6; (B) W690-700-0.6; (C) N690-800-0.3; (D) N690-800-0.6; (E) W690-800-0.3; and (F) W690-800-0.6.

element, which is applicable in coupled thermal–mechanical analysis procedure. The coupling constraint was used to simulate the end plate of steel column, and the height of hinge support was also taken into account by using MPC BEAM.

5.7.6.1 Mechanical properties

The values of specific heat, thermal expansion, and conductivity coefficients at elevated temperature adopted in this chapter were specified in EC3 (2005). Poison ratio value was taken as constant, 0.3. Elastic modulus and yield strength obtained from coupon tests were employed as shown above. Because of different chemical composition and fabricating process, HSS possesses distinct mechanical properties with mild steel at elevated temperatures, which can be observed in Fig. 5.51. To avoid the inaccuracy caused by these discrepancies, the Q690 HSS mechanical properties reduction factors obtained from test results by Wang were employed. Likewise, the constitutive relations of Q690 HSS at elevated temperatures were also derived from test results by Wang et al. (2018), as shown in Fig. 5.52.

5.7.6.2 Creep model

The Fields and Fields creep model (Fields and Fields, 1989), as shown in Eq. (5.1), was adopted to fit the creep test results of Q690 steel conducted by Wang et al. (2018).

The Fields and Fields creep model can only account for the first and secondary phase of creep. By taking a derivative of Eq. (5.1) with respect to

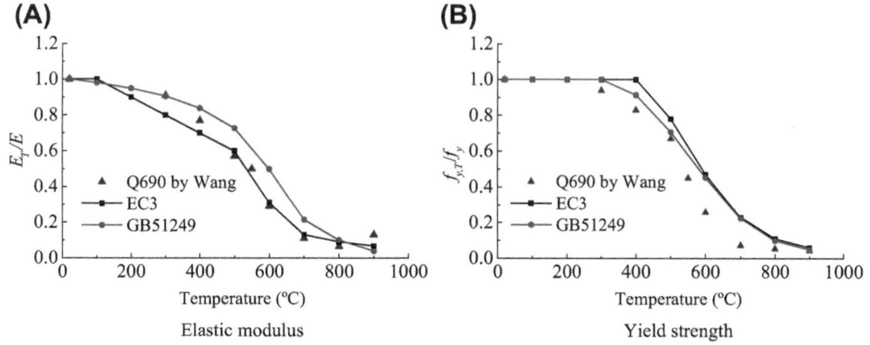

Figure 5.51 Comparison of mechanical properties reduction factors between HSS Q690 (Wang et al., 2018), EC3(2005), and GB51249(2017) at elevated temperatures. (A) Elastic modulus and (B) yield strength.

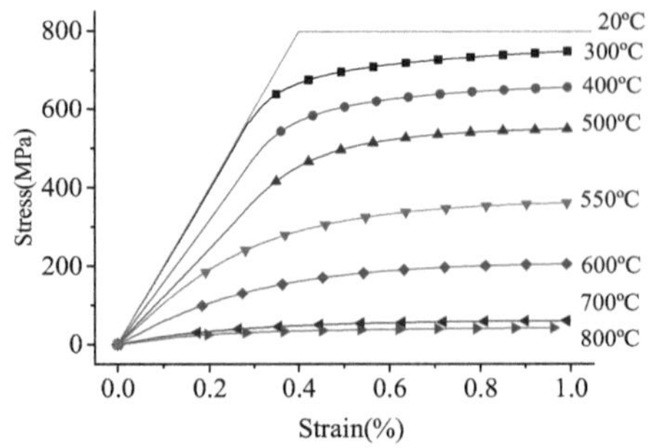

Figure 5.52 Stress–strain curves of Q690 high strength steel at high temperatures by Wang et al. (2018).

t, Eq. (5.15) can be used in the built-in, time-hardening creep law in ABAQUS.

$$\dot{\varepsilon}_{cr} = abt^{b-1}\sigma^{c} \tag{5.15}$$

where $\dot{\varepsilon}_{cr}$ is creep strain rate.

The fitted coefficients were tabulated in Table 5.17, and comparison between creep test results and fitted values are shown in Fig. 5.53, which indicates that fitted creep model has good agreement with test results and can be used in finite element analysis.

5.7.6.3 Residual stress distribution

Studies have shown that as the temperature increases, the residual stress at high temperatures will be released because of the reduction in the yield strength and elastic modulus of the steel, which will reduce the impact on the bearing capacity of the steel column at elevated temperature. However, in the high-temperature creep buckling test, because of the low stress level of the steel column, even if the residual stress is reduced, it will still have a great impact on the creep buckling failure time, which cannot be ignored.

Table 5.17 Coefficients of Fields and Fields creep model for Q690 steel.

Temperature (°C)	*a*	*b*	*c*
450	6.25E-42	14.47989	1
500	3.47E-31	11.00699	0.999,075
550	3.08E-22	8.36215	0.996,882
600	4.40E-15	5.97602	0.991,713
700	3.69E-06	2.7462	0.950,795
800	2.05E-04	2.04585	0.723,366

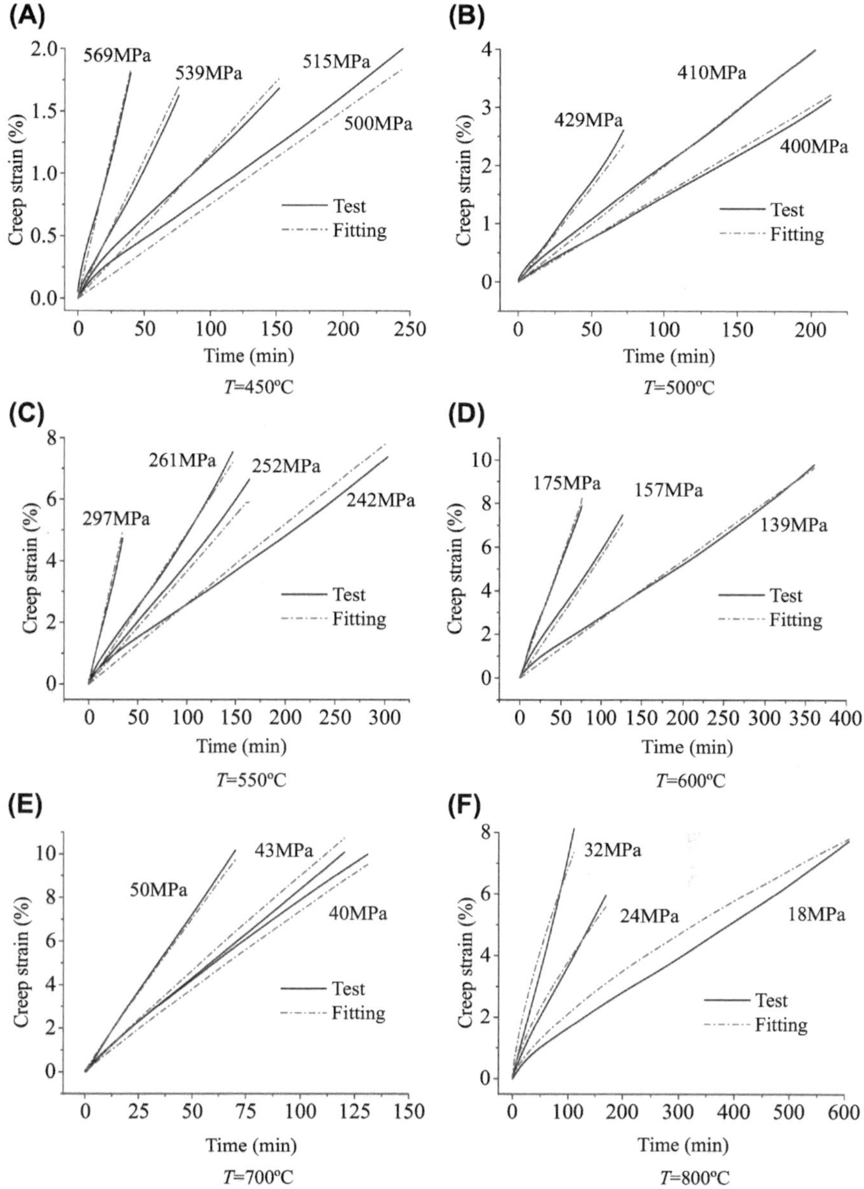

Figure 5.53 Test results and fitted values of creep strain in Q690 steel. (A) $T = 450°C$, (B) $T = 500°C$, (C) $T = 550°C$, (D) $T = 600°C$, (E) $T = 700°C$, and (F) $T = 800°C$.

The value of the residual stress at high temperature is not only related to the reduction of yield stress but also related to the stress relaxation caused by creep. Therefore, even at the same temperature level, the value of the residual stress varies from different heating process. It is necessary to consider the effect of temperature evolution of the steel column during the test on the residual stress distribution. To simulate the test, the residual stress distribution at ambient temperature will be applied on steel column at the beginning as showed in Fig. 5.54. The simplified residual stress model proposed by Wang and Qin (2016) and the maximum stress values were taken according to the measured results in literature.

5.7.6.4 Model validation

As mentioned above, the impact of heating and loading phases on creep buckling behavior of steel columns cannot be neglected. Therefore, the amplitude of temperature and axial load applied on each steel columns were input according to the measured values in the test. The first buckling mode shape of the specimen subjected to axial load was incorporated into analysis as initial imperfection, the value of which was taken as measured initial deflection at midspan.

The comparison of simulation results in terms of failure mode, axial displacement and lateral displacement are shown in Figs. 5.55–5.57 respectively. From Figs. 5.55-5.57, it indicates that the finite element analysis results show a good agreement with test results.

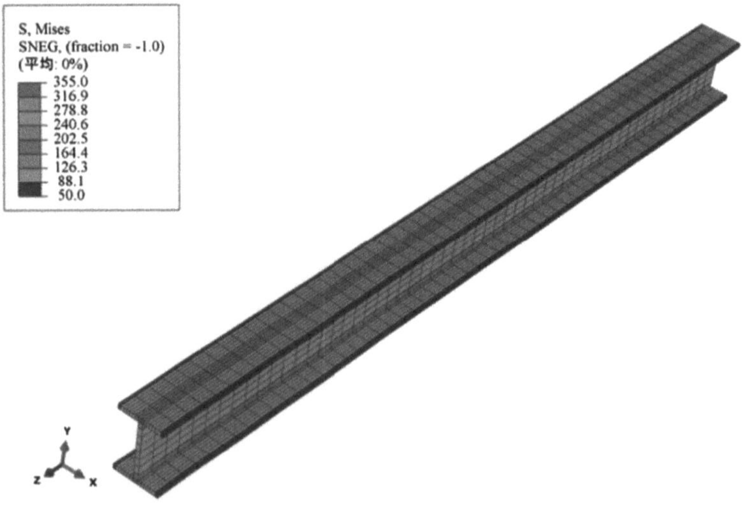

Figure 5.54 Meshing rule and initial residual stress distribution of Q690 HSS column in ABAQUS model.

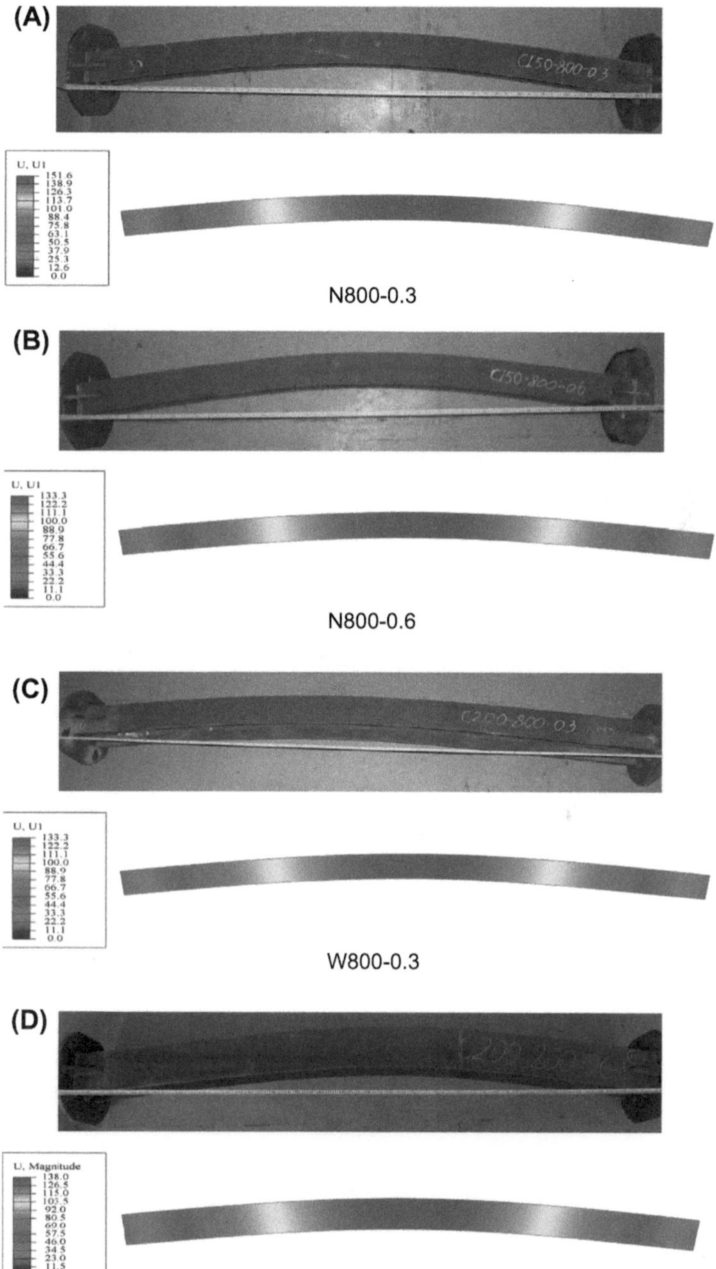

Figure 5.55 Comparison of creep buckling failure mode between FEA and test results. (A) N800-0.3, (B) N800-0.6, (C) W800-0.3, and (D) W800-0.6.

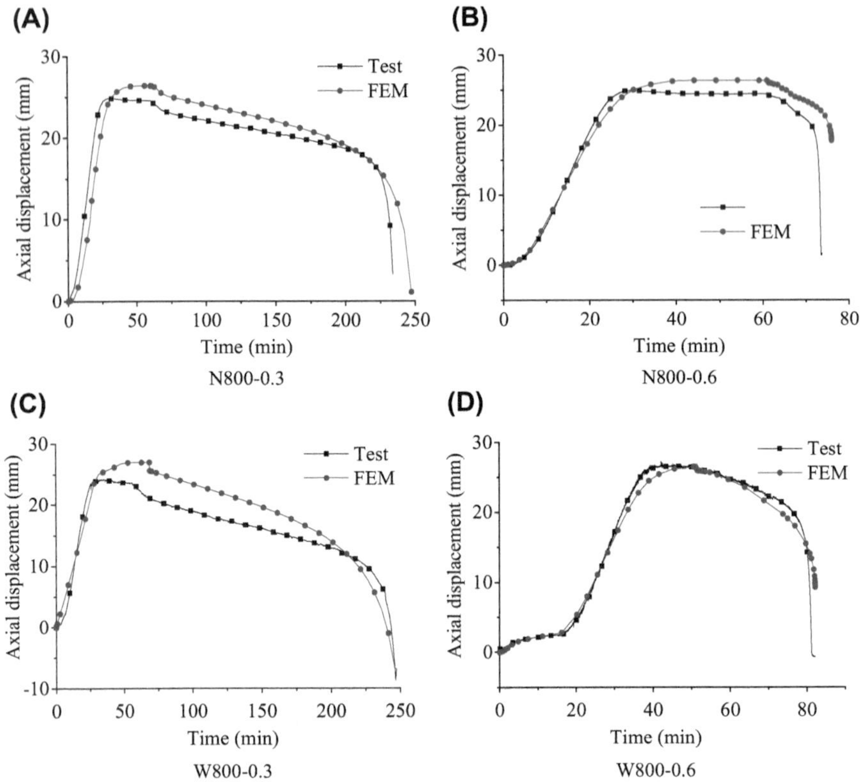

Figure 5.56 Comparison of axial displacements between FEM predictions and creep buckling test results. (A) N800-0.3, (B) N800-0.6, (C) W800-0.3, and (D) W800-0.6.

5.7.6.5 Stress analysis

Specimen W800-0.3 was chosen as example to preliminarily investigate the failure mechanism of creep buckle test. The evolution of stress in steel column was presented in Fig. 5.58, which mainly consists of the reduction process of residual stress with heating temperature. The residual tension stress decreases to 21.9 MPa at 700°C and only 6.2 MPa when temperature reaches to 800°C, then further decreases even at constant temperature level, which is caused by creep effect. Fig. 5.59 shows the evolution of stress in steel column in creep buckling phase. Because of the residual tension stress, in the initial stage, the maximum compressive stress does not occur at the edge of flange (Fig. 5.59A). With the residual stress being further eliminated, the maximum compressive stress decreases and shift to the edge of flange (Fig. 5.59B). As the lateral displacement gradually increases, the

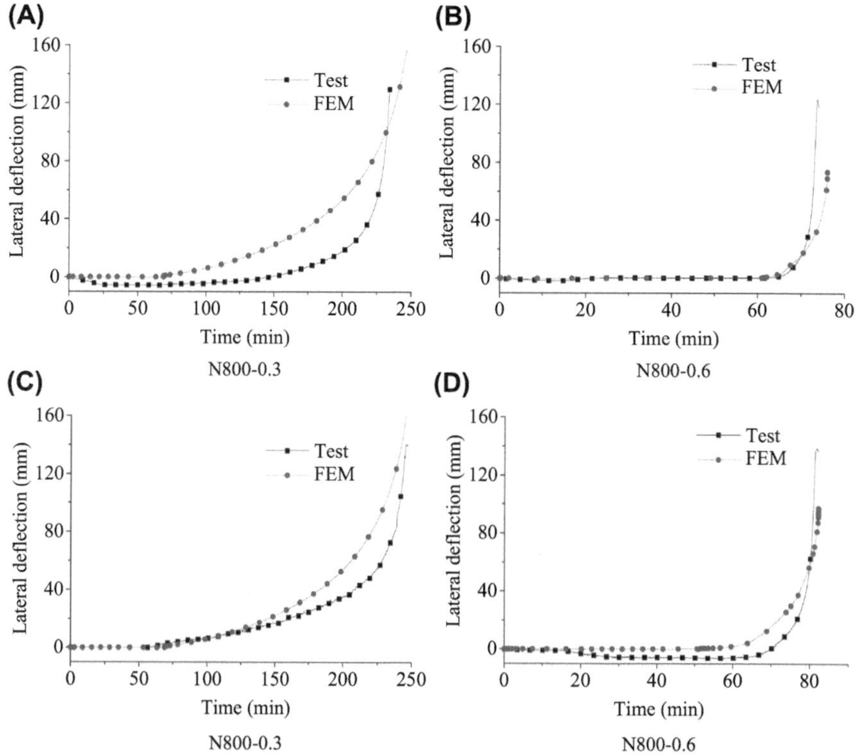

Figure 5.57 Comparison of lateral deflections between FEM predictions and creep buckling test results. (A) N800-0.3, (B) N800-0.6, (C) W800-0.3, and (D) W800-0.6.

compressive stress in flange on the other side of the steel column gradually decreases and turns into tensile stress (Fig. 5.59C). When the lateral deflection reaches to 46.4 mm (Fig. 5.59D), the internal stress level of the steel column is still within the limit of proportionality, indicating that the deformation of the steel column in the early stage is mainly composed of the creep deformation and the elastic deformation caused by the second-order effect. As the lateral deflection gradually increases, the stress level of the flange of the steel column exceeds the proportional limit (Fig. 5.59E), that is, plastic deformation occurs. Meanwhile, the deformation rate increases until column failures.

5.7.7 Parametric study

To further investigate the impacts of load ratio, temperature, and slenderness on creep buckling time of Q690 HSS columns, a series parametric

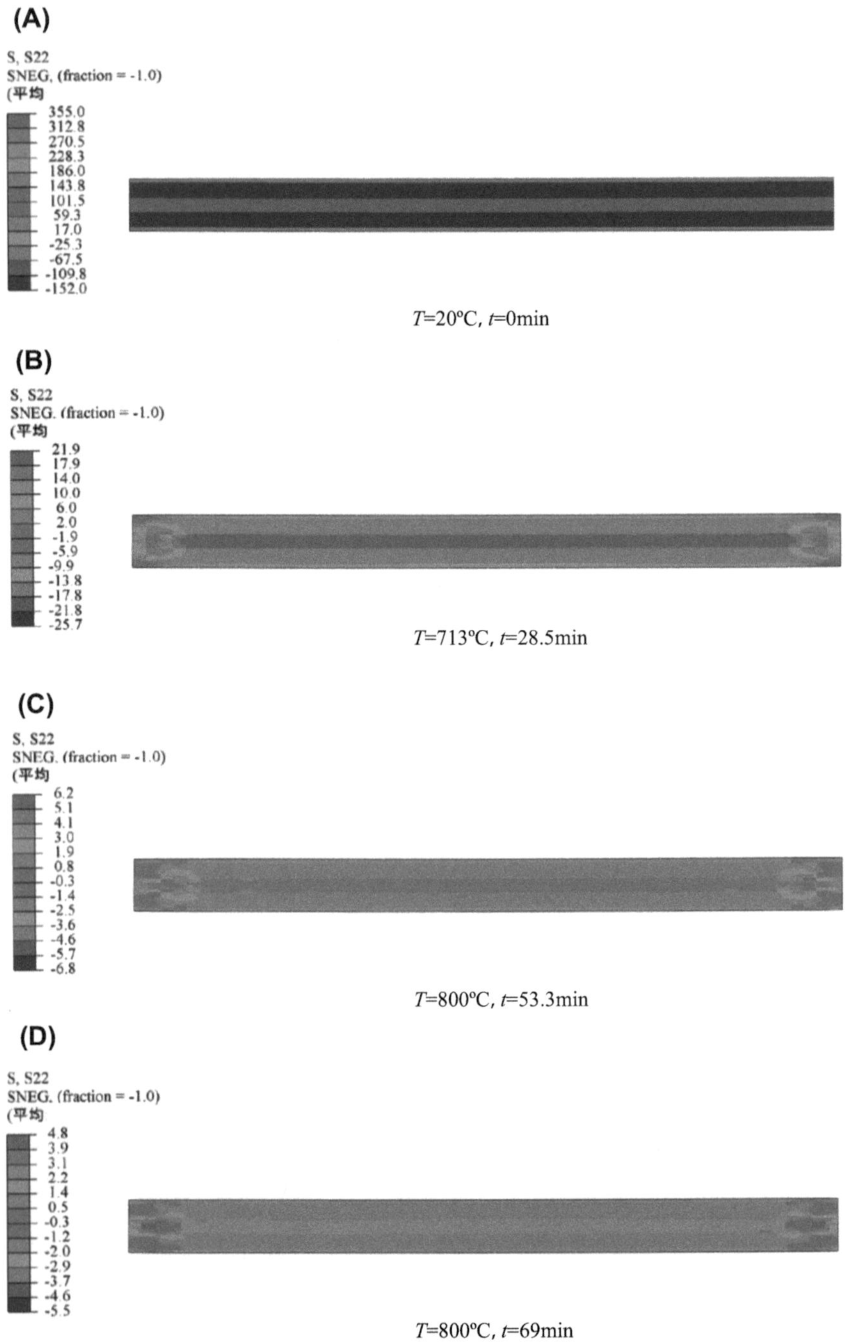

Figure 5.58 Stress distributions of steel columns at heating phase. (A) $T = 20°C$, $t = 0$ min; (B) $T = 713°C$, $t = 28.5$ min; (C) $T = 800°C$, $t = 53.3$ min; and(D) $T = 800°C$, $t = 69$ min.

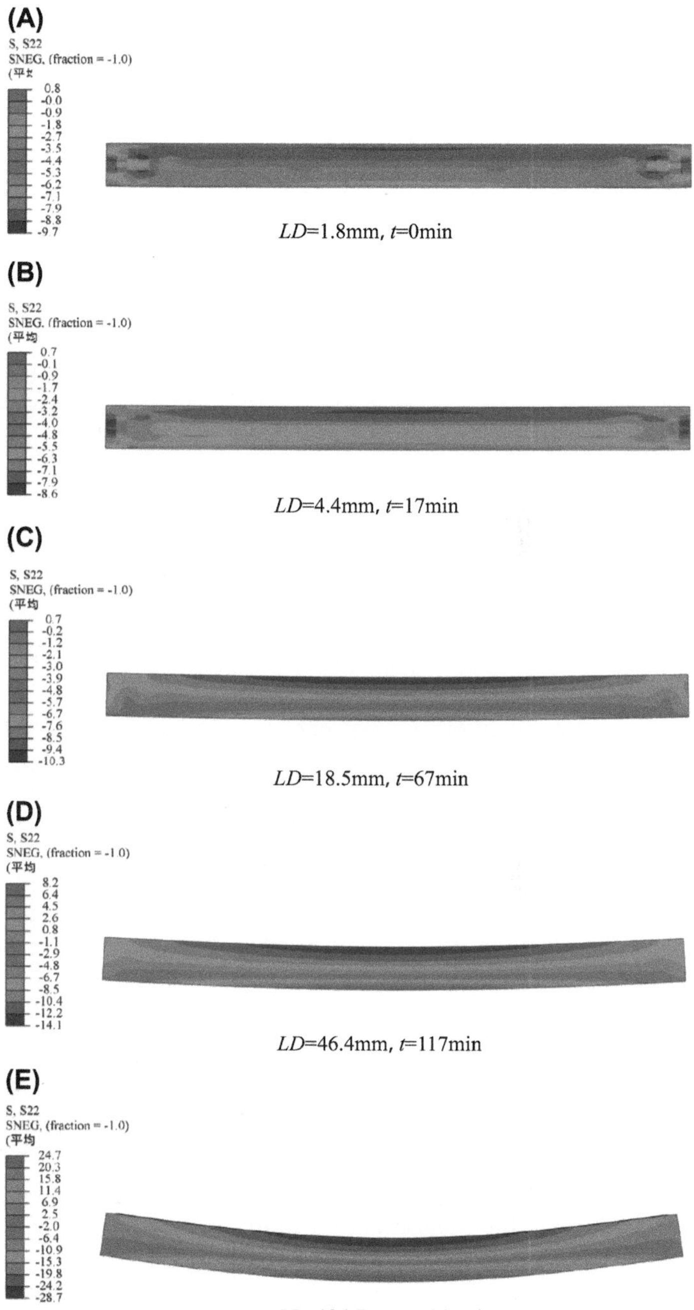

Figure 5.59 Stress distributions of steel column at constant load and temperature level. (A) $LD = 1.8$ mm, $t = 0$ min, (B) $LD = 4.4$ mm, $t = 17$ min, (C) $LD = 18.5$ mm, $t = 67$ min, (D) $LD = 46.4$ mm, $t = 117$ min, and (E) $LD = 136.7$ mm, $t = 167$ min.

studies were carried out by using the validated finite element model. H200 × 150 × 6 × 9 was chosen as the basic cross section, and the length of steel columns varies according to different slenderness. Mechanical properties at elevated temperatures, creep model, and residual stress distribution model were adopted as mentioned in finite element modeling. The initial imperfection was taken as one-thousandth of column length.

To be more accurate and simple, parametric studies only accounts for creep buckling phase but heating and loading phase, which is different with model validating. Because of neglecting the heating phase, the maximum value of residual stress at ambient temperature cannot be directly incorporated in the model because it is much greater than the yield strength of Q690 HSS at high temperature, which would lead to convergence problem. Therefore, utilizing residual stress distribution after reduction is more agreed with practical situation. The residual stress reduction factors of Q460 H-welded section at high temperatures proposed by Wang were employed, as tabulated in Table 5.18. By using Riks method in ABAQUS, the ultimate bearing capacities of specimens at different temperature levels were calculated, which were tabulated in Table 5.19.

5.7.7.1 Load ratio

When slenderness λ equals 60 and temperature T equals 800°C, the lateral deflection variations with time under various load ratios were plotted in Fig. 5.60A,B. Creep buckling time is 75 min when load ratio is 0.3 but only 1 min when load ratio increases to 0.8. On the one hand, creep strain rate is positively correlated with stress level, which is greater for larger load ratio; on the other hand, larger load ratio would lead to larger initial imperfection and greater second-order effect during creep buckling phase. The relationships between creep buckling failure time and load ratio with various slenderness were plotted in Fig. 5.60C, which indicates that the failure time decreases nonlinearly with increase of load ratio.

5.7.7.2 Temperature level

When slenderness λ equals 60 and load ratio LR equals 0.8, the lateral deflection variations at different temperature levels were plotted in Fig. 5.61A. Similar to the phenomenon of test, creep buckling failure time decreases with the increase of temperature. The reason is that under same conditions, the creep strain rate is relatively slow at lower temperature even with higher stress level. The influence of temperature level on creep buckling failure time at various load ratios was presented in Fig. 5.61B.

Table 5.18 Reduction factors of residual stress of Q460 H-welded section at elevated temperatures.

Temperature(°C)	20	100	200	300	400	500	600	700	800	900
Reduction factor	1	1.005	0.968	0.835	0.677	0.468	0.183	0.086	0.057	0.023

Table 5.19 Ultimate bearing capacity (kN) of specimens under different situations in parametric study.

Temperature(°C) / Slenderness	800	700	600	500
120	28.7	37.3	121.1	280.2
100	39.7	50.3	165.3	397
80	53.1	71.1	238.1	605.4
60	72.6	104	363.1	973.9

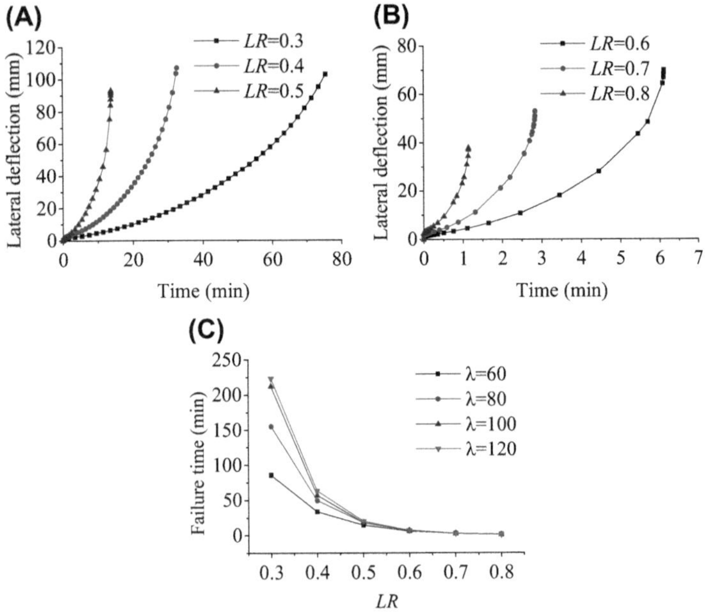

Figure 5.60 Influence of load ratio on creep buckling failure time.

Failure times were not given when $T = 500°C$ and $LR = 0.6$ and 0.7. Because the creep buckling time is too long(over 10 h) due to slow creep strain rate at low temperature and load level, which is meaningless, the third creep phase is not taken into account in Fields and Fields creep model. The analysis results in Fig. 5.61B indicate that temperature has slighter influence on failure time with larger load ratio. When $T = 500°C$ and $LR = 0.6$, steel column would fail within 7 min due to creep effect, and in this situation, the applied load is still approximately 0.6 times the ultimate bearing capacity at ambient temperature. It should be argued that creep effect cannot be neglected in fire design of steel column, otherwise leading to unsafe results.

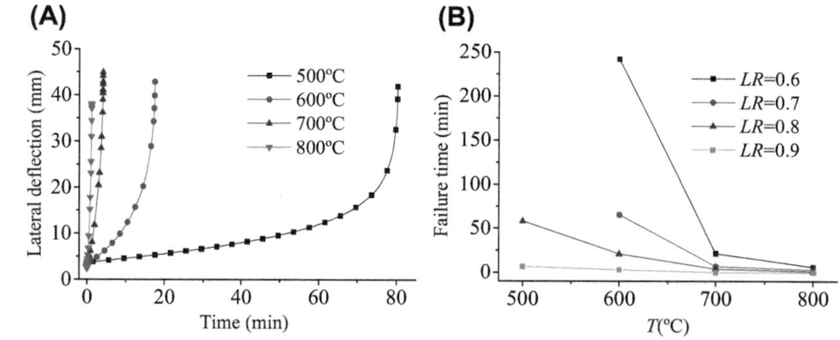

Figure 5.61 Influence of temperature level on creep buckling failure time.

5.7.7.3 Slenderness

When T is 800°C and LR is 0.4, the lateral deflection variations with different slenderness were depicted in Fig. 5.62A. It should be noted that the creep buckling failure time increases with increases of slenderness, which is diverse with the conclusion drawn from test results. Distinct from the ss in test, the section of steel columns is same in the parameter study, thus the slenderness is changed by altering the length of the steel column. Additionally, the elastic modulus of Q690 steel at 800°C is very low (13.2 GPa, which is only 0.07 times that at ambient temperature). Hence, the ultimate bearing capacity of steel column at elevated temperature is mainly controlled by stability, and ultimate bearing capacity varies greatly from different slenderness, as tabulated in Table 5.19. Consequently, with same cross section, the stress level is higher when slenderness is smaller, which results in larger creep strain rate and shorter creep buckling failure time.

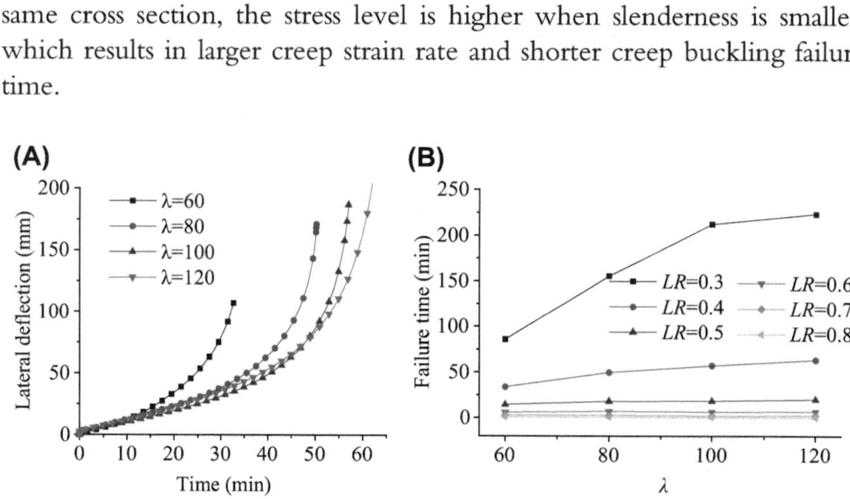

Figure 5.62 Influence of slenderness on creep buckling failure time.

It can be observed in Fig. 5.62B that when $LR = 0.3-0.5$, the greater the slenderness, the longer the creep buckling failure time. However, slenderness has a slight effect on creep buckling failure time when $LR = 0.6-0.8$.

References

American Society for Testing and Materials International (ASTM), Designation, 2009. A370-09 Standard Test Methods and Definitions for Mechanical Testing of Steel Products. West Conshohocken, PA.

Ashby, M.F., Jones, D.R., 2005. An Introduction to Properties, Applications and Design, third ed. Engineering materials I Publication, Elsevier Butterworth-Heinemann, Burlington.

ASM, 2012. International Atlas of Stress—Strain Curves, vol. 2. ASM, Material Park, OH.

Ban, H.Y., 2012. Research on the Overall Buckling Behavior and Design Method of High Strength Steel Columns under Axial Compression. Tsinghua University, Beijing.

Brnic, J., Canadija, M., Turkalj, G., Lanc, D., 2010. Structural steel ASTM A709-behavior at uniaxial tests conducted at lowered and elevated temperatures, short-time creep response, and fracture toughness calculation. J. Eng. Mech. 136 (9), 1083—1089.

Cowan, M., Khandelwal, K., 2014. Modeling of high temperature creep in ASTM A992 structural steels. Eng. Struct. 80, 426—434.

Fields, B.A., Fields, R.J., 1989. Elevated Temperature Deformation of Structural Steel. Report NISTIR, vols. 88—3899. NIST, Gaithersburg, MD.

Franssen, J., Kodur, V.K.R., Zaharia, R., 2009. Designing Steel Structures for Fire Safety. Taylor & Francis Group, London, UK.

GB 51249, 2017. Code for fire safety of steel structures in buildings. State Standard of the P.R. China. China Planning Press, Beijing, China.

GB/T 228.1-2010, 2010. Metallic Materials-Tensile Testing-Part 1: Method of Test at Room Temperature. State Standard of the P.R. China. China Standard Press, Beijing, China.

Harmathy, T.Z., 1967. A comprehensive creep model. J. Basic Eng. 89 (3), 496—502.

Kodur, V.K.R., Dwaikat, M., 2009. Response of steel beam-columns exposed to fire. Eng. Struct. 31 (2), 369—379.

Kodur, V.K.R., Aziz, E.M., 2014. Effect of temperature on creep in ASTM A572 high-strength low-alloy steels. Mater. Struct. https://doi.org/10.1617/s11527-014-0262-2.

Kodur, V.K.R., Dwaikat, M., 2010. Effect of high temperature creep on the fire response of restrained steel beams. Mater. Struct. 43 (10), 1327—1341.

Kodur, V.K.R., Dwaikat, M.M.S., Fike, R.S., 2010. High-temperature properties of steel for fire resistance modeling of structures. J. Mater. Civ. Eng. 22 (5), 423—434.

Morovat, M.A., Engelhardt, M.D., Helwig, T.A., Taleff, E.M., 2014. High-temperature creep buckling phenomenon of steel columns subjected to fire. J. Fire Struct. Eng. 5 (3), 189—202.

Poh, K.W., 2001. Stress-strain-temperature relationship for structural steel. J. Mater. Civ. Eng. 13 (5), 371—379.

Shi, G., Jiang, X., Zhou, W., Chan, T.M., Zhang, Y., 2014. Experimental study on column buckling of 420 MPa high strength steel welded circular tubes. Journal of Constructional Steel Research 100, 71—81.

Toric, N., Harapin, A., Boko, I., 2013. Experimental verification of a newly developed implicit creep model for steel structures exposed to fire. Eng. Struct. 57, 116—124.

Wang, W.Y., Liu, B., Kodur, V.K.R., 2013. Effect of temperature on strength and elastic modulus of high strength steel. J. Mater. Civ. Eng. 25 (2), 174–182.
Wang, W.Y., Yan, S.H., Kodur, V.K.R., 2016. Temperature induced creep in low-alloy structural Q345 steel. J. Mater. Civ. Eng. 28 (6), 06016003.
Wang, W.Y., Qin, S.Q., 2016. Experimental investigation of residual stresses in thin-walled welded H-sections after fire exposure. Thin-walled Structures 101, 109–119.
Wang, W.Y., Yan, S.H., Liu, J.P., 2017. Studies on temperature induced creep in high strength Q460 steel. Mater. Struct. 50 (1), 68.
Wang, W., Wang, K., Kodur, V.K.R., Wang, B., 2018. Mechanical properties of high strength Q690 steel at elevated temperature. Journal of Materials in Civil Engineering 30 (5), 04018062.
Zienkiewicz, O.C., Cormeau, I.C., 1974. Visco-plasticity-plasticity and creep in elastic solids: a unified numerical solution approach. Int. J. Numer. Methods Eng. 8, 821–845.

CHAPTER 6

Application of materials properties in structural fire engineering

Contents

Material Properties of Steel in Fire Conditions
ISBN 978-0-12-813302-6
https://doi.org/10.1016/B978-0-12-813302-6.00006-0

6.1 Introduction

The research studies on fire resistance of mild steel structures are abundant. However, the outcomes on fire resistance of high strength steel structures are very limited, as mechanical properties such as reduction factors or residual factors of high strength steels are not consistent with those of mild steel at elevated temperature or after fire exposure. The methods or approaches for fire resistance design of high strength steel members are different. In this chapter, research studies on fire resistance of high strength steel members by using the corresponding material properties are presented. The content mainly includes load-bearing capacity of high strength Q460 steel columns at elevated temperature and after fire exposure, response of restrained high strength Q460 steel columns, and design of high strength Q460 steel beams with nonuniform temperature distribution, taking creep into consideration.

6.2 Load-bearing capacity of high strength Q460 steel columns

For the mild steel columns, there are some assumptions adopted to calculate the load-bearing capacity of steel columns at elevated temperatures, and for high strength Q460 steel columns, same assumptions are utilized as follows:

(1) The initial geometry imperfection of a steel column can be represented as a half sine wave.

(2) The temperature distributions across the section and along the columns are both uniform.

(3) The distribution of residual stress over cross section of the columns at high temperatures is the same as that recommended for ambient design.

Two methods, namely critical stress method (simplified method or analytical method) and inverse calculation segment length method (numerical method), are frequently used to calculate the load-bearing capacity for mild steel columns at ambient temperature. The former one is simple to use and only applicable for special type of steel and section shape. The latter one is versatile and feasible to any steel and any section shape, but it is also too complicate and needs lots of calculation. The former one is the simplified results of a great amount data obtained by the latter method for some frequently used steel type and section shape. For high strength Q460 steel, the two methods are extended by taking the high temperature properties of Q460 steel into consideration.

6.2.1 Critical stress approach

Like the calculation for critical stress of mild steel columns at room temperature, the critical stress for high strength Q460 steel columns at elevated temperatures can be expressed (Li et al., 2006):

$$\sigma_{crT} = \frac{1}{2}\left\{(1+e_0)\sigma_{ET} + f_{yT} - \sqrt{[(1+e_0)\sigma_{ET} + f_{yT}]^2 - 4f_{yT}\sigma_{ET}}\right\} \quad (6.1)$$

where f_{yT} is yield strength of high strength Q460 steel at elevated temperatures; $\sigma_{ET} = \pi^2 E_T/\lambda^2$; E_T is modulus of elasticity at elevated temperature; λ is slenderness ratio of the column; and e_0 is ratio of equivalent initial eccentricity, for welded H-shaped and box section with flame-cut flange edges, which can be obtained by $e_0 = 0.300\bar{\lambda} - 0.035$, ($\bar{\lambda} = \lambda/\pi\sqrt{f_{yT}/E_T}$).

The stable factor of high strength Q460 steel columns can be determined by

$$\varphi_T = \sigma_{crT}/f_{yT} \quad (6.2)$$

From Eq. (6.1), (6.2), the stable factor can be obtained at different temperature and different slenderness ratio as shown in Fig. 6.1. As can be seen from Fig. 6.1, with the increase of temperature, the stable factor almost keeps constant. At temperatures of 200 and 500°C, the stable factor is a little lower than that at room temperatures. However, at temperature higher than 600°C, the stable factor is a little higher than that at room temperatures. This is due to that the stable factor at high temperature has great relation with the ratio of elastic modulus to yield strength (as can be seen

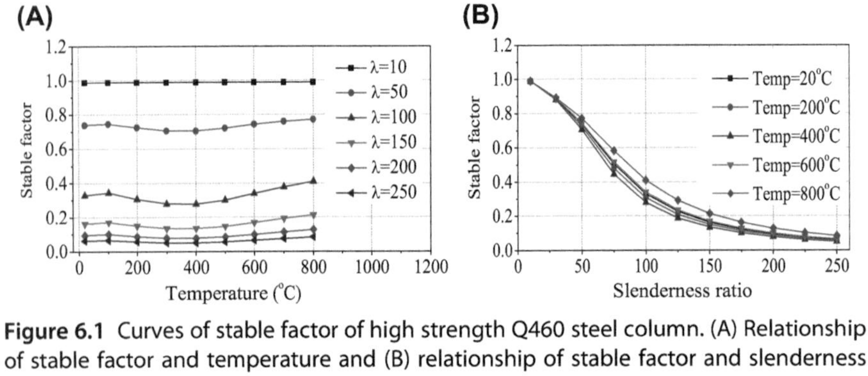

Figure 6.1 Curves of stable factor of high strength Q460 steel column. (A) Relationship of stable factor and temperature and (B) relationship of stable factor and slenderness ratio.

form the Eqs. (6.1) and (6.2)), and when the temperature is at range of 200 and 500°C, the ratio of elastic modulus to yield strength is lower than that at room temperature. For the temperature is high than 600°C, the ratio is higher (as shown in Fig. 6.2).

It is also shown that with the increase of slenderness ratio, the stable factor decreases, and during the slenderness ratio 50 through 100, the decrease of stable factor is more significant than other slenderness ratio.

To conveniently use in practice, the stable factor of Q460 steel columns at room and elevated temperature was presented in Table 6.1. After obtaining the slenderness ratio and temperature, one can find the stable factor by looking up the table.

6.2.2 Inverse calculation segment length method

In 1970s, Chen and Atsuta (1976) proposed column deflection curve CDC method to calculate the defection of steel column at a certain load and boundary conditions. In 1980s, based on the CDC method, Li and Xu (1989) proposed a numerical method named inverse calculation segment length method to calculate relationship of axial load, moment, and curvature, under consideration of initial flexure, residual stress on the section, and transverse load action. In this chapter, the inverse calculation segment length method was extended to obtain the load-bearing capacity of high strength Q460 steel columns at elevated temperatures. The flowchart of extended inverse calculation segment length method for high strength Q460 steel at elevated temperature is shown in Fig. 6.3.

Based on the steps and flowchart of inverse calculation segment length method, a computer program was designed and employed to calculate stress–slenderness relations for high strength Q460 steel columns at different temperatures.

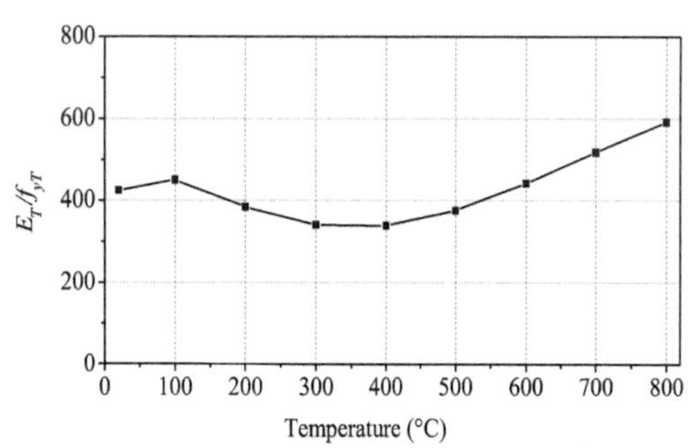

Figure 6.2 Variation of elastic modulus to yield strength ratio with temperature.

Table 6.1 Stable factor of Q460 steel columns at room and elevated temperatures.

Temperature (°C)	Slenderness ratio										
	10	30	50	75	100	125	150	175	200	225	250
20	0.988	0.885	0.738	0.505	0.330	0.225	0.162	0.122	0.095	0.076	0.062
100	0.988	0.887	0.745	0.520	0.344	0.236	0.170	0.128	0.100	0.080	0.065
200	0.988	0.883	0.725	0.479	0.307	0.208	0.149	0.112	0.087	0.069	0.057
300	0.988	0.879	0.705	0.446	0.279	0.188	0.134	0.100	0.078	0.062	0.050
400	0.988	0.879	0.705	0.444	0.278	0.187	0.133	0.100	0.077	0.062	0.050
500	0.988	0.882	0.721	0.472	0.301	0.204	0.146	0.109	0.085	0.068	0.055
600	0.988	0.886	0.743	0.516	0.340	0.233	0.168	0.126	0.098	0.079	0.064
700	0.988	0.889	0.760	0.554	0.378	0.263	0.192	0.145	0.113	0.091	0.074
800	0.988	0.892	0.772	0.582	0.409	0.290	0.213	0.162	0.127	0.102	0.084

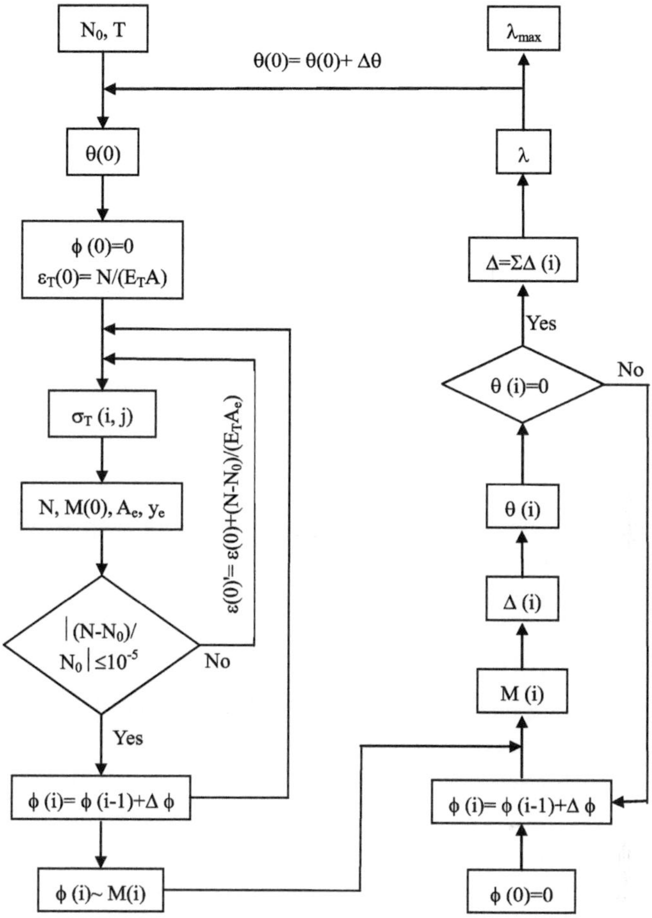

Figure 6.3 Flowchart of inverse calculation segment length method.

6.2.3 Comparison of the two methods

The curves of stable factor and slenderness ratio for high strength Q460 steel at temperatures 400 and 800°C obtained by critical stress method and extended inverse calculation segment length method were compared in Fig. 6.4. As can be seen from Fig. 6.4, the two methods agree well with each other, especially for the slenderness ratio more than 100. For the slenderness ratio less than 100, there is some difference and only about 10% departure from the average value of the results. From the comparison, we can conclude that the two methods are feasible to calculate the load-bearing capacity of high strength Q460 steel columns at elevated temperatures.

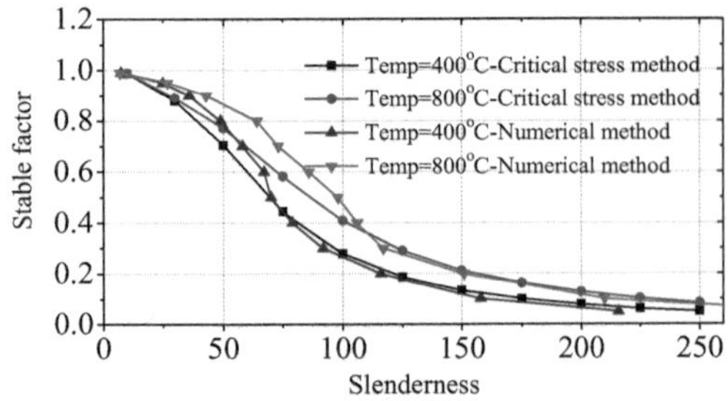

Figure 6.4 Comparison of critical stress method and numerical method.

As mentioned above, the critical stress method is simple to use and has implicit expression. The inverse calculation segment length method is versatile and feasible to any residual stress distribution and initial flexure ratio. Hence, in the following sections, the critical stress method is utilized to deduce the critical temperature of high strength Q460 steel columns, and the inverse calculation segment length method is employed to perform parametric study on the initial flexure and residual stress.

6.2.4 Critical temperature

At the elevated temperature, critical temperature is a key parameter used to represent the fire resistance for steel members. For an axially compressed steel column with a serviceability load N, the yield strength of steel reduces with the increase of temperature. When the yield strength declines to global stability critical stress, the failure occurs. The critical temperature T_{cr} is defined as the temperature at which the yield strength equals to the global stability stress (Li et al., 2006). Therefore, the critical temperature can be found by solving

$$N/\varphi_T A = f_{yT} \tag{6.3}$$

Eq. (6.3) can be rewritten in the following form:

$$\frac{N}{\varphi A f} = \gamma_R \frac{f_{yT}}{f_y} \cdot \frac{\varphi_T}{\varphi} \tag{6.4}$$

where φ is the stable factor of steel column at room temperature; f is the design strength of steel at room temperature, which is equivalent to f_y/γ_R; γ_R is the resistance factor of steel, which can be approximately adopted as 1.1.

The left item of Eq. (6.4) is the load ratio R, which is defined as the ratio of serviceability load N to the stability load-bearing capacity at room temperature and has no relation with temperature. Given a certain load ratio, the critical temperature of high strength Q460 steel column with axial load may be obtained by solving Eq. (6.4), which is a function of critical temperature T_{cr}.

Figs. 6.5 and 6.6 present the critical temperature calculated for $\lambda = 30-250$ and $R = 0.3-0.8$ for high strength Q460 steel columns. From Fig. 6.6, it can be observed that the critical temperature slightly increases with the increase of the slenderness ratio when the load ratio is lower than 0.7. However, the critical temperature decreases with the increase of the slenderness ratio when the load ratio is higher than 0.7. It is shown that in Fig. 6.6, the critical temperature reduces almost linearly with the increase of load ratio. By comparing the critical temperature with EC3 (2005) as shown in Fig. 6.6, at the same load ratio, the critical temperature of high strength Q460 steel columns is much higher than mild steel columns.

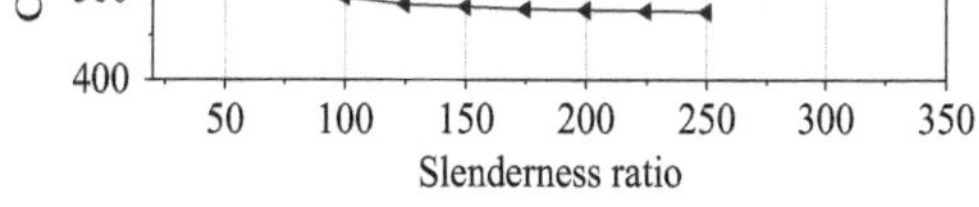

Figure 6.5 Relationship of critical temperature.

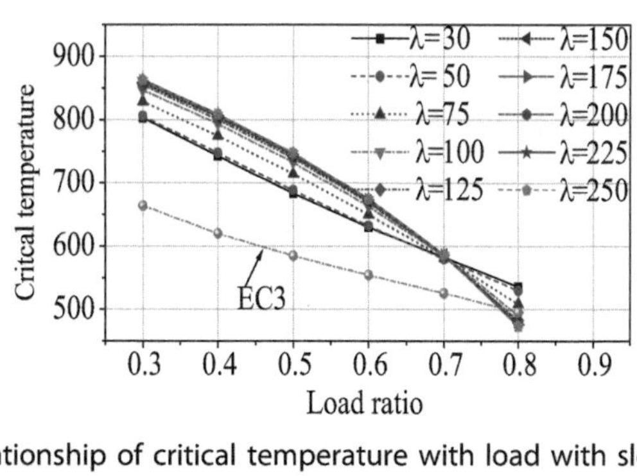

Figure 6.6 Relationship of critical temperature with load with slenderness ratio.

Therefore, the critical temperatures for mild steel columns given in EC3(2005) are not applicable for high strength Q460 steel columns.

6.2.5 Parametric study

For a compressed steel column, the load-bearing capacity mostly depends on the global instability. At room temperature, in addition to the slenderness ratio, the residual stress and geometric imperfection have obvious influence on the global instability. At elevated temperature, the influence of residual stress and geometric imperfection is unknown. To understand the influence of residual stress and geometry imperfection on load-bearing capacity of high strength Q460 steel columns at elevated temperatures, the parametric study is carried out on the initial flexure and residual stress distribution.

In the middle of 20th century, the Fritz laboratory of engineering structure in Lehigh University (Huber and Beedle, 1953) comprehensively studied the distribution of residual stress in cross section of steel members and influence of residual stress on the buckling capacity of steel columns and found that the residual stress is a key factor in determining the load-bearing capacity of steel columns. As to now, few literatures focus on the residual stress of high strength steel columns at elevated temperatures. At room temperature, the residual stress on the section of steel columns induced by weld or rolling is simplified to triangle or rectangle distribution. For high strength Q460 steel, same distribution modes are used to study the influence of residual stress value and distribution mode on the load-bearing capacity. For one distribution mode, two factor values 0.3 and 0.5 are considered. The two distribution modes and relevant factor values are shown in Fig. 6.7. The factor α means the ratio of the maximum residual stress to yield strength f_y.

By employing the computer program designed in Section 6.2.2, the load-bearing capacity for H-shaped high strength Q460 steel columns is performed, respectively, bending around strong and weak axis. The dimension of cross section for the Q460 steel column is H200 × 200 × 12 × 8 and the initial flexure is 1‰ length of column. The stable factor of high strength Q460 steel columns with distribution mode 1 and 2 and residual stress values 0.3 and 0.5 are shown in Figs. 6.8 and 6.9.

As is shown in Fig. 6.8, for distribution mode 1, the residual stress value has almost no influence on the stable factor when the slenderness ratio is bigger than 75. However, when the slenderness ratio is lower than 75, with

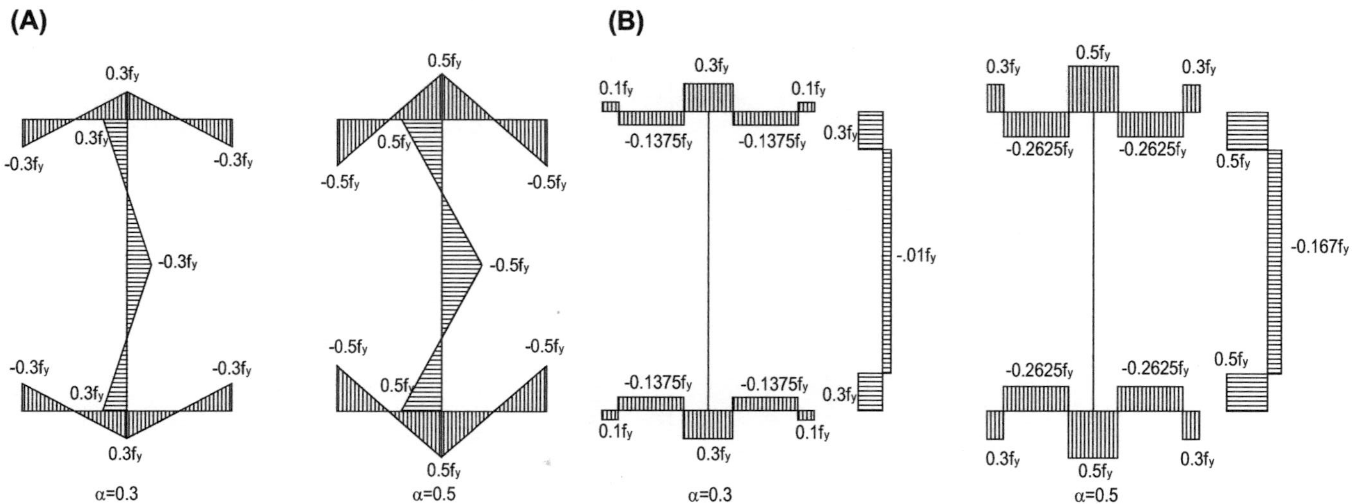

Figure 6.7 Distribution mode of initial residual stress. (A) Mode 1—triangle distribution and (B) mode 2—rectangle distribution.

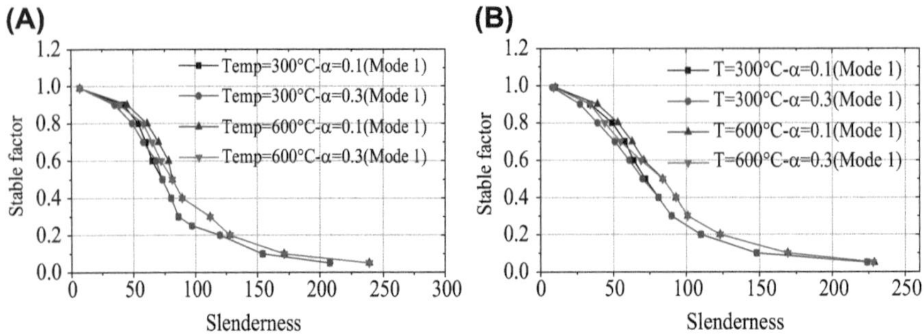

Figure 6.8 The influence of residual stress value on the stable factor at distribution mode 1. (A) Around strong axis and (B) around weak axis.

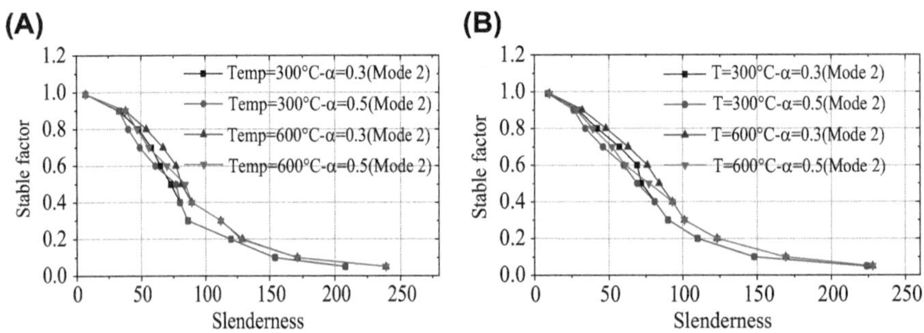

Figure 6.9 The influence of residual stress value on the stable factor at distribution mode 2. (A) Around strong axis and (B) around weak axis.

the increase of residual stress value, the stable factor decreases. The same conclusion can be drawn for both strong axis and weak axis. For distribution mode 2, the similar trend can be observed from Fig. 6.9. Therefore, the residual stress value has significant influence on the stable factor for the short and medium columns (slenderness ratio is less than 75), and for long column (slenderness ratio is more than 75), the influence is negligible.

The comparison of two distribution modes for residual stress is shown in Fig. 6.10. It is easy to find that at the same residual stress value, the distribution mode has very little influence on stable factor around strong axis. However, around weak axis, the stable factor of steel column under rectangle distribution (mode 2) is higher than triangle distribution (mode 1) for short and medium columns. For the long columns, the distribution mode has no influence on the stable factor.

In a realistic steel column, the geometric imperfection always exits, and the shape of initial flexure is various. It is proved that the initial flexure can

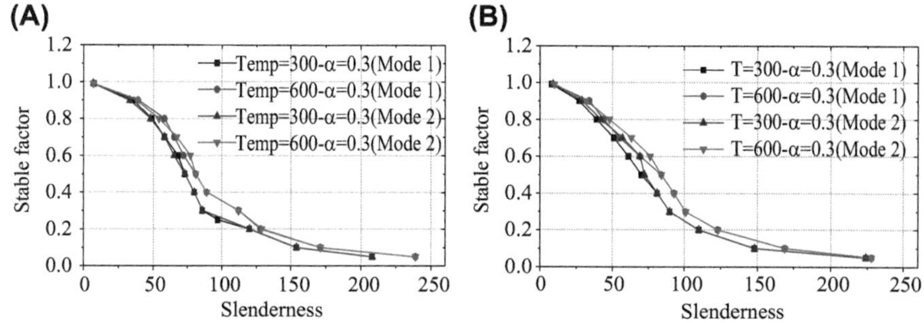

Figure 6.10 The influence of residual stress mode on the stable factor. (A) Around strong axis and (B) around weak axis.

be represented as a half sine wave (Chen, 2003). Through a large number of test data on the steel columns, the maximum initial deflection (1‰ length of column) is often adopted at the midspan of the column. For some columns, the initial flexure may be larger than 1‰ length of column. At elevated temperature, the nonuniform temperature distribution along the cross section also results in additional flexure of the column.

To study the influence of initial flexure on the stable factor for high strength Q460 steel columns, the computer program designed in Section 6.2.2 is adopted to obtain the stable factor at two different initial flexure values, 1‰ and 3‰. The stable factor for high strength Q460 steel columns with different initial flexure is shown in Fig. 6.11. As can be seen from Fig. 6.11, at high temperatures, the initial flexure has very significant influence on the stable factor. With the increase of initial flexure, the stable factor reduces obviously both for bending around strong and weak axis.

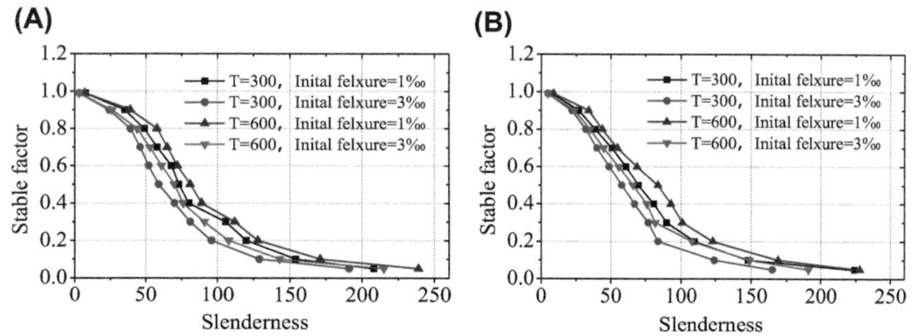

Figure 6.11 Influence of initial flexure on the stable factor. (A) Around strong axis and (B) around weak axis.

6.2.6 Finite element analysis and experiment validation

To validate the method presented above, the finite element (FE) analysis was performed on global stability. The software ANSYS was employed to conduct the analysis. The comparison between global stability analysis result from ANSYS and predicted results by critical stress method was made to validate the critical stress method.

The ultimate load-bearing capacity for six high strength Q460 steel columns at elevated temperatures was performed by using the element BEAM188. The section dimension of the column is H200 × 200 × 12 × 8, and the initial flexure is 1‰ length of column. The distribution of residual stress is shown in Fig. 6.12A. Three kinds of slenderness ratio 50, 100 and 150 and two temperatures 300 and 600°C were considered in the analysis.

Fig. 6.12A shows a typical stress distribution along high strength Q460 steel columns at temperature 600°C. Fig. 6.12B shows a typical load–displacement curve at temperature 300°C, and the load-bearing capacity is defined as the peak point of the curve. The analysis results were compared with the critical stress method and are shown in Fig. 6.13. From the comparison, we can find that the finite element analysis results agree well with critical stress results. Therefore, the method proposed in this chapter is validated to predict the load-bearing capacity for high strength Q460 steel columns at elevated temperatures.

As to now, the test result on load-bearing capacity of Q460 steel column at elevated temperature was not found. However, there is some data for Q460 steel column at room temperature (Wang et al., 2012c). As presented above, the critical stress approach can also be used to calculate the stable factor of Q460 steel columns at room temperature. To validate

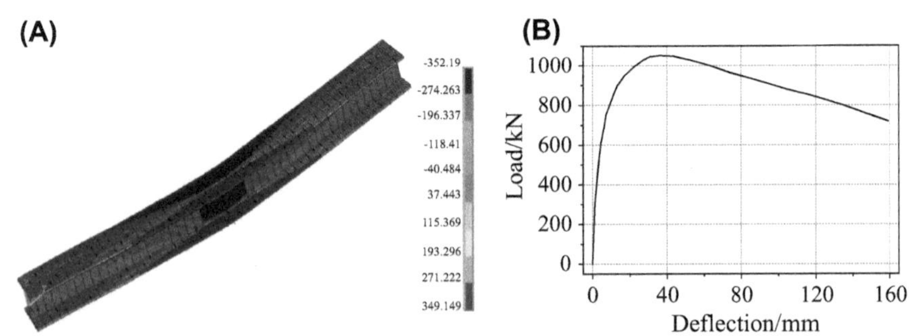

Figure 6.12 Global stability analysis results for high strength Q460 steels. (A) Stress distribution (600°C) and (B) load–displacement curve (300°C)

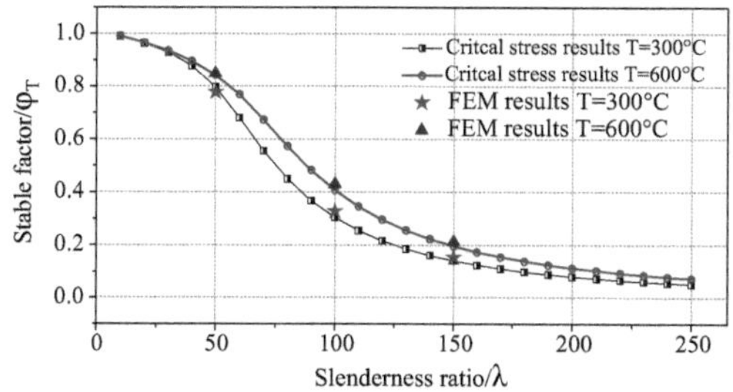

Figure 6.13 Comparison of FEM and critical stress method.

Table 6.2 Comparison of experimental results and analytical result

Specimen No.	Slenderness ratio	Stable factor: Test results	Stable factor: Analytical results	Ratio of analysis to test	Average
S-1	82.5	0.449	0.444	0.99	0.93
S-2	81.9	0.496	0.449	0.91	
S-3	56.2	0.765	0.682	0.89	
S-4	56	0.708	0.684	0.97	
S-5	41.5	0.881	0.809	0.92	
S-6	41.6	0.869	0.808	0.93	

the method by experiment, the comparison of stable factor was made between test results and analytical results (as shown in Table 6.2). Form the comparison, it is shown that the results agree well and only 7% difference between them.

6.3 Response of restrained high strength Q460 steel columns

To investigate the fire response of restrained high strength steel columns, fire tests were carried out for the high strength steel columns with axial and rotational end restraints in a fire furnace.

6.3.1 Specimen preparation

Eight column specimens were made of Q460 steel plate welded to an H-shaped section of H200 × 195 × 8 × 8, in which four specimens were designed with axial end restraints with the length of 4.3 m, whereas the others are designed with both axial and rotational end restraints with the length of 4.48 m. The restraining stiffness at each end was provided by two H-shaped steel beams made of Q235 steel with the length of 3.2 m. Two cross sections, namely H200 × 150 × 6 × 9 and H300 × 150 × 6.5 × 9, were fabricated for the beam to generate two different restraining stiffness. The mechanical properties of the test specimens and the restraining beams are obtained by the standard tension coupon test according to the ASTM A370 test protocol (2009). The test results are tabulated in Table 6.3.

For the specimens with the axial end restraints, the end conditions are hinge connected to the restraining beam to ensure the specimen ends can freely rotate about its weak axis. The end conditions of strong axis are seen as fixed and cannot rotate. For the specimens with both axial and rotational end restraints, extended end plate connections were used to connect the column ends to the restraining beam in such a way both axial and rotational deformation at column ends are prevented. The aforementioned two types of end connections are illustrated in Fig. 6.14.

Different magnitudes of the applied load and restraining stiffness were considered in the tests. The axial restraint ratio β_a is defined as

$$\beta_a = \frac{K_b}{K_c} = \frac{48E_bI_b}{l_b^3} / \frac{E_cA_c}{l_c} \tag{6.5}$$

Table 6.3 Material properties of steels by coupon tests.

Steel	Thickness	Yield strength	Ultimate strength	Elastic modulus
Q235	9 mm	285 MPa	415 MPa	2.10×10^5 MPa
Q460	8 mm	585 MPa	660 MPa	2.12×10^5 MPa

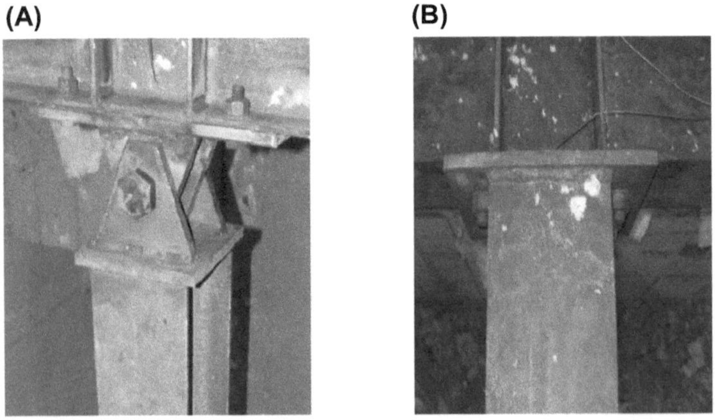

Figure 6.14 Connection between column specimen and restraining beam. (A) Hinged connection and (B) extended end-plate connection.

where K_b is the flexural stiffness associated with the midspan deflection of the restraining beam; K_c is the axial stiffness of the column; I_b is the moment of inertia of the beam; A_c is the column cross-sectional area; and l_b and l_c are the length of the beam and column, respectively.

The rotational restraint ratio is defined as

$$\beta_r = \frac{K_{rb}}{K_{rc}} = \frac{12E_bI_b}{l_b} / \frac{3E_cI_c}{l_c} = \frac{4E_bI_bl_c}{E_cI_cl_b} \tag{6.6}$$

where K_{rb} is the rotational stiffness associated with the midspan rotation of the restraining beam and K_{rc} is the end rotational stiffness of the column.

The load ratio R is expressed as

$$R = N/N_{cr} \tag{6.7}$$

where N is the applied load placed on the column top end and N_{cr} is the ultimate load capacity of the column evaluated based on GB50017-2017(2017) at ambient temperature.

The detail information about the specimens is tabulated in Table 6.4.

6.3.2 Test setup and measurements

The test specimens are heated in a fire furnace. The dimension of the furnace is 3.6 m wide, 4.6 m long, and 3.3 m high. The maximum heat power generated by the furnace is 5 MW. Eight natural gas burners are installed in the furnace, and the furnace temperature was recorded by 10 thermocouples placed in the test chamber over a fire test. The plan view of the furnace is shown in Fig. 6.15. During the fire test, the temperature readings in thermocouples (noted as FT1 ~ FT8) are compared with that of ISO-834 heating curve, and the control system automatically adjusts corresponding fuel supply to maintain the furnace temperature with that of the heating curve.

Table 6.4 Parameters of the specimens.

Specimen number	End restraint	Load (ratio)	β_a	β_r
S-1	Axial	0.25	0.45	0
S-2		0.40	0.45	0
S-3		0.25	0.17	0
S-4		0.40	0.17	0
S-5	Axial and rotational	0.20	0.45	36
S-6		0.20	0.17	14
S-7		0.36	0.45	36
S-8		0.38	0.17	14

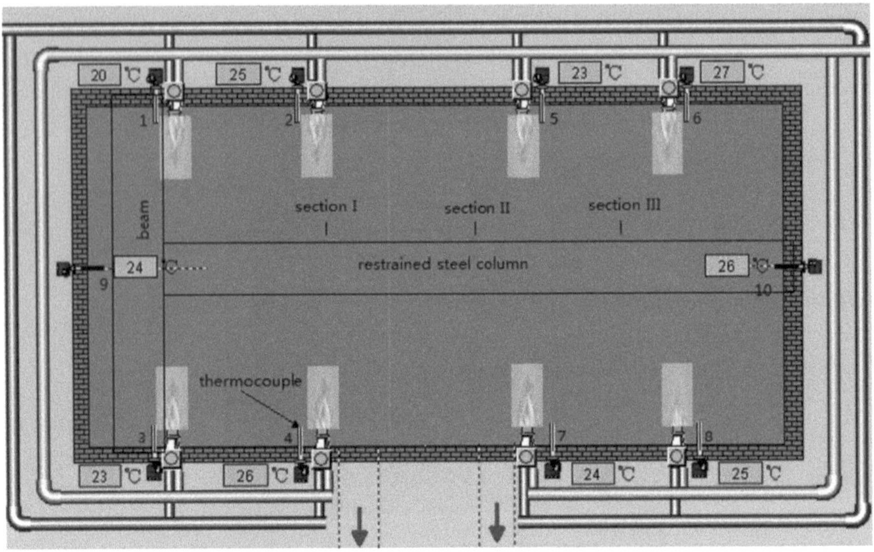

Figure 6.15 Plan view of the fire furnace

A horizontal self-reaction loading system, consisting of a steel frame and two steel restraining beams (top beam and bottom beam), was designed to apply loading on the test specimen and provided desirable boundary conditions for specimen. The frame was horizontally placed on the top of furnace. Axial load was applied on the test specimen by a horizontal hydraulic jack with capacity of 1000 kN. For the axially restrained columns, the top steel restraining beam provides an axial restraint on the test column as the temperature increases. For the axially and rotationally retrained columns, the top restraining beam provides not only the axial restraint but also the rotational restraint, and the bottom restraining beam provides only the rotational restraint to the column end. The axial restraint at the bottom of the specimen is provided by the stub column that supports the bottom restraining beam. The bottom beam hinged at the two ends is connected to the frame with use of high strength steel bolts before the installation of the test column. After the installation of test specimen and top restraining beam, the bolts at the ends of top beam keep loose to apply the load on the restrained column. Until finishing applying load, the top restraining beam will be firmly connected to the frame by tightening the high strength bolts. As the top beam was placed inside the furnace, it was protected with fire insulation to maintain the restraining stiffness during the test. Shown in Fig. 6.16 is the layout of the test frame and the specimen setup for the specimen with extended end-plate connections.

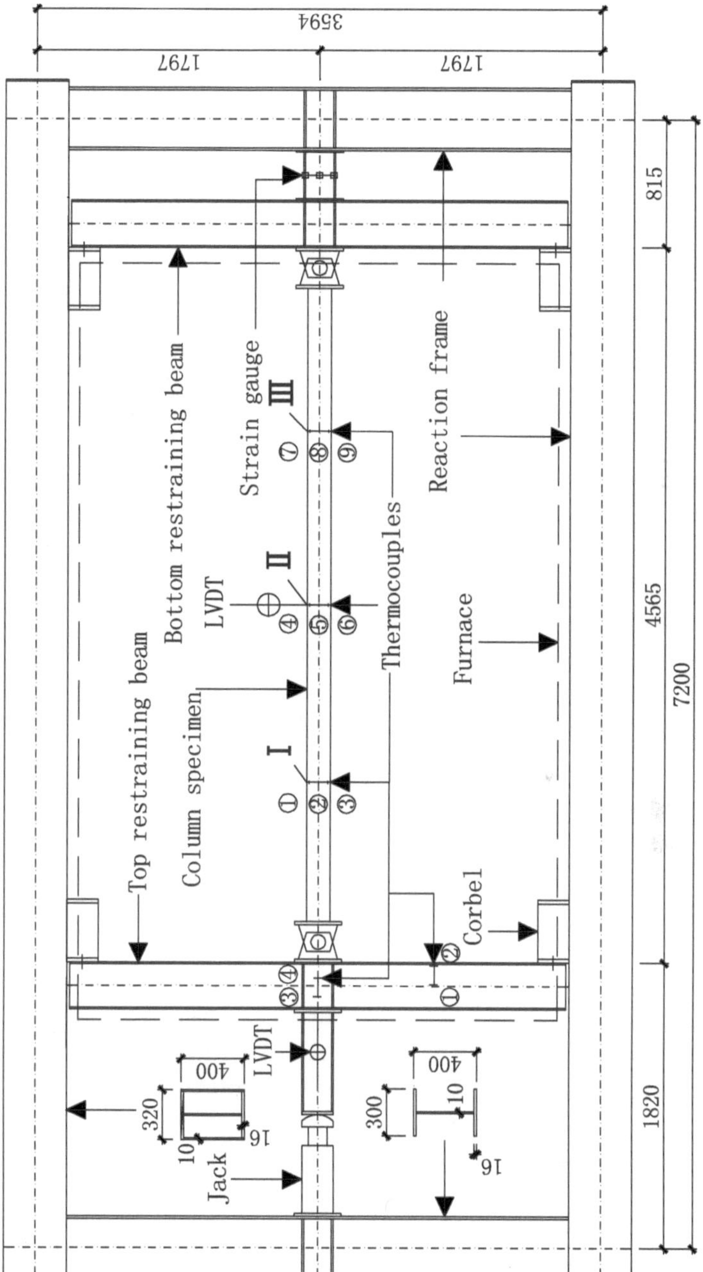

Figure 6.16 Experimental loading system layout and specimen setup.

Thermocouples, strain gages, and linear variable differential transformers (LVDTs) were used to record the thermal and structural responses of the test columns during the fire tests. The instrumentation of the test is shown in Fig. 6.17. Temperatures of the specimen were recorded by nine type-*K* thermocouples, 2.0 mm in diameter. The thermocouples were located at the 1/3, 1/2, and 2/3 of the specimen length. Three strain gages were placed to the short stub section connected the bottom restraining beam to evaluate the axial force generated in the restrained specimen. At locations of the top and midheight of the specimen, two LVDTS were installed to measure the axial displacement and lateral deflection of the specimen, respectively. The applied load on the test specimen was measured by a load cell mounted on the hydraulic jack.

6.3.3 Test procedure

The test process comprises of the following steps:

(1) The specimen was first placed into the loading frame, and instrumentations on the specimen, including the thermocouple, strain gages, and LVDTs were subsequently installed.

(2) After all the test devices and instrumentation are checked and zeroed out, the applied loading from the hydraulic jack was increased to 10% of target load and kept for 5 min. The load was released to zero. This preloading process was repeated twice to ensure that all instruments exhibit linear variation with the load and all the readings returned to their starting value after the load was completely released. After preloading, the applied load was increased with a loading rate of 20 kN/min until it reached to the target load. At this point, the top restraining beam will be firmly connected to the frame, and the magnitude of the applied load was maintained throughout the duration of the fire test.

(3) The furnace was turned on, and the furnace temperature was controlled in consistent with ISO-834 heating curve. If the axial displacement at the top end of specimen or the lateral deflection at the midheight of the specimen reached the maximum range ($l_c/100$) or failure limit ($l_b/20$), respectively, the furnace was turned off and the applied load was released, and the test is terminated.

6.3.4 Test results

The test data obtained from the test were utilized to investigate the behavior of restrained high strength Q460 steel column subjected to elevated

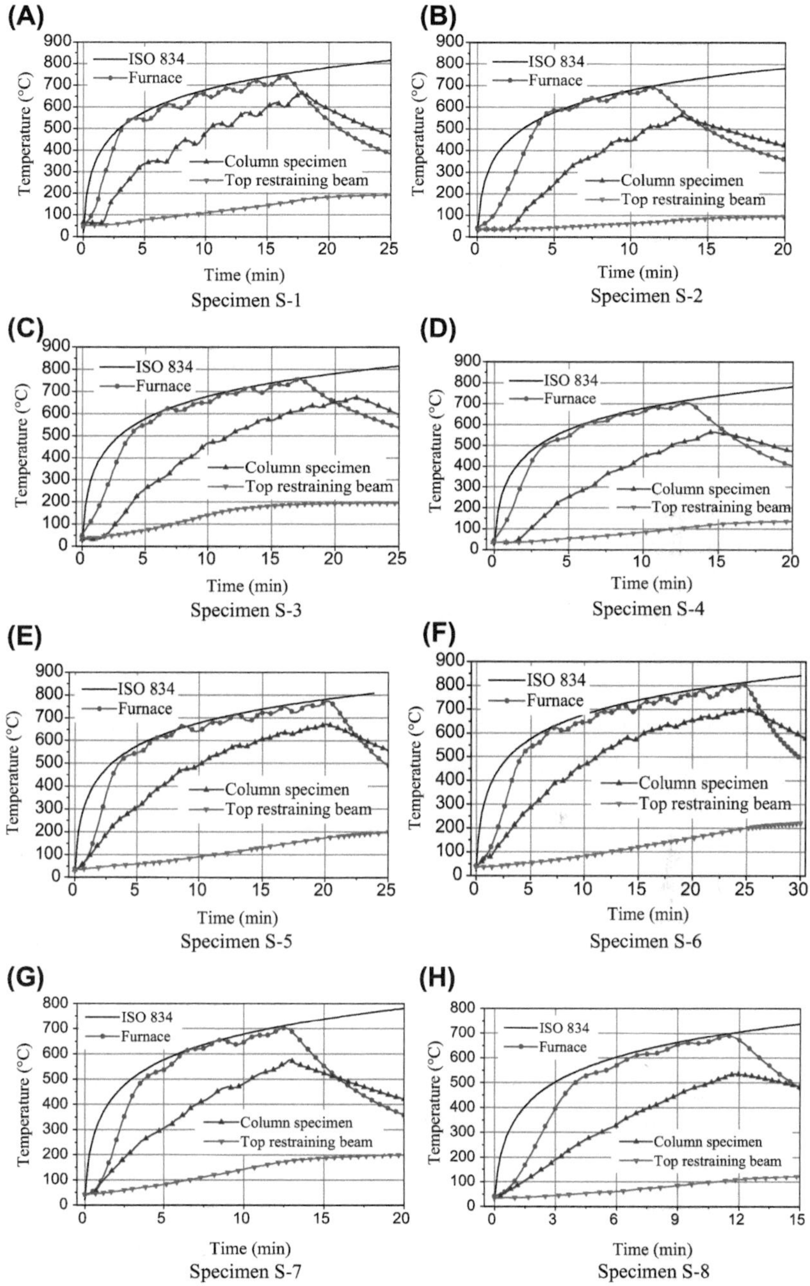

Figure 6.17 Temperature distributions of furnace, specimen, and top restraining beam.

temperature. The effects of the applied load and axial and rotational restraints on the behavior of high strength Q460 steel columns were evaluated through the comparison with thermal responses, structural responses, and failure patterns of the specimens.

6.3.4.1 Temperature evolution

The temperature evolution of the furnace, column specimens, and top beams is plotted in Fig. 6.17. It can be seen that the furnace temperature was unable to match with that of ISO-834 curve initially due to the limitation of the maximum heat power of the furnace. However, a few (3−5) min later, the furnace temperature matched to ISO-834 curve closely. The column specimen was directly exposed to fire, and its temperature increased quickly, which was attributed to the fact that the flange and web are thin and the thermal conductivity of steel is very high. The dispersion of temperatures across the column cross section recorded by thermocouples showed that it was less than 50°C. Therefore, the effect of nonuniform distribution of the temperature in the column was negligible, and the average temperature of the column was employed to represent the temperature of the specimen in the discussion of structural responses of specimens. During the test, the top restraining beam was wrapped with fire insulation and the corresponding temperature is relatively slow with the maximum temperature being only approximately 220°C.

6.3.4.2 Axial displacement and lateral deflection

According to the previous research (Wang et al., 2010), for a restrained mild steel column in fire condition, the temperature at which the axial force reaches the maximum magnitude is defined as buckling temperature, whereas the temperature associated with that when the axial force returns to its initial magnitude is defined as the failure temperature or critical temperature. When the axial displacement induced by thermal expansion approaches the maximum value, the axial force generated because of the axial restraint also reaches the maximum value.

Illustrated in Fig. 6.18 are the variations of the axial displacements of the specimens versus the time duration and specimen's temperature. As shown in the figure, the axial displacements increased gradually with an abrupt drop when the maximum displacements are reached. From the fact that all the specimens failed at elevated temperature within 25 min, it can be concluded that the unprotected high strength steel columns are quite sensitive to fire. The failure temperature (T_1) of the specimens based on the

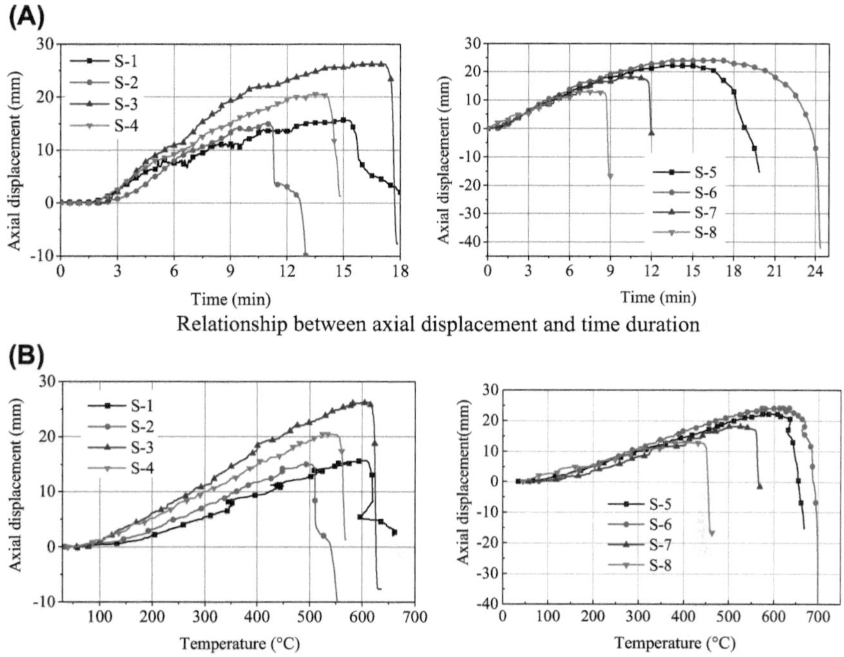

Figure 6.18 Axial displacements of test specimens.

abrupt change in axial displacements can be determined from Fig. 6.18, and they are tabulated in Table 6.5. It can also be found from Fig. 6.18 that the load ratio is also a critical factor to influence the fire resistance of the restrained column. At the same applied load ratio, taking an example of specimens S-1 and S-3, the specimen with higher axial restraining stiffness yields smaller axial displacement. It can be concluded that at the same applied load ratio, larger axial restraining stiffness would result in lower failure temperature of the columns. This can be attributed to the fact that the higher axial restraining stiffness would produce larger thermal force in the column. Consequently, the higher the axial force in the column would result in a lower the failure temperature.

Table 6.5 Critical temperature and failure time of the specimens.

Specimen number	S-1	S-2	S-3	S-4	S-5	S-6	S-7	S-8
Buckling temperature/°C	606	493	582	530	607	617	512	412
Failure temperature T_1/°C	620	510	625	564	655	688	564	454
Failure temperature T_2/°C	652	550	603	558	669	699	560	356

It was observed that specimens S-1 and S-2 reached to a metastable state in postbuckling phase after the column specimens lost their capacities of resisting applied load due to buckling. In this phase, the applied load was resisted by the top restraining beam instead of the column. It should point out that this phenomenon was only observed for the column specimens with high axial restraining stiffness and relatively lower load level, such as shown in S-1 and S-2. This phenomenon was not observed in columns with low restraining stiffness. In fact, this phenomenon was also observed in investigations of the restrained mild steel column by Franssen (2000) and Wang (2004).

The relationship between the lateral deflection at the midheight of the specimen and the time duration and the temperature at midheight of specimens is presented in Figs. 6.19A,B, respectively. It can be seen from the figures that at initial stage of heating, the lateral deflection remains quite stable before reaching the failure state. Abrupt increase of the lateral deflection was observed for all the specimens. The transition period from

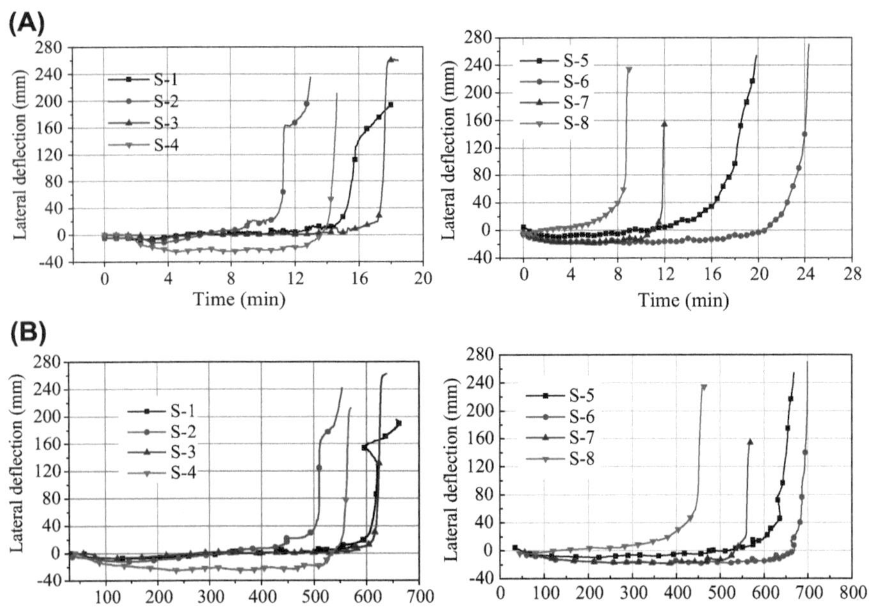

Figure 6.19 Lateral deflection of test specimens. (A) Relationship between the lateral deflection and time duration and (B) relationship between the lateral deflection and specimen's temperature.

stable to failure state took a few seconds, and the corresponding increase of the lateral flection was approximately about 200 mm. It is also observed that some specimens experienced negative deflections in the initial stage of heating, such as specimen S-4. This may result from the fact that the measurement of the deflection is located at the flange of the specimen; the negative deflection can be attributed to possible flange curling and torsion of the specimen in the initial heating stage.

In the case that the specimens are only subjected to the axial restraint, it is noticed from Fig. 6.19 that it took longer for specimen S-6 to reach the failure than that of specimen S-5 even though both specimens were subjected to the same applied load. This is primarily because the top restraining beam connected to S-5 was stiffer, which yields a higher value of axial restraint ratio. Consequently, the higher value of axial restraint ratio results in a higher axial force associated with thermal expansion in the specimen and illustrated in Fig. 6.19. However, in the case that columns were subjected to both axial and rotational restraints, different from what was observed from specimens S-7 and S-8, it took a longer time for specimen S-7 to failure even though the axial restraint ratio of S-7 was greater than that of S-8. The reason of that is attributed to the higher value of the rotational restraint ratio associated with S-7, which enhanced the capacity of the specimen against the flexural torsional buckling comparing with that of specimen S-8.

6.3.4.3 Axial compressive force in the specimen

The variations of the axial force ratio, P/P_0, with the time duration and specimens' temperature are showed in Fig. 6.20, where P denotes the axial force calculated based on the elastic modulus and strain of the stub column connected to the bottom end of the test specimen measured from the test and P_0 is the applied load on top of the specimen. The axial force induced by thermal expansion in the axially restrained column is greater for the column with the higher value of the axial restraint ratio and the lower value of the applied load ratio, which consequently leads to a higher value of P/P_0, such as the case of S-1. The similar phenomenon is also observed for the specimen with both axial and rotational restraints and is clearly illustrated by specimen S-5 in the figure. The axial force in the column declined quickly after large lateral deformation was observed, which signified the failure of the column.

Based on the aforementioned discussion, in addition to the failure temperature (T_1), defined based on the axial displacement of the specimen,

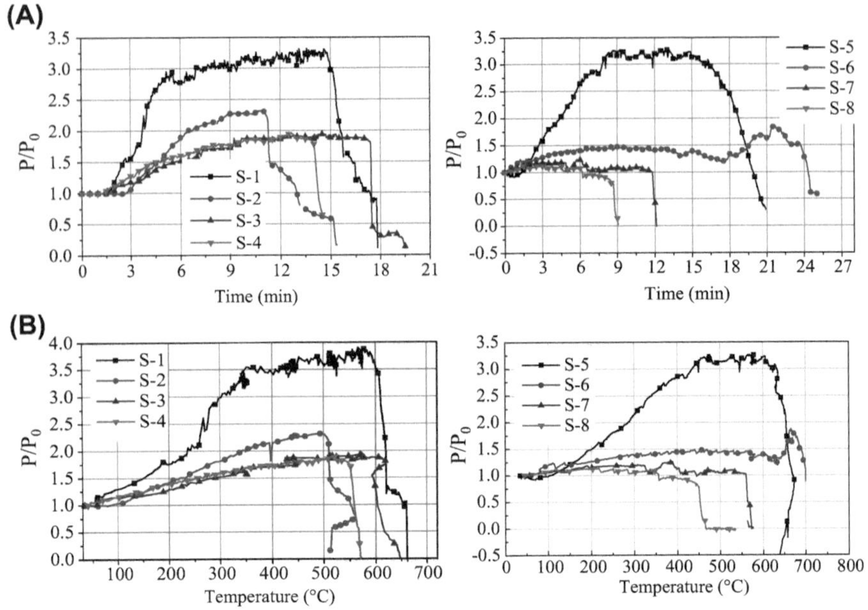

Figure 6.20 P/P_0 of test specimens. (A) Relationship between the axial force ratio P/P_0 and time duration and (B) relationship between the axial force ratio P/P_0 and specimen's temperature.

the axial force of the specimen can be alternatively adopted to define the failure temperature (T_2) of the specimen. For the reason of comparison, in the failure, temperatures T_1 and T_2 are presented in Table 6.5. It can be seen from the table that there are some discrepancies among the failure temperatures T_1 and T_2, particularly for specimen S-8. The discrepancy can be attributed to the fact that the calculated axial force may not always be accurate by using strain obtained from the stub at the top of the specimen as the bottom beam could resist the axial force of the specimen in some degree.

6.3.4.4 Failure mode

Fig. 6.21 shows the failure modes of the test specimens. It is clearly shown that the failure modes of the specimens are influenced by the type of end restraints and magnitude of applied load. For the specimens S1—S4, as the rotation about the weak axis of the specimens is not restrained, the failure mode is the flexural buckling about the weak axis of the specimens. For specimens with both axial and rotational restraints, in case of low magnitude

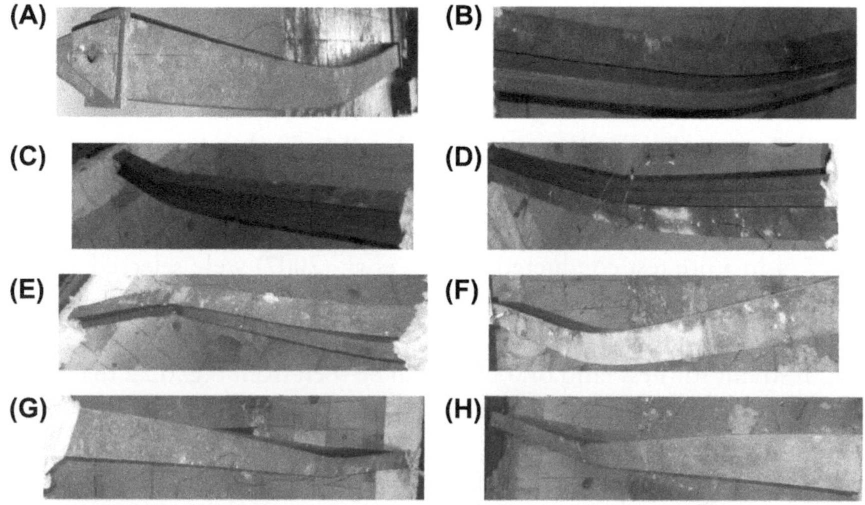

Figure 6.21 Failure modes in test specimens. (A) Specimen S-1, (B) specimen S-2, (C) specimen S-3, (D) specimen S-4, (E) specimen S-5, (F) specimen S-6, (G) specimen S-7, and (H) specimen S-8.

of applied load, specimens S-5 and S-6, flexural buckling about the weak axis of the specimens was observed, no matter what the magnitudes of axial and rotational restraints are. However, for the specimens subjected to high magnitude of applied load and with both axial and rotational restraints, S-7 and S-8, flexural torsional buckling was observed as indicated in Fig. 6.21G,H. The flexural torsional buckling can be attributed to the fact that the temperature distribution in these two specimens at the failure time may not be uniform. Considering the fact, the two specimens were subjected to a higher applied load ratio, and the corresponding flexural torsional buckling failure occurred at an earlier stage comparing with that of other specimens that were subjected to a lower applied load ratio. In the early stage of the test, the temperature increases quickly, and the effects of nonuniform distribution can be more significant than that in the later stage. The nonuniform temperature distribution in the specimen resulting in the stiffness distribution of the specimen is no longer double symmetrical. As for the other specimens, the flexural buckling failure happened in a longer fire exposure time due to the low applied load ratios, and the corresponding temperature distribution at this stage is relatively uniform. Therefore, it appears that the failure mode of the specimens is more sensitive to the magnitude of applied load other than the magnitudes of axial and rotational restraints. Further investigations on the behavior of the columns with the high applied load ratio and effect of nonuniform temperature distribution on the column failure mode are needed in future research.

6.3.5 Comparison with restrained mild steel columns

To investigate the difference of the fire responses between the restrained columns with high strength steel and mild steel, test results of specimen RS97_4 in a fire test conducted by Tan et al. (2007) were selected for the comparison with that of specimen S-2 in this investigation as both specimens have identical slenderness ratio of 96, applied load ratio of 0.5, and axial restraint stiffness ratio of 0.16. It can be seen from Fig. 6.22 that there are significant differences between the maximum axial displacement and P/P_0 ratio because the length of RS97_4 is only 1.5 m, which is considerably shorter than that of S-2. However, it is noticed that P/P_0 ratios of the two specimens follow a close evolution in the early stage of the elevated temperature. It should point out that with the same test conditions, axially restrained column of Q460 steel generally exhibits much better fire resistance than that of mild steel. This conclusion also can be drawn by comparing the test results of this investigation with that of tests performed by Ali et al. (1998). The failure temperatures of axially restrained mild steel columns with slenderness of 98 ranged from 333 to 410°C for applied load ratios and axial restraint stiffness ratios within a range of 0.4–0.6 and 0.1–0.2, respectively. Studies (Wang et al., 2010) on fire resistance of restrained steel columns have shown that increase of either the applied load or axial restraint stiffness would lead to a greater

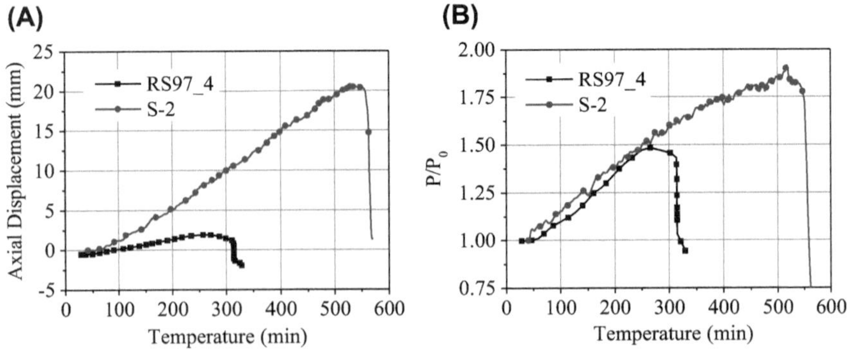

Figure 6.22 Fire resistance comparison between mild and high strength restrained steel columns. (A) Axial displacement and (B) axial applied load ratio.

reduction in the failure temperature. For a column with the applied load ratio of 0.3, the axial restraint stiffness ratio of 0.34, and a slenderness of 51 similar to that of specimen S-7 in this investigation, the failure temperature of column HEA200-K128-L30 was approximately 515°C as reported in Wang et al. (2013a,b), which is less than 564°C obtained from this investigation for specimen S-7. Such observation can explain that reduction factors of material properties for high strength steel Q460 at elevated temperature are greater than that of mild steel given in EC3 (2005).

6.3.6 Finite element simulation

Fire test of full-scale specimen is costly and time consuming. Finite element simulation has been widely adopted as it can generate reasonable predictions for both thermal and structural responses of structures. The finite element software ANSYS was employed to perform fire response analysis of restrained high strength Q460 steel columns. Temperature-dependent thermal and mechanical properties of the steel were adopted in the analysis, and the restrained high strength Q460 steel columns with different magnitudes of the applied loads and boundary conditions were exposed to four-side fires in the simulation. Thermal and structural model of the column was established for finite element thermal and structural analysis, respectively.

6.3.6.1 Thermal analysis model

The analysis results from the three-dimensional (3D) finite element analysis of steel column subjected to standard ISO-834 fire indicated that the temperature in the steel column was uniformly distributed along the column length (Kodur et al., 2009). Therefore, a simple two-dimensional (2D) finite element analysis model to simulate the temperature distribution in cross section was established. The steel column section was modeled by using element PLANE55 in ANSYS, which has four nodes with a single degree of freedom (temperature) at each node. Heat conduction can be simulated by specifying appropriate property for the thermal conductivity of the steel. In this investigation, the property specified in EC3 was adopted. To simulate the effect of heat convection and thermal radiation, the surface of PLANE55 2D thermal solid element was covered by SURF151 element, as depicted in Fig. 6.23. In the thermal analysis, the recorded furnace temperature from the test was assigned to the exposed

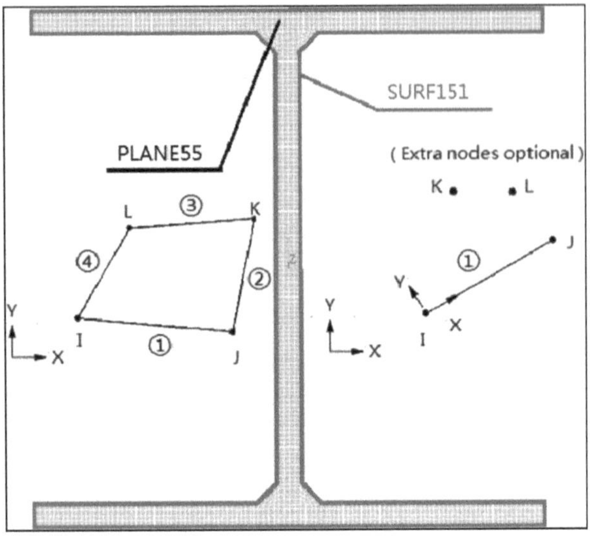

Figure 6.23 FE model for thermal analysis.

surface of SURF151 element. The value of convection coefficient was given in Chen (2010), and the Stefan–Boltzmann radiation constant of $\sigma = 5.67 \times 10^{-8}\ \mathrm{W/(m^2\ K^4)}$ was used in the analysis. The thermal conductivity, convection coefficient, steel density, and heat specific, as per EC3, were plotted in Fig. 6.24A

6.3.6.2 Structural analysis model

The mechanical model of the restrained steel column with two different column end boundary conditions and the applied load were shown in Fig. 6.25. Because of the large values of slenderness ratios, BEAM188

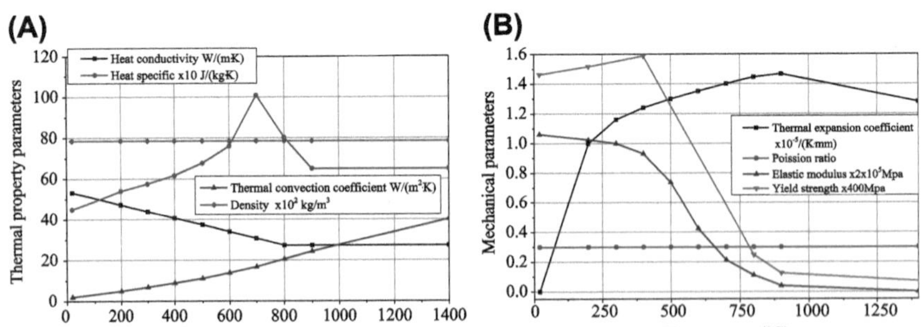

Figure 6.24 Physics parameters of Q460 steel. (A) Thermal property parameters of Q460 steel and (B) mechanical parameters of Q460 steel.

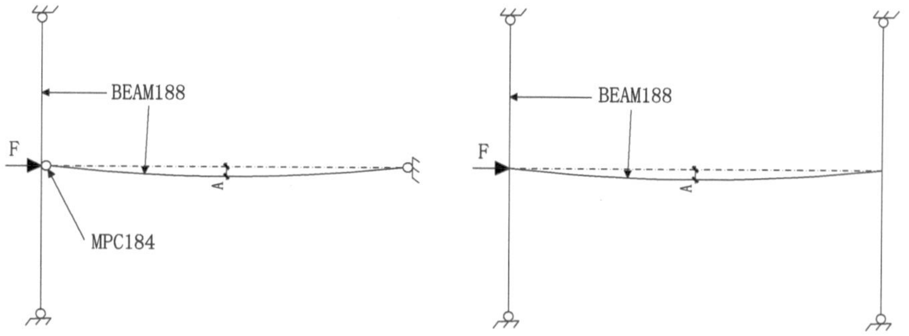

Figure 6.25 Analytical model of ANSYS.

element was adopted to model the columns and connected restraining beams. BEAM188 element is Timoshenko beam element with both effects of shear deformation and large deformation being taken into accounted (Zhang et al., 2018). The pin-ended joint between the restraining beam and the column was simulated by using multipoint constraint element MPC184. Such element can also simulate a rigid joint element by using two nodes at same location and is capable of considering large deformation and nonlinear material behavior. With using PLANET82 element, the cross-sectional mesh of the column specimen and restraining beam was shown in Fig. 6.26. The column and beam were both divided into 50 segments in their longitudinal direction. The constitutive models of Q460 and Q235 steel are bilinear and ideal elastoplastic model, respectively. The corresponding mechanical properties at elevated temperature are presented in Fig. 6.24B, where the Poisson's ratio and thermal expansion coefficient are adopted from EC3. Elastic modulus and yield strength were determined according to test results

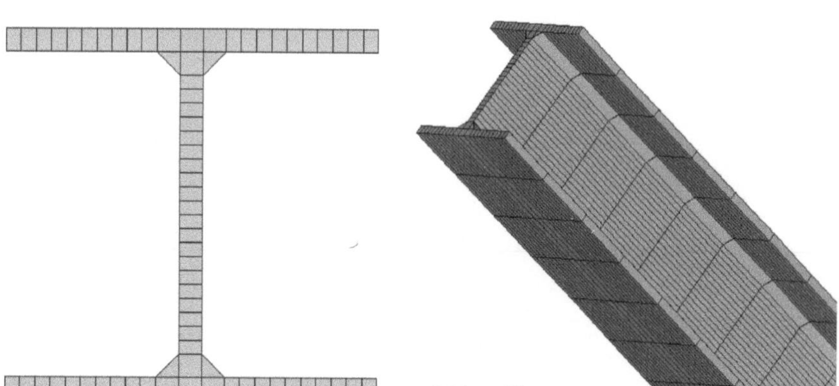

Figure 6.26 FE model mesh.

presented in Section 3.2 of Chapter 3. The variation of residual stress as a function of temperature proposed by Wang et al. (2015) was adopted to calculate residual stress at elevated temperature. The geometric imperfection was incorporated into the model as $A = \frac{l}{1000}\sin\left(\frac{\pi x}{l}\right)$.

6.3.6.3 Model validation

The thermal analysis indicated that the maximum temperature located at the flange edges of the column section and the minimum temperature located at the intersection between the column web and flange. The maximum and minimum temperatures are presented in Fig. 6.27. It can be seen that the difference between these two temperatures is insignificant as the difference is less than 36°C. Such trivial difference can be neglected because resulted variations of the yield strength and elastic modulus are very small. Therefore, the temperature in the steel column can be considered as uniformly distributed. For the eight fire tests presented in this study, the temperature evolution and structural response of the specimens were all simulated with the finite element analysis. The only differences in the model of each specimen were the applied load level and restraint ratio. Good agreements have been achieved between the numerical and experimental results in terms of axial displacement and deflection at the mid-height. For the purpose of the demonstration, thermal and structural analysis models were validated by comparison of the numerical results with the test results of specimen S-5 and S-6. Fig. 6.28 shows the column temperatures obtained by the finite element model and the test for cross section I, II, and III shown in Fig. 6.16. It can be seen from the figure the numerical and experimental results are in good agreement.

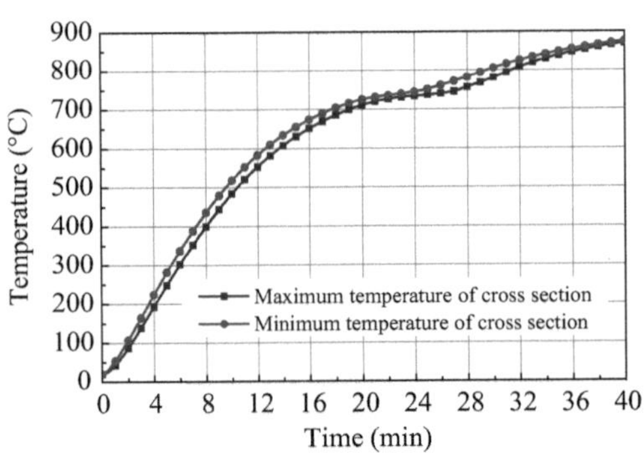

Figure 6.27 Maximum and minimum temperatures of cross section by thermal analysis.

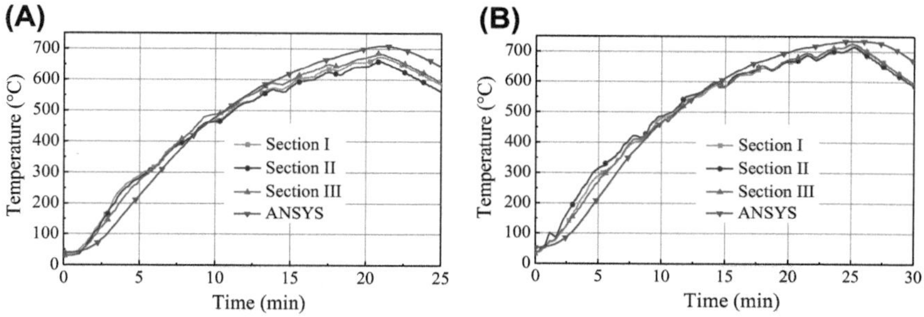

Figure 6.28 Comparison of numerical and experimental results. (A) Specimen S-5 and (B) specimen S-6.

In Fig. 6.29, predictions on the axial displacements and lateral deflections of specimens S-5 and S-6 obtained by ANSYS are compared with the corresponding test results. In the test, if the column specimen experienced a large deformation, the test would be stopped immediately for the safety reason and avoidance of potential damages to the furnace. However, in the finite element analysis, the simulation would keep on going as long as the result convergence could be achieved. Thus, the predicted final deformation could exceed that was obtained from the test. Generally, the results obtained from the finite element modeling are less conservative than the tests. It can be found that the predicted axial displacements agree well with the test results for the temperature up to 520°C. For the lateral deflection at midheight of the column, the predicted results agree well with test ones for the temperature up to 550°C. The discrepancies of the predicted and test results for both the axial displacement and

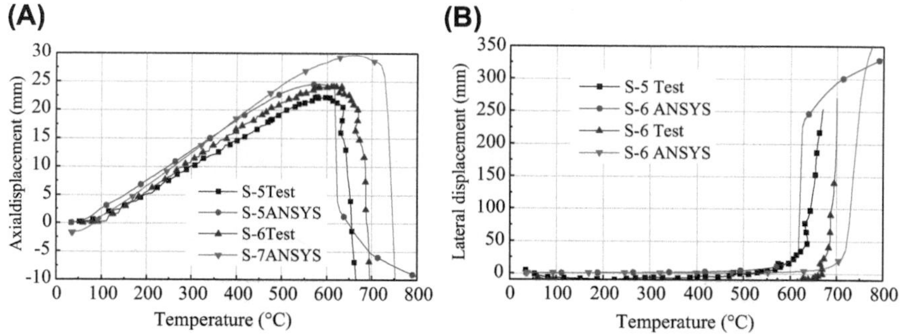

Figure 6.29 Comparison of column responses between test and prediction. (A) Axial displacement and (B) deflection at midheight.

lateral deflection become to increase when the temperatures exceed the aforementioned values. The maximum difference of buckling temperature obtained from the prediction and test is approximately 50°C. The difference can be attributed to three possible reasons: first, that the local geometric imperfection was not considered in the analysis; second, the creep deformation of the steel column was not taken into consideration in the analysis; and third, the reason for the discrepancy could be the variation in material properties at elevated temperatures, given that the assumptions for the material models are adopted from other test and not from a direct material testing of the specimen.

As for the failure mode, the results of finite element modeling indicated that flexural buckling about the weak axis is predominate failure mode for all eight specimens. This is true for all the specimens except S-7 and S-8 in which flexural torsional buckling was observed in the tests. As previously explained, the flexural torsional buckling associated with S-7 and S-8 may result from the high applied load ratio and possible nonuniform temperature distribution in the specimens at the early stage of fire exposure. In the finite element modeling, the possible nonuniform temperature distribution was not accounted. Consequently, the failure mode obtained from the finite element modeling for S-7 and S-8 is flexural buckling about the weak axis.

6.3.7 Parametric study

To facilitate parametric studies, a simplified finite model of the column was established by replacing the restraining beam at each end of the column with an elastic spring as shown in Fig. 6.30. The spring element is COMBIN14 element, which has ability to simulating elastic springs associated with axial and torsional deformation in 1D, 2D, or 3D models. The axial stiffness and torsional stiffness of COMBIN14 element were determined by that of the restraining beam.

The simplified finite model shown in Fig. 6.26 and analytical model shown in Fig. 6.30 were employed to simulate the fire test of specimen S-5. The results obtained from the both models are presented in Fig. 6.31 for the

Figure 6.30 FE model of restrained steel column for parametric analysis.

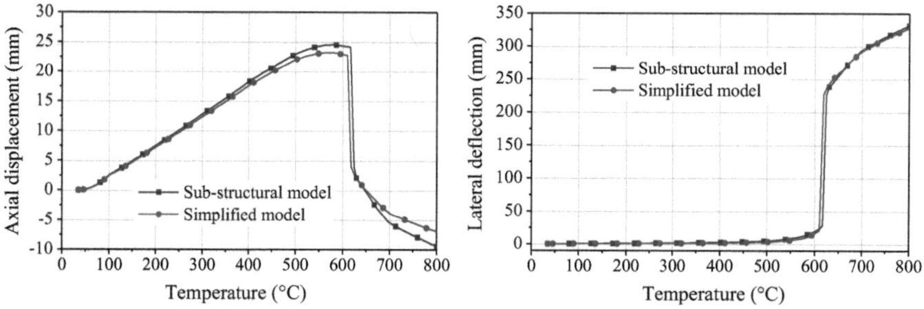

Figure 6.31 Results comparison of analytical model and simplified model.

reason of comparison. In general, good agreements was found on both of the lateral deflection and axial displacement except minor deviations on the axial displacement when the temperature exceeds 700°C. The analytical model is capable of simulating the fire test with both constant and variable end restraining stiffnesses, which correspond the elastic and inelastic behaviors of the restraining beam, respectively. The simplified model, however, is only applicable to the steel column with the constant end restraining stiffness. In the parametric study, the validated simplified finite element model was employed to perform the analysis of investigating effects of axial and rotational restraints as well as the applied load and column slenderness ratios on the buckling and failure temperatures. The effect of residual stress and global geometric imperfections were taken into account in the parametric study.

6.3.7.1 Axial restraint stiffness

The influence of the axial restraining stiffness on fire resistance of restrained mild steel column has been studied by many researchers. Fire tests on restrained steel column conducted by (Rodrigues et al., 2000) showed that the axial restraint could decrease the buckling temperature of the steel column, and for the column with the larger axial restraining stiffness, the axial force would drop more slowly after the buckling of the column. The realistic values of the axial restraint ratio for structural steel frames in practice are approximately ranged from 0.004 to 0.05 (Wang et al., 1997).

The relationships between the axial force and temperature for different axial restraining stiffness were plotted in Fig. 6.32, in which the rotational restraint ratio β_r was taken as 0 and the applied load ratio is 0.5. Analysis results indicated that the buckling temperature and failure temperature of Q460 steel column reduced with the increase of axial restraining stiffness and the axial force decreased more slowly after the buckling for columns

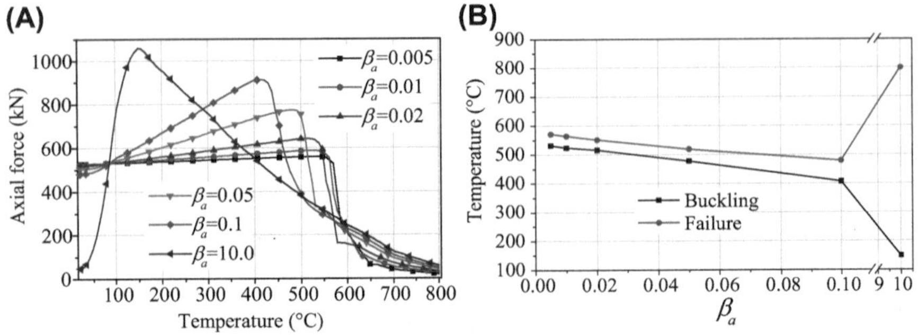

Figure 6.32 The effect of axial restraint on fire behavior of the column. (A) Curves of axial force and temperature (B) Buckling and failure temperature

with larger axial restraining stiffness. When axial restraint ratio β_a reached to 10, the axial restraining spring undertook the majority of the applied load after the buckling of the column, which results in a higher failure temperature. It can be seen from Fig. 6.32 that the difference between buckling and failure temperature is greater for columns with larger axial restraining stiffness. Consequently, the postbuckling stage stretches a longer range, and the column can survive longer time duration in fire because of the lager axial restraining stiffness.

6.3.7.2 Rotational restraining stiffness

The FE analysis carried out by Valente and Neves (1999) indicated that the failure temperature of steel column is higher for columns with larger rotational restraining stiffness until it reached to a certain magnitude. Wang et al. (2010) reported that the axial force–temperature curves were similar to that with rotational restraint ratio of 2.0 when the rotational restraint ratio (β_r) exceeded 2.0. In structural steel frames, the rotational restraint ratio for a column is not only related to the stiffness of the connected beams but also to that of beam-to-column connections, which is often within the range of 0–1.0 (Qiang et al., 2018). The influence of the rotational restraint on Q460 steel column was presented in Fig. 6.33, where $\beta_a = 0.05$, $\rho_N = 0.5$, and $\lambda = 90$. From the analysis results, the limit of rotational restraint ratio on the mild steel column is also applicable for high strength Q460 steel column, and the critical value of the rotational restraint ratio for Q460 steel column can be taken as 1.0. That is, if the rotational restraint ratio is less than 1.0, the higher rotational restraint ratio would result in the higher buckling temperature. When the rotational restraint ratio exceeds

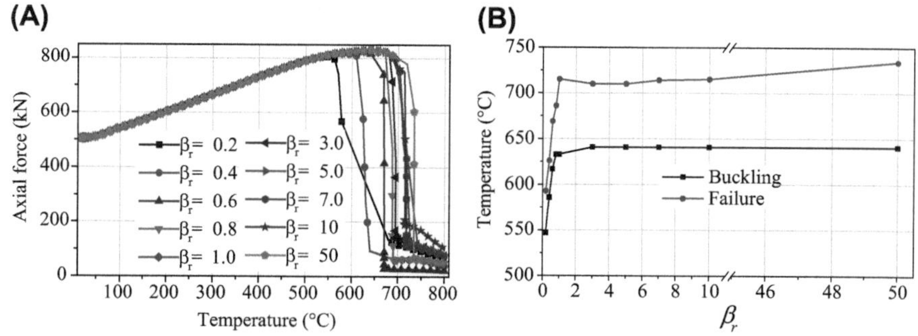

Figure 6.33 The effect of rotational restraint on fire behavior of column. (A) Relationship of axial force and temperature (B) Buckling and failure temperature

1.0, the effect of the rotational restraint ratio on buckling temperature is not obvious.

6.3.7.3 Applied load ratio

Based on a study by Wang (2004) on the fire resistance of restrained mild steel columns, it was found that the axial force in the column would drop slowly when the applied load ratio is less than 0.5. However, the rate of the axial load decrease becomes more rapid if the applied load ratio is greater than 0.5. The effects of applied load ratio on fire resistance of an axially restrained Q460 steel column from the parametric study were shown in Fig. 6.34, where $\beta_a = 0.05$, $\beta_r = 0$, and $\lambda = 90$. Both analysis and test results showed a similar trend that supports the aforementioned conclusion. It was also observed that from the figure that the increase of the applied load ratio decreases both the buckling and failure temperatures. This indicates that for higher applied load ratio, the deterioration of the fire resistant of Q460 steel column would be more severe. In addition, the drop of the axial force after column buckling becomes more rapid, and the difference between the buckling and failure temperatures comes smaller with higher applied load ratio. This may explain the reason of the axial restraint delays and the failure of Q460 steel column when the applied axial load is relatively small.

6.3.7.4 Column slenderness ratio

Fig. 6.35 shows the effects of column slenderness ratio on the fire resistance of restrained Q460 steel column with $\beta_a = 0.05$, $\beta_r = 0$, and $\rho_N = 0.5$. It is evidenced that both of the buckling and failure temperatures decrease with

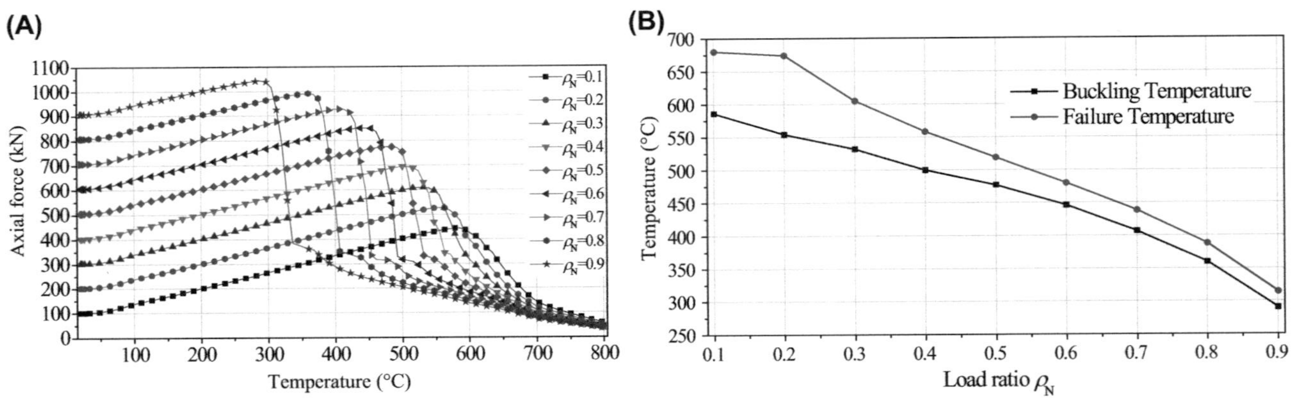

Figure 6.34 Effects of applied axial load on fire resistance of the restrained Q460 steel column. (A) Curves of axial force and temperature and (B) buckling and failure temperature.

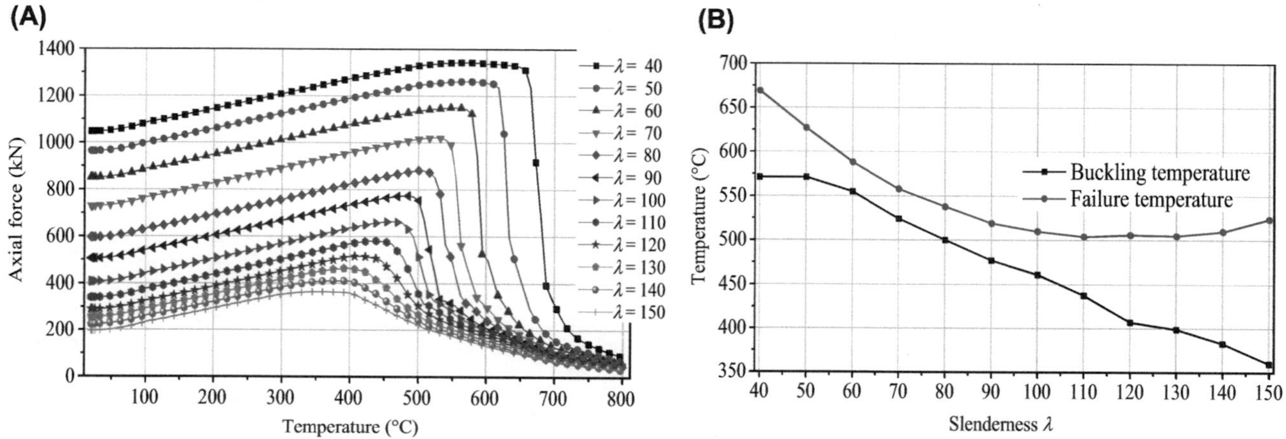

Figure 6.35 Effects of slenderness ratio on fire resistance of restrained 460 steel column. (A) Curves of axial force and temperature and (B) buckling and failure temperature.

the increase of the slenderness ratio, and the slenderness ratio has a minor influence on the failure temperature when the ratio is less than 60. When the slenderness ratio is less than 60, the drop rate of the axial force becomes slower as the decrease of the slenderness ratio, which results in larger differences between the buckling and failure temperatures. When the slenderness ratio ranges between 60 and 90, the effect of the slenderness ratio on the difference between the buckling and failure temperatures is very minor. However, the difference increases as the increase of the slenderness ratio when the ratio exceeds 90.

6.4 Load-bearing capacity evaluation of steel columns after fire exposure

To investigate the postfire behavior of high-strength Q460 steel, a comprehensive test program was designed. The test program consisted of two stages: the first stage is fire exposure of the test specimens to ISO-834 standard fire, and the second one is compressive test on the specimen after fire exposure.

6.4.1 Specimens

A total of four columns fabricated with Q460 steel were tested. The web plates were connected to the flange plates with a pair of fillet welds with 8 mm leg size. A gas shielded arc welding process was used with CO_2 shielding gas. The voltage of the welding gun was 25 V, the ampitudes were 230 A, and the welding speed was about 35 cm/min. The key parameter considered in the specimen was cross-sectional dimension. The measured geometric properties of the four columns are tabulated in Table 6.6, and the cross-sectional dimensions are illustrated in Fig. 6.36. The sectional area (A), second moment of area (I), and radius of gyration (i) of the cross sections in Table 6.4 were all calculated based on the measured actual cross-sectional dimensions. The length of specimens is L, with L_t being the effective length taken as L plus 298 mm (the distance between the centers of rotation of both pin-ended supports). An example of the test specimen identification symbol is as follows: B1-460, where B denotes the box section, the next digit represents specimen label, and the following three digits refer to the steel nominal yield strength in N/mm^2. The design length was 2540 mm for all specimens.

Table 6.6 Details of the specimens.

Specimen label	H/mm	B/ mm	$t_f = t_w$/ mm	A/mm^2	I/mm^4	i/ mm	L/ mm	L_t/ mm
B1-460	150.2	150.2	8.25	4683.4	1.577E + 07	58.0	2528	2826
B2-460	181.7	181.7	8.25	5723.9	2.877E + 07	70.9	2538	2836
H1-460	249.7	125.7	8.27	4005.8	2.748E + 06	26.2	2532	2830
H2-460	200.9	101.2	8.24	3188.1	1.431E + 06	21.2	2528	2826

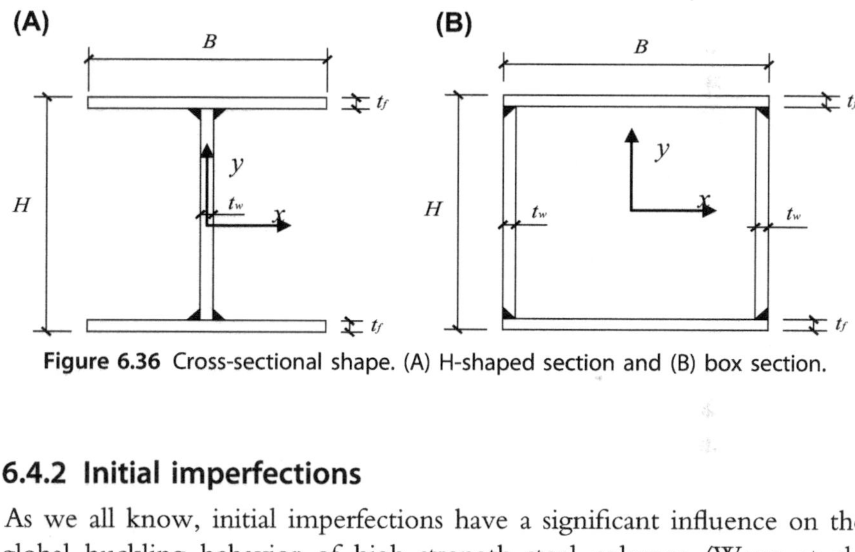

Figure 6.36 Cross-sectional shape. (A) H-shaped section and (B) box section.

6.4.2 Initial imperfections

As we all know, initial imperfections have a significant influence on the global buckling behavior of high strength steel columns (Wang et al., 2014b). The initial bending, and residual stress along cross section were all determined before the compressive test. The recorded deviations of the column along the weak axis from the straight line connect centroids at two ends, tabulated in Table 6.7, including the results v_1, v_2, and v_3 at the three quarter points along the column length and the maximum deviation $v_0 = \max(v_1, v_2, v_3)$. It should be noted that all the imperfection for H-shaped section is about week axis of the section, namely y axis as shown in Fig. 6.36. For box section, the weak axis means the axis around which the imperfection is bigger.

Table 6.7 Measured results of initial geometric imperfections.

	Geometric imperfection (mm)			
Specimen label	v_1	v_2	v_3	v_0
B1-460	−6.05	−8.25	−6.32	−8.25
B2-460	2.77	1.74	1.66	2.77
H1-460	3.31	4.20	3.99	4.20
H2-460	−0.23	−0.35	−0.53	−0.53

Residual stress measurements at room temperature and after fire exposure were carried out with 10 H-section columns and 10 box section columns by Qin (2015). All the columns were fabricated with the same parent plates. The H-section columns and box section columns had the same nominal cross-sectional dimensions respectively as the test specimens. A simplified residual stress distribution for H-sections and box section after fire exposure is reproduced here, as shown in Fig. 6.37, in which f_y is the steel nominal yield strength in N/mm^2 (460 N/mm^2); α is residual stress reduction coefficient (the ratio of residual stress after fire exposure to that at ambient condition) as shown in Table 6.8; and η is the ratio of maximum residual stress value at ambient condition to nominal yield strength of steel as shown in Table 6.9.

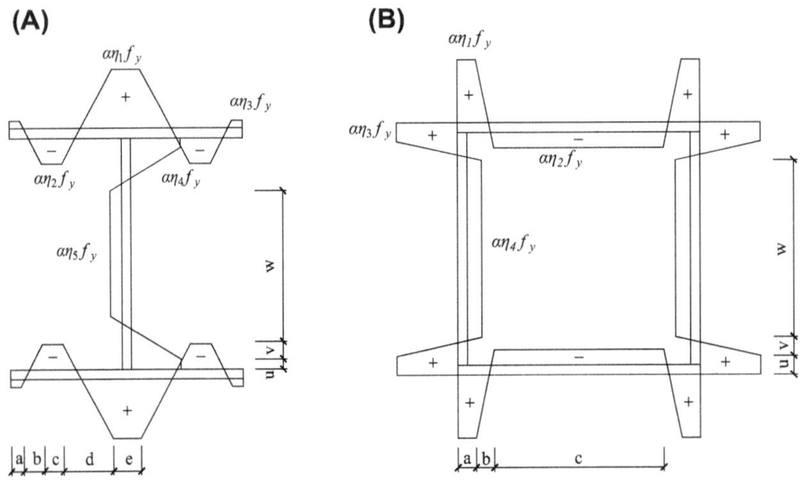

Figure 6.37 Simplified residual stress distribution of Q460 steel section after fire exposure. For (A) H-shaped section and (B) box section.

Table 6.8 Residual stress reduction coefficient α.

Cross-sectional shape	Temperature for exposure/oC				
	20	200	400	600	800
H-section	1	0.95	0.75	0.25	0.1
Box section	1	0.97	0.65	0.20	0.1

Table 6.9 Values of η_1-η_5.

Cross-sectional shape	η_1	η_2	η_3	η_4	η_5
H-section	0.65	0.4	0.15	0.65	*
Box section	0.9	0.25	1.0	0.2	—

Note: Value of * can be obtained by calculating the section balance equation.

6.4.3 Test setup and procedure

The fire exposure experiments were carried out in a furnace that was specially designed for testing structural members at Tongji University. The dimension of furnace is 3.6 m wide, 4.6 m long, and 3.3 m high. The maximum heat power that the furnace can produce is 5 MW. Eight natural gas burners located within the furnace provide thermal energy, while 10 thermocouples, distributed throughout the test chamber, monitor the furnace temperature during a fire test. The temperature in the furnace was controlled automatically and was set according to the ISO-834 standard fire and air cooling. The specimens were put horizontally (as shown in Fig. 6.38) in the furnace in an unstressed condition to expose to a relatively uniform temperature zone.

Figure 6.38 Photo of steel columns after fire exposure.

A thermocouple was placed on each specimen by drilling a small hole at the midheight of column. The temperatures of the furnace and steel columns were measured during both the heating and cooling phases as shown in. The temperatures in four steel columns are relatively uniform, but the actual temperatures response of steel column is slightly lower than the furnace temperatures as shown in Fig. 6.39. After fire exposure, the tests were paused while all the specimens were cooled to approximately room temperature.

All the specimens were tested by applying a concentric loading with a 5000 kN hydraulic compression machine at Chongqing University. The photos of the test setup are illustrated in Fig. 6.40. Both the top and bottom supports of the specimen were set to be fixed around strong axis and pin-supported around weak axis to assure that the buckling occurs around weak axis. In Fig. 6.40, symbols (1), (2), (3), and (4) refer to hydraulic compression machine, test specimen, pin support, and force transducers, respectively.

6.4.4 Test instruments

The arrangement of displacement transducers, dial indicators, and strain gages used in the test is shown in Fig. 6.41. The axial deformations of the specimens were recorded by dial indicators *V*01 and *V*02. Displacement transducers *H*01−*H*03 were placed at the midheight of the column to measure the in plane lateral deflection. Transducers *V*03, *V*04, *V*05, and *V*06 were attached on both sides of the pin-ended supports to generate the end rotation of the column (θ), which was calculated by using Eq. (6.8) for the top end.

$$\theta = \frac{R_{V3} - R_{V4}}{d} \times \frac{180}{\pi} \tag{6.8}$$

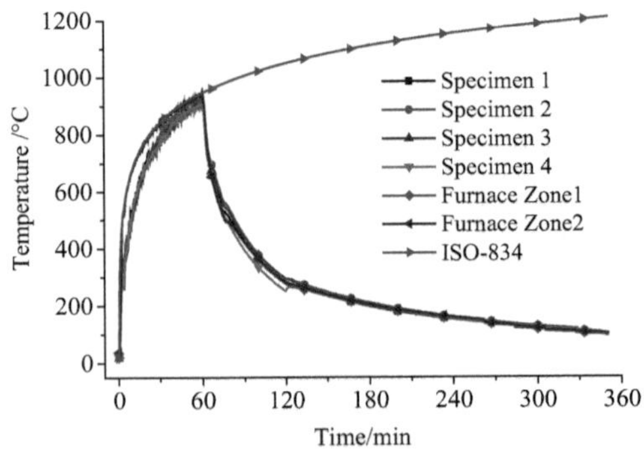

Figure 6.39 Temperatures of furnace and specimen.

(A) (B)

Figure 6.40 5000 kN hydraulic compressive test setup. (A) Photo of compressive test setup and (B) view of connection at the ends of specimen.

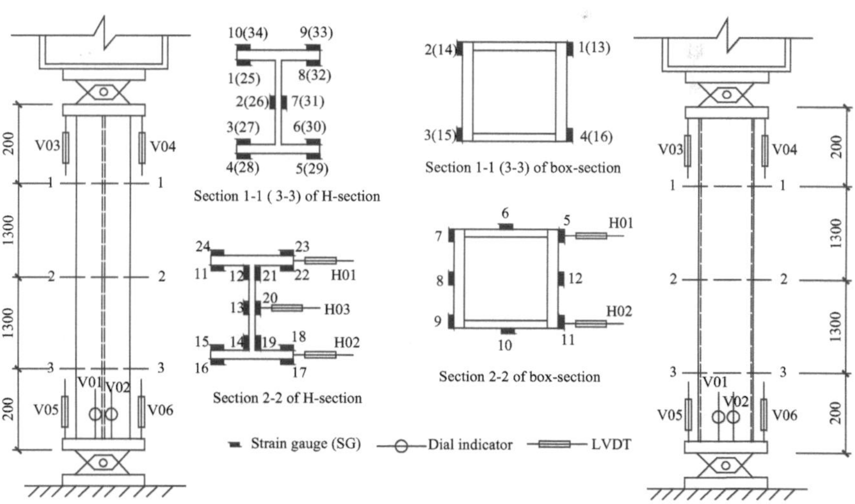

Figure 6.41 Arrangement of test instruments.

where R_{V3} and R_{V4} are readings of transducers *V*03 and *V*04, respectively, and d is the distance between these two measurement points.

Strain gages were placed at the midheight cross section and sections 200 mm away from both ends of the specimens to record the loading force and validate the measured geometric imperfections. Before the formal compressive test, a preload with 10% of the predicted maximum load-bearing capacity was used to verify the test instrumentations and then unloaded. In the formal test, the loading rate was 10 kN for each grade load up to approximately 85% of the expected load-bearing capacity, after which the rate was controlled by a displacement. During the test, the axial load, axial displacement, lateral deflection, and strains in the steel columns were recorded and saved.

6.4.5 Test results and discussions

Data generated from compressive tests were utilized to obtain the residual compressive load-bearing capacity and failure characteristics of Q460 steel columns after fire exposure.

6.4.5.1 Structural response

The applied load versus midheight deflection curves of all specimens are plotted in Fig. 6.42. It should be noted that the test results of specimen H2-460 is not saved due to mechanical problem of the test setup during the test. In the future, more tests will be considered to be carried out on postfire behavior of high strength steel columns, and the results will reported in the further study. Therefore, only the test data of the other three specimens were reported herein. From Fig. 6.42, it can be seen that the specimens were sensitive to initial geometric imperfections and in the early stage of

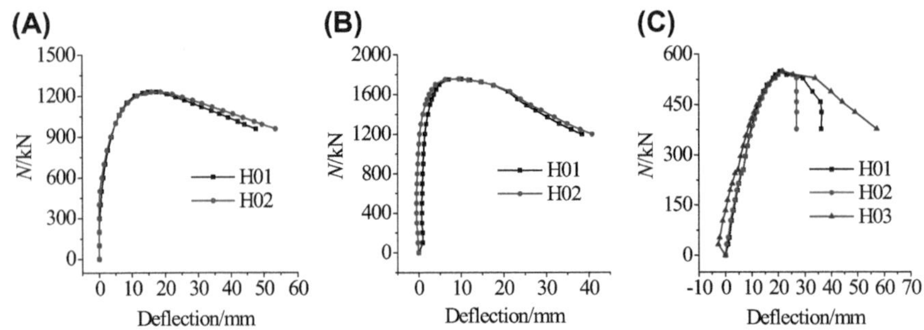

Figure 6.42 Load-deflection curves. (A) Specimen B1-460, (B) specimen B2-460, and (C) specimen H1-460.

load-deflection curves will exhibit reduction of stiffness. For box section specimens, the data generated from the LVDTs arranged on the same specimen are similar, which indicates that only flexural buckling occurs on the specimen and almost no torsion deformation was seen. However, for H-section specimen, there are some discrepancies of the three measured deflections. This may be attributed to local buckling on the flange or web at the midheight of column. The load—axial displacement curves measured at the end of the columns are shown in Fig. 6.43. Unfortunately, dial indicators *V*02 in specimen B1-460 did not work due to mechanical problems. It can be seen that the axial displacements obtained from the dial indicators *V*01 and *V*02 are quite consistent with each other. The typical axial load—strain curves recorded at the midheight of the steel columns are presented in Fig. 6.44. It is convenient to determine the occurrence of bending and easy to identify the yield status of the midheight cross section

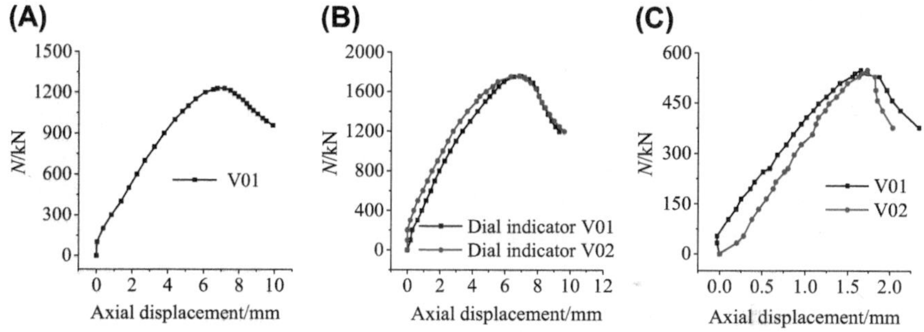

Figure 6.43 Load—axial displacement curves. (A) Specimen B1-460, (B) specimen B2-460, and (C) specimen H1-460.

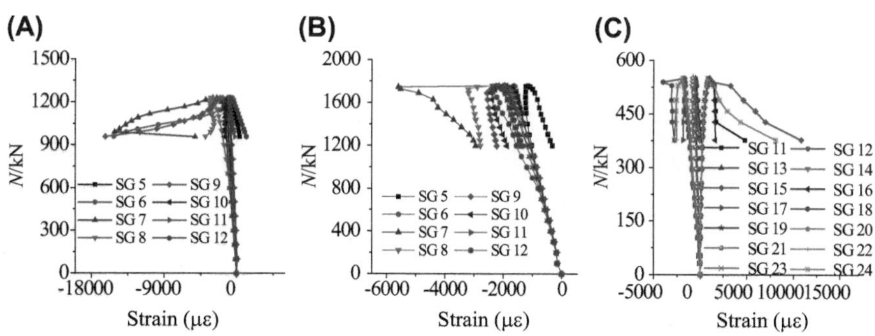

Figure 6.44 Load—strain curves. (A) Specimen B1-460, (B) specimen B2-460 and (C) specimen H1-460.

by using the load–strain curves. Although all specimens were designed as axial concentrically loaded columns, the bending of specimen H1-460 occurred at the beginning of loading due to geometrical imperfection.

Fig. 6.45 summarizes the load–rotation curves at two ends of each column obtained from the experimental results, with the end rotation being determined from Eq. (6.1). It can be found that the end rotation develops smoothly with an increase of compression load, and so the rotation performance of the specially designed pin-ended support system was very effective. Consequently, the end supports of the test specimens could be regarded as perfectly pin-ended ones.

6.4.5.2 Failure mode and load-bearing capacity

The failure modes of all specimens are shown in Fig. 6.46. From the figure, it can be seen that the failure mode of the specimens was global flexural buckling along the weak axis. This phenomenon agrees well with what we identify from the section shape of column and boundary condition at the both ends.

From the load–displacement curves shown in Fig. 6.43, the load-bearing capacities of columns are obtained and summarized in Table 6.10, where P_{cr} is the measured load-bearing capacity of columns; φ_T' is the stability factor of steel columns after fire exposure; A is the area of cross section; and f_{yT}' is the yield strength after fire exposure. Table 6.10 indicates that the global stability factors for box-shaped specimen are much higher than that for H-shaped section as the slenderness ratio of former is lower that the latter.

6.4.6 Finite element analysis

Results from the above tests were utilized to develop and validate a finite element model to predict the structural response and load-bearing capacity of Q460 steel column after fire exposure.

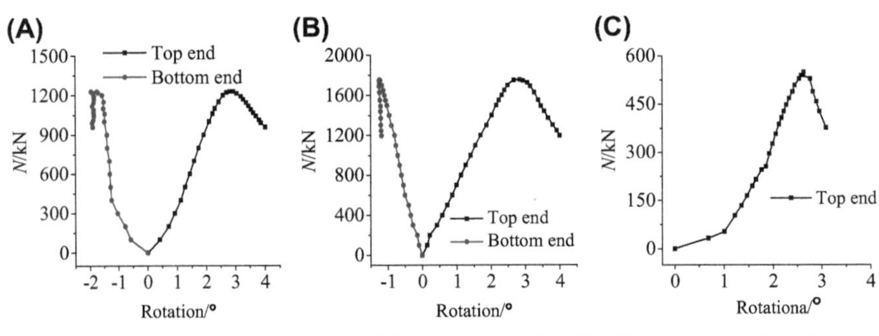

Figure 6.45 Load–rotation curves. (A) Specimen B1-460, (B) specimen B2-460, and (C) specimen H1-460.

Figure 6.46 Photos of specimens after failure. (A) Specimen B1-460, (B) specimen B2-460, and (C) specimen H1-460.

Table 6.10 Measured ultimate load capacities of specimens.

Specimen label	B1-460	B2-460	H1-460
P_{cr}/kN	1229	1756	549
$\varphi'_T = P_{cr} / Af'_{yT}$	0.697	0.814	0.364

6.4.6.1 Strain–stress relationship

Previous studies (Tao et al., 2012; Li and Guo, 1993) have shown that the mechanical properties of steel after fire exposure highly depend on the maximum temperature experienced during the heating phases. Hence, the stress–strain relationships of steel after fire exposure can be obtained by substituting the key mechanical properties, such as residual elastic modulus, ultimate strength, and corresponding strain, after fire exposure for those of the unexposed steel. The multilinear kinematic hardening model using von Mises yielding criterion as shown in Fig. 6.47 was utilized to simulate the stress–strain relationship for the Q460 steel with air cooling method. The mechanical properties after fire exposure, including the strength and strain, were determined by the tension coupon test presented in Section 4.3 of Chapter 4. The results of mechanical

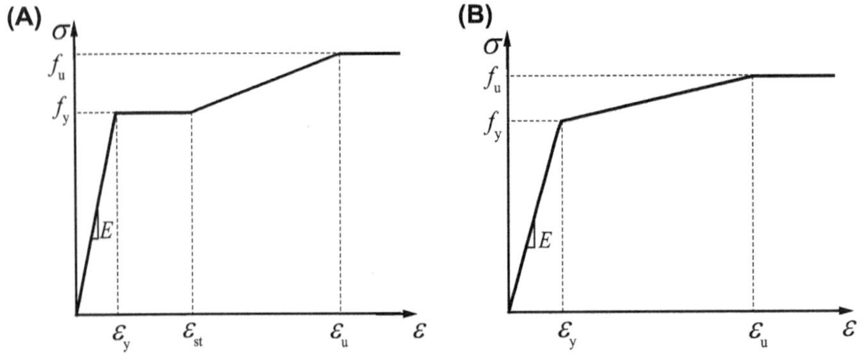

Figure 6.47 Stress–strain relationship of Q460 steel after fire exposure. For steel under (A) natural cooling (B) and water cooling.

properties at room temperature are shown in Table 6.3. As Q460 steel had no yield plateau after exposing temperature up to 800°C under water cooling condition, the stress–strain relationship was followed as Fig. 6.47B. Poisson's ratio was taken as 0.3.

6.4.6.2 Model and boundary conditions

A 3D finite element model (shown in Fig. 6.48) was established by employing the general finite element software ANSYS. Four-noded shell element SHELL181 was selected to discretize the steel columns with 50

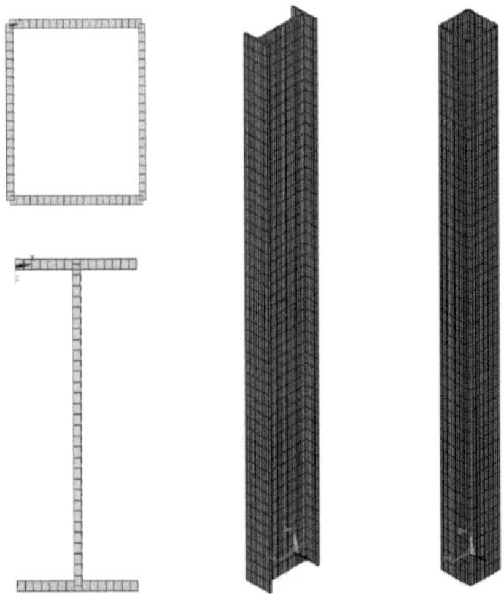

Figure 6.48 The mesh of cross section and model of columns.

elements along the column length. The mesh dimension for the cross section is adopted 8 mm through a mesh sensitivity study.

Initial geometrical imperfections were considered in the finite element simulation. The cross-sectional residual stress after fire exposure was applied as an initial stress at the element integration points, and the magnitude and distribution model was determined according to the relevant experimental results as mentioned in Section 6.4.2.

Both ends of all the specimens were set to fix around the strong axis and pin-supported along the weak axis. All displacements at both top and bottom ends were restrained except the axial displacement of the loading point. To capture the peak load, the arc-length solution method was selected, with the load, deflection, end rotation, and stress and strain distributions being recorded during the analysis.

6.4.6.3 Validation of the FE model

The established finite element model (FEM) was validated by comparing its results with the experimental data on Q460 steel columns after fire exposure. The comparisons for axial displacement and midheight deflection are presented in Fig. 6.49 and 6.50, respectively. Good agreement for both load—deflection curve and load—axial displacement curve was seen from the comparison except load—axial displacement curve for specimen H1-460. The axial stiffness tested for specimen H1-460 is higher than that in analysis. It may attribute to real imperfection is lower than tested imperfection. From Fig. 6.49C, it is shown that the deflection at the midheight of the specimen H1-460 is also lower than analysis. Generally, the finite element models are verified.

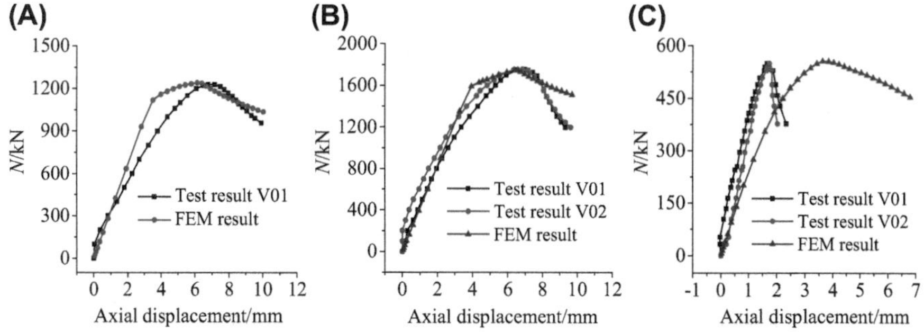

Figure 6.49 Comparisons of axial displacement between FEM and test results. (A) Specimen B1-460 and (B) specimen B2-460 and (C) specimen H1-460.

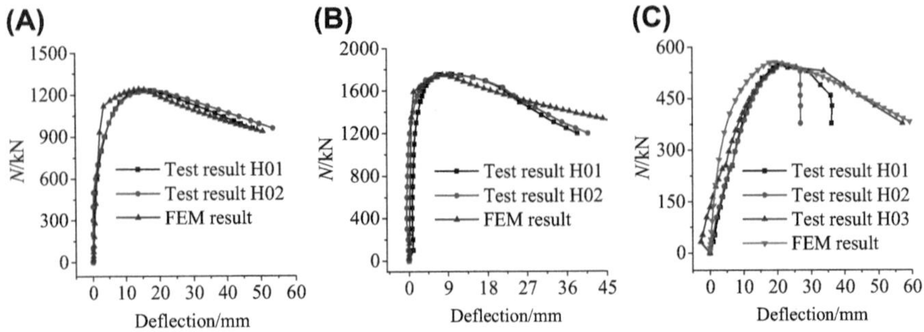

Figure 6.50 Comparisons of deflection between FEM and test results. (A) Specimen B1-460, (B) specimen B2-460, and (C) specimen H1-460.

Table 6.11 Comparison of the FE results and test results.

Specimen label	Experimental/kN	FEM/kN	FEM/experimental
B1-460	1229	1240	1.01
B2-460	1756	1746	0.99
H1-460	549	555	1.01

The predicted residual load-bearing capacity and the experimental results were compared, as shown in Table 6.11. It can be seen that the FE results are in good agreement with the experimental load-bearing capacity.

To further validate the finite element model, comparisons were also made against additional reported tests on Q460 steel columns at ambient temperature (Ban, 2012; Wang et al., 2012a,b; 2013b). Comparisons of the test and finite element results, together with the finite element results from the authors of the original investigations, are presented in Fig. 6.51.

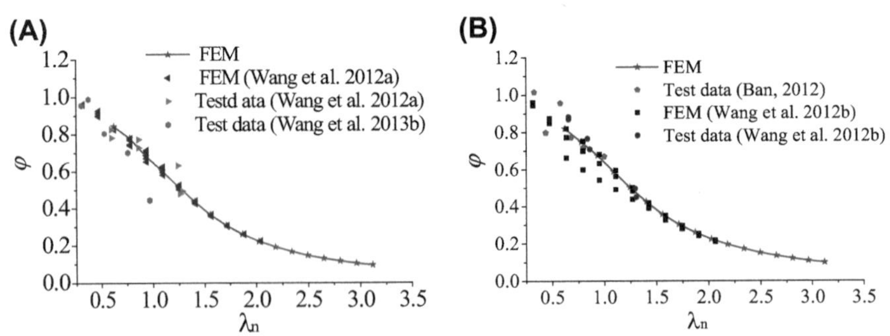

Figure 6.51 The comparison of FEM and test results on Q460 steel column at ambient temperature. (A) H-shaped section column, and (B) box section column.

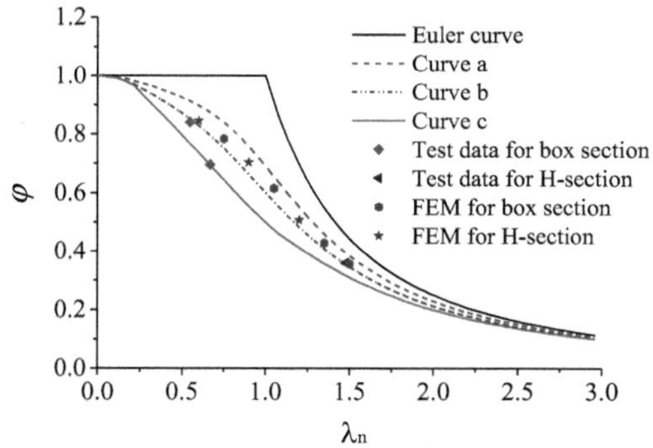

Figure 6.52 Comparison of test results and design code of GB 50017-2003.

Acceptable agreement may be seen, and it can be concluded that the generated finite element model is capable of predicting accurately the behavior of Q460 steel columns at ambient and after fire exposure.

The stability factors φ (the ratio of buckling capacity of column to the strength of cross section) obtained from the experiments and FEM analysis are compared with the multiple buckling curves of GB 50017-2017 (2017) in Fig. 6.52. For H-shaped section columns that buckled about their weak axis in the test, the column curve "*b*" in GB50017-2017 agrees better with the test results and FEM analysis, while the curve "*c*" in GB50017-2017 is relatively conservative with the test results of box section specimens, which in accordance with GB 50017-2017 for columns at ambient condition. Therefore, GB50017-2017 design curves can be extended to postfire-welded Q460 high strength steel columns.

6.4.7 Parametric study

To extend the ranges of the key parameters to investigate their effects on the residual load capacity of Q460 steel columns after fire, parametric studies were undertaken. The investigated parameters include heating temperatures, slenderness ratio, and cooling methods (air cooling and water cooling). A total of 272 pin-ended columns subjected to axial compression and buckling about weak axes were simulated in the parametric study. The investigated columns were divided into eight groups with different temperatures of each section type. At each temperature, there were 17 columns with slenderness ratio ranging from 40 to 200. The dimensions of columns for parametric study are listed in Table 6.12.

In the parametric studies, the mechanical properties of Q460 steel were taken as Table 4.5 according to different cooling methods. The initial

Table 6.12 Dimensions of simulated columns of H-section and box section.

Section shape	*B*/mm	*H*/mm	t_w/mm	t_f/mm	*A*/mm2	I_y/cm4	i_y/mm
H-section	200	250	11	21	10,688	2802	51.2
Box section	240	240	12	12	10,944	9508	93.2

geometric imperfection was considered by updating the geometry according to the first-order eigenvalue buckling mode, with the maximum deviation taken as 1‰ of the column length. The residual stress was considered at the initial load step. Based on the nonlinear finite element model, the peak load for each column was captured to determine the global buckling strength of Q460 steel column. The influences of these key parameters on the stability factor φ of Q460 steel columns are shown in Fig. 6.53–6.55, in which φ_{nT} and φ_{wT} are the stability factors under natural and water cooling methods, respectively. The influence of cooling method is not obvious after fire exposure, except at the temperature of 800°C. The comparison shows that the stability factor with water cooling method at 800°C is higher than that of air cooling method by about 12%. This phenomenon can be attributed to the fact that the yield strength of steel under water cooling (361.7 MPa) is much lower than that under air cooling (411.7 MPa) after exposure to 800°C. The stability factors may be seen to reduce with increasing slenderness ratio and increase with rise of temperature due to decrease of yield strength with the elevation of temperature exposed.

6.4.8 Proposed simplified design approach

As is mentioned above, the GB50017-2003 design curves can be extended to postfire welded Q460 high strength steel columns, but they are

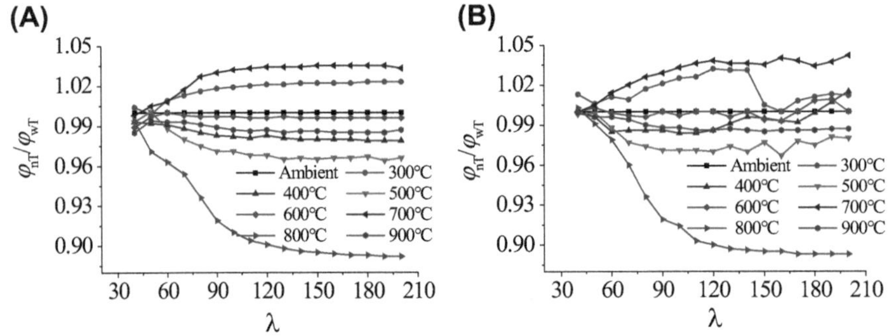

Figure 6.53 Influences of cooling method on stability coefficient. (A) H-shaped section and (B) box section.

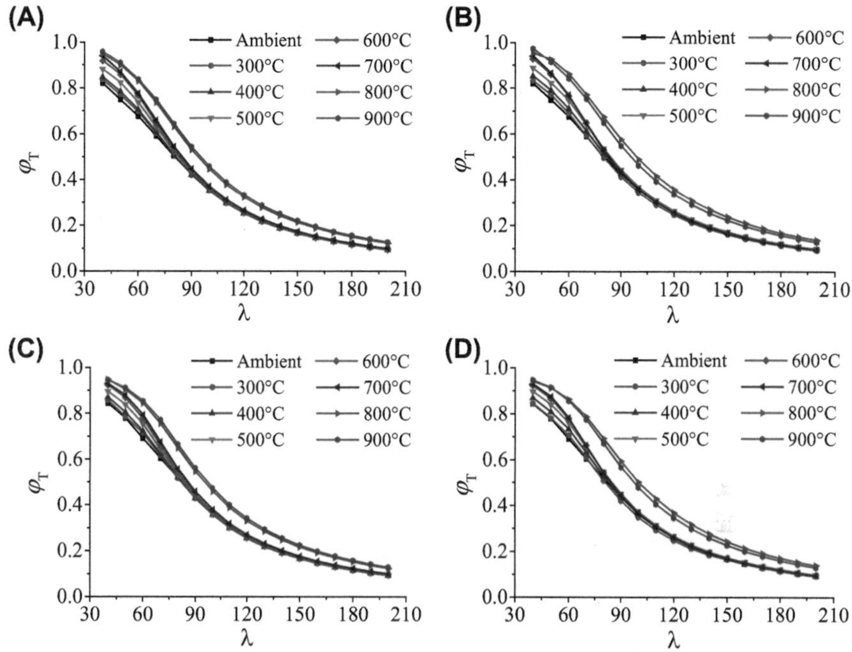

Figure 6.54 Influences of cooling methods on stability coefficient. (A) H-shaped section under natural cooling, (B) H-shaped section under water cooling, (C) box section under natural cooling, and (D) box section under water cooling.

conservative. To evaluate the load-bearing capacity of Q460 steel columns after fire exposure, a new design approach is proposed based on the large number of numerical analysis.

Li et al. (2006) proposed a simple approach for predicting the load-bearing capacity of steel columns in fire, given as Eq. (6.9):

$$N = \varphi_T A f_{yT} \tag{6.9}$$

where φ_T is the stability factor of steel column at elevated temperature and f_{yT} is the yield strength of steel at elevated temperature.

To keep consistency with the above design approach for steel columns in fire, a design approach is proposed for predicting the residual load-bearing capacities of Q460 steel columns after being exposed to ISO-834 standard fire.

Eq. (6.10) is intended to determine the residual load-bearing capacity of Q460 steel columns after exposure to ISO-834 fires in the investigated parameter ranges and may be used to evaluate an estimate of the postfire

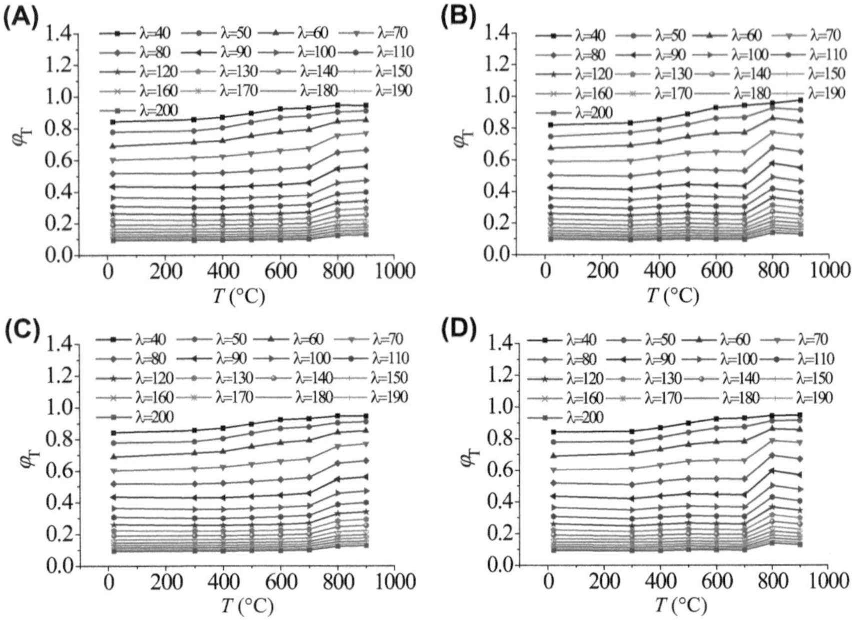

Figure 6.55 Influences of slenderness ratio on the stability factor. (A) H-shaped section under natural cooling, (B) H-shaped section under water cooling, (C) box section under natural cooling, and (D) box section under water cooling.

residual load-bearing capacity for Q460 steel columns at the initial stage of assessment of fire damage.

$$N = \varphi'_T A f'_{yT} \tag{6.10}$$

where φ'_T is the stability factor after fire exposure and f'_{yT} is the yield strength after fire exposure.

To obtain the residual load-bearing capacity, the key problem is to determine the stability factor of steel columns after exposure to fire. The stability calculation parameter after cooling down from high temperatures was calculated as the ratio of postfire stability factor to that at ambient condition, which was shown as Eq. (6.11).

$$\alpha_c = \varphi'_T / \varphi \tag{6.11}$$

According to the influences of different parameters on the stability calculation parameter, the FE parametric analysis results were obtained, which were shown in Tables 6.13–6.16 for H-shaped section and box section, respectively.

Table 6.13 α_c for H-section Q460 steel column after fire under natural cooling.

	Temperature/°C						
λ	**300**	**400**	**500**	**600**	**700**	**800**	**900**
40	1.019	1.033	1.075	1.119	1.146	1.158	1.170
60	1.030	1.044	1.092	1.135	1.149	1.229	1.239
80	1.005	1.011	1.041	1.056	1.082	1.256	1.280
100	0.984	0.984	1.003	1.012	1.039	1.248	1.281
120	0.975	0.974	0.986	0.994	1.022	1.245	1.284
140	0.969	0.967	0.978	0.987	1.015	1.246	1.286
160	0.969	0.966	0.977	0.986	1.016	1.251	1.293
180	0.968	0.963	0.974	0.983	1.013	1.251	1.294
200	0.968	0.962	0.973	0.983	1.014	1.252	1.298

Table 6.14 α_c for box-section Q460 steel column after fire under natural cooling.

	Temperature/°C						
λ	**300**	**400**	**500**	**600**	**700**	**800**	**900**
40	1.015	1.032	1.062	1.095	1.103	1.123	1.122
60	1.033	1.047	1.092	1.127	1.149	1.223	1.236
80	0.997	1.004	1.025	1.047	1.076	1.252	1.283
100	0.978	0.978	0.993	1.008	1.038	1.255	1.294
120	0.98	0.977	0.990	1.003	1.034	1.266	1.307
140	0.99	0.991	1.003	1.016	1.047	1.291	1.332
160	0.993	0.996	1.011	1.024	1.056	1.305	1.348
180	0.994	0.998	1.016	1.028	1.061	1.314	1.357
200	0.995	0.998	1.017	1.030	1.063	1.318	1.361

Table 6.15 α_c for H-section Q460 steel column after fire under water cooling.

	Temperature/°C						
λ	**300**	**400**	**500**	**600**	**700**	**800**	**900**
40	1.015	1.040	1.083	1.131	1.148	1.165	1.187
60	1.021	1.053	1.105	1.135	1.140	1.276	1.248
80	0.989	1.026	1.067	1.059	1.054	1.342	1.292
100	0.966	1.002	1.033	1.015	1.007	1.372	1.298
120	0.955	0.991	1.018	0.998	0.989	1.383	1.301
140	0.949	0.986	1.012	0.990	0.982	1.390	1.304
160	0.949	0.986	1.011	0.990	0.982	1.399	1.312
180	0.947	0.984	1.008	0.987	0.979	1.401	1.313
200	0.946	0.983	1.008	0.986	0.982	1.404	1.315

Table 6.16 α_c for box-section Q460 steel column after fire under water cooling.

	Temperature/°C						
λ	300	400	500	600	700	800	900
40	1.002	1.030	1.064	1.096	1.102	1.119	1.123
60	1.022	1.063	1.104	1.129	1.134	1.249	1.241
80	0.980	1.019	1.053	1.050	1.048	1.338	1.296
100	0.955	0.994	1.022	1.011	1.006	1.374	1.310
120	0.949	0.990	1.019	1.004	0.999	1.408	1.324
140	0.960	0.994	1.032	1.017	1.010	1.440	1.350
160	0.993	1.003	1.043	1.026	1.018	1.459	1.366
180	0.983	0.999	1.042	1.018	1.022	1.470	1.376
200	0.982	0.983	1.043	1.030	1.022	1.476	1.379

6.5 Fire resistance of high strength Q460 steel beam

6.5.1 Temperature distribution across the section

In most design codes, the effect of nonuniform temperature distribution on fire response of steel beams was not considered due to complexity. In practical application, a steel beam is most likely to be exposed to fire from three sides as the top flange of the beam is contacted with the concrete floor slab, which results in the nonuniform temperature distribution of the cross section. To analyze temperature of the steel beam, the following assumptions are made: (1) temperature remains as a constant along the longitudinal axis of the steel beam, (2) thermal transfer between web and flange of the steel beam is not taken into account, and (3) thermal transfer between the top flange and the concrete slab is ignored, thus the upper surface of the top flange is considered in adiabatic condition.

The first aforementioned assumption is based on the fact that the difference of temperature along the longitudinal axis of the steel beam is not significant if the beam is relatively long. The heat transfer between flange and web of steel beam is very limited as the difference between their temperatures is negligible. From the first assumption, the temperature variation along the length of beam is neglected. According to the second and third assumptions, the steel beam can be divided into three components along the cross-sectional depth of the beam: top flange with three-side fire exposure, web, and bottom flange with four-side fire exposure. Therefore, the calculation of temperature in each component of the steel beam can be performed respectively based on the shape coefficient of the component according to the simplified method presented in Li et al. (2006).

A typical temperature distribution obtained from the simplified method by Li et al. (2006) for a steel beam with section H250 × 250 × 9 × 14 exposed to the ISO-834 standard fire is shown in Fig. 6.56. In practice, the more realistic parametric fire or nonuniform fires are often used. Attempts have been made to establish relationship between realistic and standard fires through use of equivalent time. The equivalent time for a realistic fire is defined as the time of exposing a structural element to the standard fire that would produce the same effect as that of the realistic fire. It can be noted that the temperature in top flange is much lower than the average temperature, and the temperature in web and bottom flange is considerably higher than average temperature. Furthermore, there is a short plateau on temperature profile when the temperature of web and bottom flange rise to 730°C. This can be attributed to the fact that the specific heat recommended in EC3 reaches a maximum of 5000 J/(kg·K) when the steel temperature is around 730°C. To obtain the temperature relationship between different components of the steel beam, the temperature distribution for a large number of steel beam sections with section height ranging from 250 to 500 mm and flange width between175–350 mm was computed first. Then the average temperature of top flange (T_{tf}), bottom flange (T_{bf}), and web (T_{wb}) was subsequently computed for different sections. Finally, the average temperature (T_s) of the steel beam exposed to standard fire is calculated with the assumption that the beam is exposed to the fire on four sides.

The temperature of different component in a steel beam can be taken as a function of average temperature (T_s) in the form of relative value, which is convenient to demonstrate the cross-sectional temperature of the steel beam. By employing linear least squares regression, Eqs. (6.12)–(6.14) are obtained to determine the relative temperature coefficients for different components of steel beams subjected to elevated temperature.

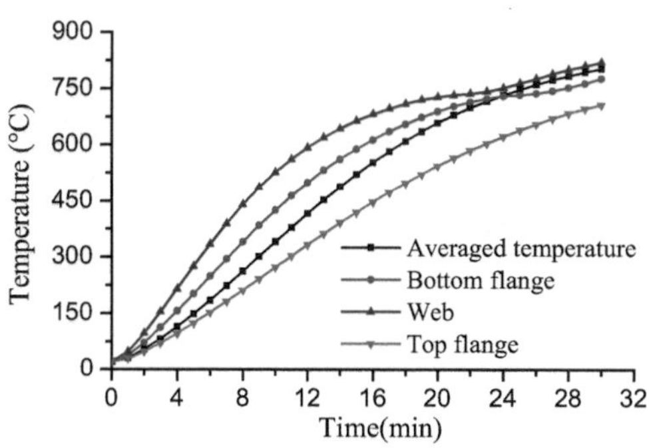

Figure 6.56 Typical temperature distribution in steel beam.

Bottom flange of H-shaped steel beam:

$$\eta_{bf} = \begin{cases} 3.125 \times 10^{-3} T_s + 0.9375 & 20^\circ\mathrm{C} \le T_s \le 100^\circ\mathrm{C} \\ -4.39 \times 10^{-4} T_s + 1.2939 & 100^\circ\mathrm{C} < T_s \le 800^\circ\mathrm{C} \end{cases} \tag{6.12}$$

Web of H-shaped steel beam:

$$\eta_{wb} = \begin{cases} 0.01 T_s + 0.8 & 20^\circ\mathrm{C} \le T_s \le 100^\circ\mathrm{C} \\ -1.36 \times 10^{-3} T_s + 1.936 & 100^\circ\mathrm{C} < T_s \le 500^\circ\mathrm{C} \\ -1.024 \times 10^{-3} T_s + 1.768 & 500^\circ\mathrm{C} < T_s \le 750^\circ\mathrm{C} \\ 1 & 750^\circ\mathrm{C} < T_s \le 800^\circ\mathrm{C} \end{cases} \tag{6.13}$$

Top flange of H-shaped steel beam:

$$\eta_{tf} = \begin{cases} -1.538 \times 10^{-3} T_s + 1.0307 & 20^\circ\mathrm{C} < T_s \le 150^\circ\mathrm{C} \\ 0.8 & 150^\circ\mathrm{C} < T_s \le 600^\circ\mathrm{C} \\ 4 \times 10^{-4} T_s + 0.56 & 600^\circ\mathrm{C} < T_s \le 800^\circ\mathrm{C} \end{cases} \tag{6.14}$$

where η_{bf}, η_{wb}, and η_{tf} are the relative temperature coefficients of bottom flange, web, and top flange, respectively. T_s is the average temperature of a steel beam if it exposed to fire on four sides.

To evaluate the reliability of the fitting equations (Eqs. 6.12–6.14), the average temperatures of different dimension of the components in steel sections were calculated and compared with the temperatures determined by Eqs. (6.12)–(6.14), which are shown in Fig. 6.57. The temperature on the horizontal axis is the average temperature of steel beam with four-side fire exposure. As can be seen from this figure, the maximum error of the fitting equations is 10.3% when the temperature ranges from 150–700°C. So Eqs. (6.12)–(6.14) can be used with an acceptable accuracy to represent

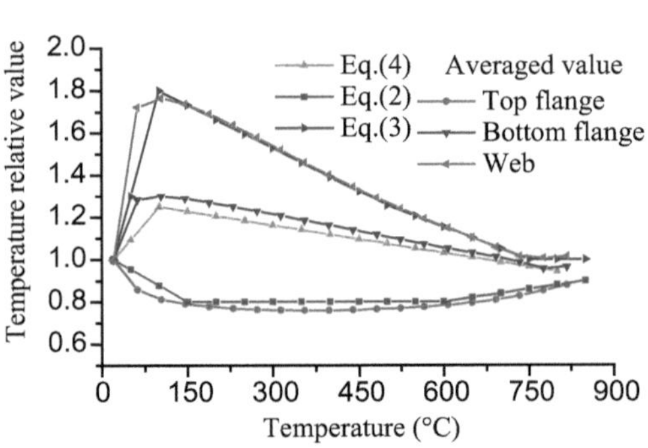

Figure 6.57 Fitting curves of relative temperature coefficients.

the temperature of each component. Based on the mechanical properties of Q460 steel at elevated temperatures, the temperature of Q460 steel ranges from 150 to 700°C when the magnitude of yield strength reduction factor of the steel is between 0.96 and 0.40. Therefore, the temperature in each component subjected to ISO-834 standard fire can be obtained by

$$T_{bf} = \eta_{bf} T_s \tag{6.15a}$$

$$T_{wb} = \eta_{wb} T_s \tag{6.15b}$$

$$T_{tf} = \eta_{tf} T_s \tag{6.15c}$$

where T_{bf} is the representative temperature of bottom flange; T_{wb} is the representative temperature of web; and T_{tf} is the representative temperature of top flange.

6.5.2 Bending bearing capacity

Flexural strength failure and overall buckling failure are two primary failure modes of steel beams. The evaluation of flexural strength of a steel beam at elevated temperature is simple, and this chapter focuses on the strength associated with the overall buckling failure, namely, lateral torsional buckling strength. According to the beam stability theory, there are several parameters that affect the critical moment resistance of the beam, including lateral flexural stiffness, torsional stiffness, and load type, etc. Furthermore, the flexural and torsional stiffness are only associated with the elastic modulus, shear modulus, and cross-sectional dimensions of the steel beam. Therefore, the procedure of evaluation of lateral torsional buckling strength for the conventional mild steel beam is also applicable to high strength Q460 steel beam. Hence, the critical moment capacity of high strength Q460 steel beam with a uniform temperature distribution can be obtained by substituting the elastic modulus at ambient temperature with that at elevated temperature.

When the temperature of the steel beam is nonuniformly distributed, the representative temperatures of top flange, bottom flange, and web of steel beam provided by Eq. (6.15) can provide a reasonable approximation for the cross-sectional stiffness of the beam by applying equivalent stiffness principle of the cross section.

Assuming T is the time-dependent temperature of web of the steel beam, the representative temperature of steel beam T_s, the top flange temperature T_1, and the bottom flange temperature T_2 can be calculated based on Eq. (6.15). By employing the elastic modulus reduction factor of high strength Q460 steel in Eq. (6.1), the elastic modulus of different component of a steel beam at the corresponding temperature can then be obtained as

$$E_{T1} = E_T\psi(T_1)/\psi(T) = f(T_1, T)E_T \tag{6.16a}$$

$$E_{T2} = E_T\psi(T_2)/\psi(T) = f(T_2, T)E_T \tag{6.16b}$$

where E_T is the elastic modulus of web at temperature T and E_{T1} and E_{T2} are the elastic modulus of top flange and bottom flange at temperature T_1 and T_2, respectively.

Using the equivalent stiffness principle, if the flexural stiffness of each flange with respect to their individual axis of symmetry is identical as shown in Fig. 6.58, the equivalent flexural stiffness of top and bottom flanges can be determined as follow:

Top flange of H-shaped steel beam:

$$\begin{aligned} E_{T1}I_1 &= f(T_1, T)E_T I_1 = f(T_1, T)E_T \cdot bt_f^3/12 = E_T \cdot [f(T_1, T)b]t_f^3/12 \\ &= E_T \cdot b_1 t_f^3/12 \end{aligned} \tag{6.17}$$

Bottom flange of H-shaped steel beam:

$$\begin{aligned} E_{T2}I_2 &= f(T_2, T)E_T I_2 = f(T_2, T)E_T \cdot bt_f^3/12 = E_T \cdot [f(T_2, T)b]t_f^3/12 \\ &= E_T \cdot b_2 t_f^3/12 \end{aligned} \tag{6.18}$$

where I_1 and I_2 are the moment of inertia of the top and bottom flange with respect to their horizontal axis, respectively and b_1 and b_2 are the equivalent width of the top flange and bottom flange resulting from the different elevated temperature in the flanges, respectively.

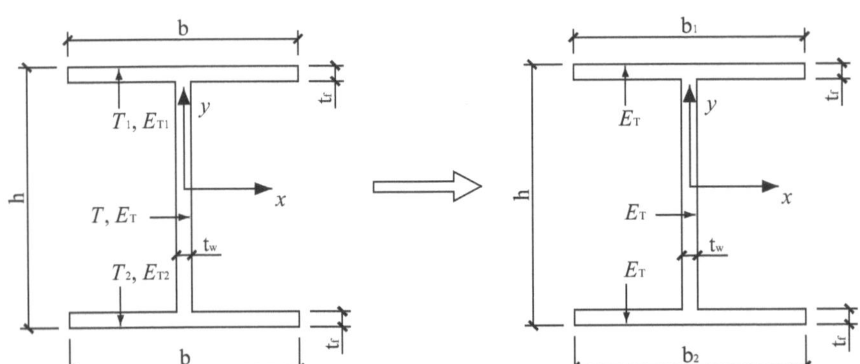

Figure 6.58 Dimensions of original and equivalent cross sections of the steel beam.

After the replacement of width of flanges by using equivalent stiffness principle, the cross section of the beam becomes single symmetric section from double symmetric section. To distinguish that, the moment inertial of the section is denoted with superscript "*eq*," and the critical moment capacity M_{crT}^{eq} of the steel beam subjected to a nonuniform temperature distribution can be expressed as

$$M_{crT}^{eq} = C_1 \frac{\pi^2 E_T I_y^{eq}}{l^2}\left[C_2 a^{eq} + C_3 \beta^{eq} + \sqrt{(C_2 a^{eq} + C_3 \beta^{eq})^2 + \frac{I_\omega^{eq}}{I_y^{eq}}\left(1 + \frac{G_T I_t^{eq} l^2}{\pi^2 E_T I_\omega^{eq}}\right)}\right] \tag{6.19}$$

where, C_1, C_2, and C_3 are the parameters related to load type (uniform load or concentrated load), respectively; G_T is the shear modulus of steel at temperature T; E_T is the elastic modulus of steel at temperature T; I_y^{eq} is the equivalent weak axis inertia moment of the section; I_ω^{eq} is the equivalent warping moment of inertia of the steel beam; I_t^{eq} is the equivalent torsional moment of inertia of the steel beam; a^{eq} is the eccentricity of the applied load from the sectional shearing force center; and β^{eq} is the equivalent factor representing the degree of sectional asymmetry and is defined as follows:

$$\beta^{eq} = \frac{1}{2I_x^{eq}} \int_A y(x^2 + y^2)dA - y_0 \tag{6.20}$$

The critical bending capacity M_{crT}^{eq} of the steel beam subjected nonuniform elevated temperature can also be written as

$$M_{crT}^{eq} = \varphi_{bT}^{eq} W^{eq} f_{yT} \tag{6.21}$$

where W^{eq} is the equivalent gross section modulus of the steel beam.

Therefore, the overall stability coefficient of a steel beam with a nonuniform temperature distribution can be evaluated by the following equation:

$$\varphi_{bT}^{eq} = C_1 \frac{\pi^2 E_T I_y^{eq}}{l^2 W^{eq} f_{yT}}\left[C_2 a^{eq} + C_3 \beta^{eq} + \sqrt{(C_2 a^{eq} + C_3 \beta^{eq})^2 + \frac{I_\omega^{eq}}{I_y^{eq}}\left(1 + \frac{G_T I_t^{eq} l^2}{\pi^2 E_T I_\omega^{eq}}\right)}\right] \tag{6.22}$$

To obtain a simple expression for the overall stability coefficient (φ_{bT}^{eq}) of a steel beam subjected to pure bending, the integral in Eq. (6.20) can be neglected compared with y_0 for commonly used cross sections, and then the factor β^{eq} is approximated as

$$\beta^{eq} \approx -y_0 \approx \frac{h}{2} \cdot \frac{I_1^{eq} - I_2^{eq}}{I_y^{eq}} = \frac{h}{2}\left(2\alpha_b^{eq} - 1\right) = 0.5\eta_b^{eq} h \tag{6.23}$$

where I_1^{eq} and I_2^{eq} are the inertia moment of top flange and bottom flange around y axis of the steel beam, respectively, in which the factors α_b^{eq} and η_b^{eq} can be computed by

$$\alpha_b^{eq} = \frac{I_1^{eq}}{I_1^{eq} + I_2^{eq}} \approx \frac{I_1^{eq}}{I_y^{eq}} \tag{6.24}$$

$$\eta_b^{eq} = 2\alpha_b^{eq} - 1 = \frac{I_1^{eq} - I_2^{eq}}{I_y^{eq}} \tag{6.25}$$

Practically, the temperature of top flange is lower than that of bottom flange and web of the steel beam because the top flange is partially protected by concrete slab. Based on the numerical analysis, the factor β^{eq} is approximated as

$$\beta^{eq} \approx 0.4\eta_b^{eq} h \tag{6.26}$$

Then the factor η_b^{eq} can be rewritten as

$$\eta_b^{eq} = 0.8\left(2\alpha_b^{eq} - 1\right) \tag{6.27}$$

The equivalent torsional moment of inertia of a steel beam can be simplified as

$$I_t^{eq} = \frac{1.25}{3}\left(b_{f1}t_f^3 + b_{f2}t_f^3 + h_0 t_{weq}^3\right) \approx \frac{1}{3}\left(b_{f1}t_f + b_{f2}t_f + h_0 t_{weq}\right)t_f^2 = \frac{1}{3}At_f^2 \tag{6.28}$$

where A is the sectional area of the steel beam.

The equivalent warping moment of inertia of the steel beam can be evaluated by

$$I_\omega = \frac{I_1 I_2}{I_y}h^2 = \alpha_b(1 - \alpha_b)I_y h^2 \tag{6.29}$$

By substituting Eq. (6.23)–(6.29) into Eq. (6.22), the following relationship is obtained for the overall stability coefficient $\varphi_{bT,pb}^{eq}$ of a steel beam with a nonuniform temperature distribution under pure bending:

$$\varphi_{bT,pb}^{eq} = \frac{\pi^2 E_T}{2f_{yT}} \cdot \frac{A^{eq}h}{\lambda_y^{eq^2} W^{eq}}\left[\sqrt{1 + \left(\frac{\lambda_y^{eq} t_f}{4.4h}\right)^2} + \eta_b^{eq}\right] \tag{6.30}$$

where λ_y^{eq} is the equivalent slenderness ratio about y axial of a steel beam and W^{eq} is the equivalent section modulus of a steel beam.

The above expressions can be utilized to analysis the overall stability of a steel beam with a nonuniform temperature distribution subjected to pure bending. In addition, the overall stability coefficient of a steel beam bearing any type transverse load at elevated temperature can be evaluated as per Eq. (6.22).

As the effect of temperature on the Poisson's ratio of steel can be neglected, the following equations are obtained:

$$G_T/E_T = G/E \tag{6.31}$$

$$\lambda_y = l/i_y \tag{6.32}$$

where G is the shear modulus of steel; E is the elasticity modulus of steel at ambient temperature; λ_y and i_y are the slenderness ratio and gyration radius with regard to the weak axis of the section, respectively. Therefore, the factor β_b^{eq} is obtained as the ratio of Eq. (6.22) and Eq. (6.30) as

$$\beta_b^{eq} = \frac{2C_1}{h} \cdot \frac{C_2 a^{eq} + C_3 \beta^{eq} + \sqrt{\left(C_2 a^{eq} + C_3 \beta^{eq}\right)^2 + \frac{I_\omega^{eq}}{I_y^{eq}}\left(1 + \frac{G I_t^{eq} l^2}{\pi^2 E I_\omega^{eq}}\right)}}{\sqrt{1 + \left(\frac{\lambda_y^{eq} t_f}{4.4h}\right)^2} + \eta_b^{eq}} \tag{6.33}$$

where β_b^{eq} is the equivalent critical moment factor.

Therefore, the overall stability coefficient of a steel beam with a nonuniform temperature distribution subjected to any load can be written as

$$\varphi_{bT}^{eq} = \beta_b^{eq} \frac{\pi^2 E_T}{2 f_{yT}} \cdot \frac{A^{eq} h}{\lambda_y^{eq^2} W^{eq}} \left[\sqrt{1 + \left(\frac{\lambda_y^{eq} t_f}{4.4h}\right)^2} + \eta_b^{eq}\right] \tag{6.34}$$

It should be noted that the overall stability coefficient above is only applicable for steel beam in elastic stage. The overall stability critical stress decreases apparently as the effect of plasticity on the overall stability critical stress is pronounced when the steel beam works in an elastic–plastic stage. A modified method is proposed in *Code for Design of Steel Structures* (GB50017-2003) for evaluating the overall stability coefficient at elevated temperature as

$$\varphi'_{bT} = \begin{cases} \varphi_{bT} & \varphi_{bT} \le 0.6 \\ 1.07 - 0.282/\varphi_{bT} \le 1.0 & \varphi_{bT} > 0.6 \end{cases} \tag{6.35}$$

Then overall stability coefficient of the steel beam with nonuniform temperature distribution can be computed by substituting φ_{bT}^{eq} into Eq. (6.35).

$$\varphi_{bT}^{eq'} = \begin{cases} \varphi_{bT}^{eq} & \varphi_{bT}^{eq} \leq 0.6 \\ 1.07 - 0.282/\varphi_{bT}^{eq} \leq 1.0 & \varphi_{bT}^{eq} > 0.6 \end{cases} \tag{6.36}$$

Thus, the criterion of overall stability of the steel beam with nonuniform temperature distribution can be written as

$$M/\varphi_{bT}^{eq'} W^{eq} \leq \eta_T \gamma_R f \tag{6.37}$$

where γ_R is the resistance partial coefficients and can be adopted as 1.1 (CECS200:2006) and η_T is the strength reduction factor of steel at elevated temperature.

Based on the aforementioned approach, the strength and elastic modulus reduction factors of high strength Q460 steel are recommended for the analysis of the overall stability of high strength Q460 steel beam subjected to a nonuniform elevated temperature distribution. In addition, the overall stability coefficient can be evaluated by substituting Eq. (6.1) into Eq. (6.34) and thus Eqs. (6.36) and (6.37) can be used to evaluate overall stability of a high strength Q460 steel beam subjected to a nonuniform elevated temperature distribution.

6.5.3 Critical temperature

As the flexural capacity of the steel beam decreases as the increase of temperature, the temperature in steel reaches a certain value defined as critical temperature when the flexural capacity reaches moment induced by applied load on the steel beam. Load response ratio R is determined as the ratio of maximum bending moment in the steel beam to the bending capacity of the beam, and the expression is

$$R = M/\left(\varphi_b' W f\right) \tag{6.38}$$

where R is the load response ratio of steel beam; M is the maximum bending moment induced by applied load of the steel beam; φ_b' is the modified overall stability coefficient of steel beam in elastic stage at room temperature; W and f are the cross section modulus and design strength of steel beam, respectively.

Alternatively, Eq. (6.38) can be rewritten as

$$M/(Wf) = \varphi_b' R \tag{6.39}$$

The criterion of overall stability of the steel beam at elevated temperature is written as

$$M/\left(\varphi'_{bT}W\right) \leq \eta_T\gamma_R f \tag{6.40}$$

where φ'_{bT} is the modified overall stability coefficient of a steel beam at temperature T.

Then the load response ratio of a steel beam can be calculated by substituting Eq. (6.39) into Eq. (6.40):

$$R=\varphi'_{bT}/\varphi'_b\eta_T\gamma_R \tag{6.41}$$

Based on the aforementioned equation, the critical temperature of a high strength Q460 steel beam can be derived in accordance with the overall stability coefficient and load response ratio. The relationship between the critical temperature the load response ratio obtained from the proposed method for the case of uniform temperature distribution is plotted in Fig. 6.59A for different modified overall stability coefficients. As can be seen from the figure, the critical temperature decreases linearly with the increase of load ratio when the load response ratio is greater than 0.6. In addition, the overall stability coefficient has considerable influence on the critical temperature at a given load ratio. The lower the overall stability coefficient is, the higher the critical temperature is.

By substituting Eq. (6.38) into Eq. (6.37), the load response ratio of a steel beam with a nonuniform temperature distribution under fire condition can be obtained as

$$R=\eta_T\gamma_R\varphi_{bT}^{eq'}W^{eq}/\left(\varphi'_b W\right) \tag{6.42}$$

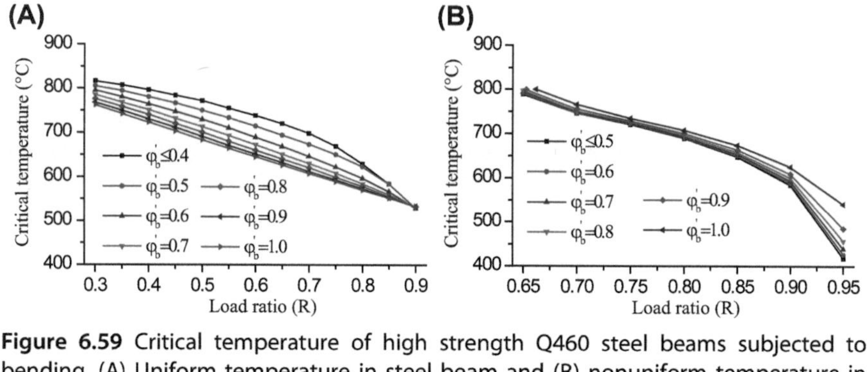

Figure 6.59 Critical temperature of high strength Q460 steel beams subjected to bending. (A) Uniform temperature in steel beam and (B) nonuniform temperature in steel beam.

Thus, the ratio of section modulus of the steel beam at elevated and ambient temperature (W^{eq}/W) almost kept constant at temperatures below 780°C, and the ratio ranges from 1.0 to 1.08 for different sections. Therefore, the value W^{eq}/W can be adopted as 1.0 for the reason of simplicity. Therefore, the load response ratio of a steel beam with a nonuniform temperature distribution under fire condition is approximated as

$$R = \eta_T \gamma_R \varphi_{bT}^{eq'} / \varphi' \tag{6.43}$$

The critical temperature of high strength Q460 steel beam T_d^{eq} can be obtained based on the overall stability coefficient and load response ratio of the steel beam subjected to the nonuniform temperature distribution under fire condition as that shown in Fig. 6.59B. It can be seen from the figure that the critical temperature of high strength Q460 steel beam decreases gradually as the increase of the temperature when the load response ratio is less than 0.8. The decrease of the critical temperature becomes more rapid when the ratio is greater than 0.8. On the other hand, the critical temperature increases gradually as the increase of the magnitude of the stability coefficient. The influence of the overall stability coefficient on the critical temperature becomes substantial when the load response ratio exceeds 0.9. For example, the difference of the critical temperature between the stability coefficients being 0.5 and 1.0 reaches to 120°C for the load response ratio of 0.95.

6.5.4 Validation by finite element analysis

The proposed equivalent stiffness method (ESM) is verified by comparing its predictions with the results from rigorous 3D finite element analysis carried out using the ANSYS. The comparison covers a wide range of steel beams with variation of different parameters, such as beam span length, applied load magnitude, and temperature distribution mode. The 3D finite element model is capable of tracing thermal and structural response of fire exposed high strength Q460 steel beams from preloading stage till failure of the beam. In thermal analysis, the results from the finite element analysis are validated by test data. In structural analysis, the critical bending moment of high strength Q460 steel beams are compared with the results predicted by the ESM.

6.5.4.1 Thermal analysis

A model of test specimen, made of high strength Q460 steel with a section of H400 × 200 × 8 × 13, is subjected to ISO-834 standard fire. SOLID70 element is selected to simulate steady and transient heat transfer between the steel beam and fire source. SOLID70 is an eight-node cubic thermal element with conduction capability, and it can be meshed into tetrahedron or prisms. The thermal analysis was carried out by applying the ISO-834 standard time—temperature curve in the finite element thermal model. The initial temperature is assumed as 20°C, and the relevant thermal boundary conditions of steel beam are established based on the realistic conditions, 4-side fire exposure or 3-side fire exposure. As there is a lack of temperature-dependent thermal properties of high strength Q460 steel, temperature-dependent thermal properties of steel as specified in EC3 (2005) are adopted in the thermal analysis for specific heat, thermal conductivity, thermal radiation coefficient, thermal convection coefficient, density, and so on.

Test result of a mild Q235B steel (nominal yields strength is 235 MPa) beam exposed to fire in three sides, subjected to heating and cooling (Li and Guo, 2008), was selected to validate the established thermal model due to lack of similar test result for high strength Q460 steel. The test specimen was exposed to fire with three sides by protecting top flange with fire insulation. From the layout of the fire test, the temperatures of different component in the beam were measured. The thermal analysis was performed by using recorded furnace temperature. The comparison of temperature in different components, namely, top flange, web, and bottom flange, between analysis and test are shown in Fig. 6.60. It can be seen that the measured temperature is in good agreement with the finite element predictions.

6.5.4.2 Structural analysis

Flexural-torsional buckling is one of the critical failure modes of an unbraced steel beam at ambient temperature. SHELL181 element is utilized to evaluate overall structural stability capacity of high strength Q460 steel

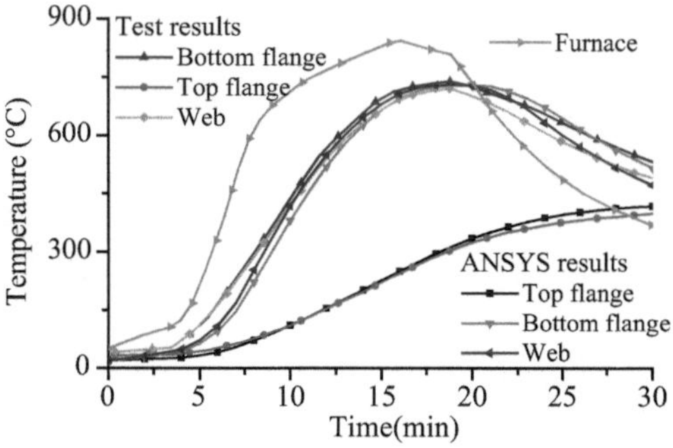

Figure 6.60 Comparison of test results with that of the finite element analysis.

beam 3D finite element analysis. SHELL181, which is used to model the steel beam, has four nodes with six degrees of freedom (three translations and three rotations) per node. This element can capture local buckling of flanges and web and also flexural—torsional buckling of the steel beam and is well-suited for large rotation, large strain, and nonlinear problems. The material properties of steel strength and elastic modulus under fire condition utilized in structural analysis are selected from test results in Chapter 3. A bilinear elastic—plastic model for stress—strain relationship of steel was adopted in the structural analysis for the reason of simplicity. The element mesh of the developed model used in structural analysis is shown in Fig. 6.61A. However, local bucking and distortional buckling may arise in flange and web of the steel beam before the occurring of overall lateral torsional buckling. To avoid the local bucking and distortional buckling, stiffeners are incorporated in the finite element model. Five to seven stiffeners are arranged in the longitudinal direction of the steel beam depending on the span length. The stiffeners are connected with web of steel beam and coupled displacement with flange in the in-plane of the stiffeners. In such way, the possible local bucking and distortional buckling in flange and web are prevented, and there is no effect on the torsional and flexural stiffness of the steel beam. The ends of the steel beam are restrained as clamped condition in which restrains in-plane displacements of all nodes. The stiffeners and constraints of the member model are shown in Fig. 6.61B.

There are two stages for simulating the fire response of the steel beam in finite element analysis. In the first stage of static analysis, the specimen is preloaded in ambient temperature of 20°C. Then, temperatures obtained from the thermal analysis are imposed on nodes of the shell elements. Finally, transverse load in the form of either concentrated or uniform load is applied on the steel beam. In the second stage of structural eigenvalue

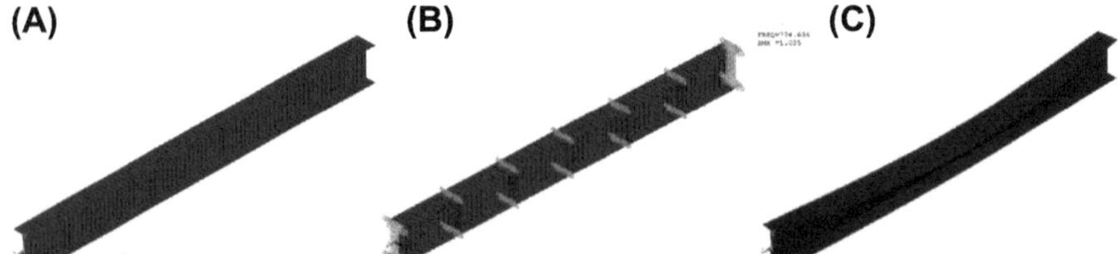

Figure 6.61 Finite element model of steel beam. (A) Meshing of beam, (B) boundary condition and stiffeners, and (C) buckling mode.

buckling analysis, the bucking model and eigenvalue extraction method are set up, and the finite element models are solved. The flexural–torsional buckling mode of steel beam is shown in Fig. 6.61C. The critical bending moment can be obtained by multiplying eigenvalue and applied load. With the increase of temperature, the critical bending moment decrease due to the deterioration of mechanical properties, and the critical temperature of steel beam can be defined as the temperature at which the critical bending moment reaches maximum bending moment of the beam.

To verify the proposed method in Section 6.5.2, several high strength Q460 steel beams of section H400 × 200 × 8 × 13 are analyzed with different span lengths (4, 6, and 8 m) and different temperature distribution patterns. The beams with 4-side and 3-side fire exposure are analyzed to simulate the scenarios of uniform temperature distribution and nonuniform temperature distribution. As there is a lack of fire test data on high strength Q460 steel beams under fire, predictions from the proposed approach (EMS) are compared with results from finite element analysis (ANSYS) in Fig. 6.62. The temperature in vertical axis is average temperature of the

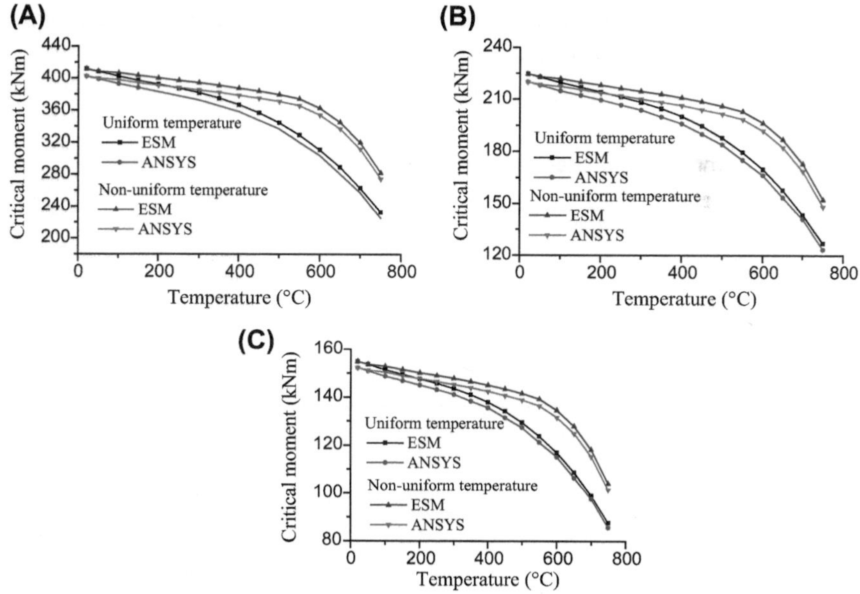

Figure 6.62 Results comparison of equivalent stiffness method and finite element analysis. (A) 4 m span, (B) 6 m span, and (C) 8 m span.

steel beam with 4-side fire exposure or the web temperature of the steel beam with 3-side fire exposure. It can be seen that the calculated critical bending moment by the ESM is in good agreement with the predicted results by the finite element modeling. Therefore, the presented ESM appears to provide a good prediction of the behavior of high strength Q460 steel beams subjected to uniform and nonuniform elevated temperature distributions.

To evaluate the accuracy in quantity for critical bending moment obtained from the ESM, the results are also compared to those obtained from finite element analysis for Q460 steel beams under different web temperatures (300 and 600°C) and beam span lengths (4, 6, and 8 m) as shown in Table 6.17. It can be seen from the table that the critical bending moment obtained from the ESM compares well with those obtained from finite element analysis with only approximate 3% difference.

Critical temperature obtained from the ESM is also compared with those obtained from the finite element analysis for beams under different load ratios (0.7 and 0.9) and beam span lengths (4, 6, and 8 m) with nonuniform temperature distribution pattern in Table 6.18. It can be seen that the critical temperature obtained from the ESM is also in good agreement with those obtained from the finite element predictions with maximum difference being only 6%.

Table 6.17 Comparison of lateral torsional buckling resistance at different temperatures.

	T = 300°C			T = 600°C		
Span	**ESM**	**ANSYS**	**Errors**	**ESM**	**ANSYS**	**Errors**
4 m	394 kN m	385 kN m	2.24%	362 kN m	354 kN m	2.28%
6 m	215 kN m	210 kN m	2.06%	196 kN m	192 kN m	2.24%
8 m	148 kN m	145 kN m	1.69%	135 kN m	132 kN m	2.12%

Table 6.18 Comparison of critical temperature under different load ratios.

	$R = 0.7$			$R = 0.9$		
Span	**ESM**	**ANSYS**	**Errors**	**ESM**	**ANSYS**	**Errors**
4 m	750°C	744°C	0.80%	601°C	575°C	4.33%
6 m	747°C	732°C	2.01%	589°C	557°C	5.43%
8 m	742°C	728°C	1.89%	575°C	542°C	5.74%

6.5.5 Parametric study

The validated ESM was utilized to investigate the effects of steel grade and temperature distribution pattern on the critical bending moment and critical temperature of steel beams subjected to fire. The beam section is H400 × 200 × 8 × 13, which is exposed to fire with three sides as the upper surface of top flange is insulated with concrete slab. The sectional area of the beam is 85.76 cm^2, the moment of inertia about major axis is 24,300 cm^4, and the corresponding section modulus is 1215 cm^3. Both concentrated load and uniformed distributed load are considered. All specimens are exposed to the IOS-834 standard fire scenario. For the case of convenient comparison, the temperature evaluated in the following investigation is denoted by the average temperature and web temperature of the steel beam for the uniform temperature distribution and the nonuniform temperature distribution, respectively.

6.5.5.1 Steel grade

To investigate the effect of steel grade on critical bending moment of steel beams subjected to the nonuniform temperature distribution, both high strength Q460 steel and mild Q235 steel beams with different span lengths (4, 6, and 8 m) and subjected to uniform distributed load and concentrated load are investigated using the above developed numerical analysis procedure. Shown in Fig. 6.63 is the degradation of critical bending moment of steel beams with the increase of temperature for different span lengths. It is clearly seen that the critical bending moment of high strength Q460 steel beam is greater than that of mild Q235 steel beam. It is also found that the influence of steel strength on the critical moment of the beam is not significant before the temperature reaching 400°C as the decrease of the

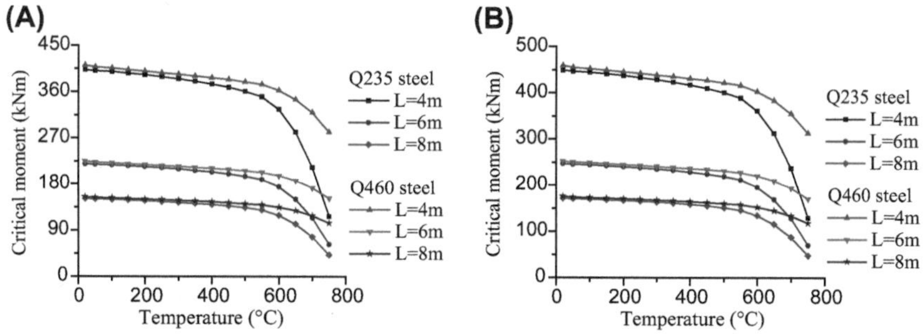

Figure 6.63 Critical bending moment of high strength Q460 steel and mild Q235 steel beams. (A) Subjected to distributed load and (B) subjected to concentrated load.

critical bending moment is primarily resulted from the degradation of mechanical properties of steel. When the temperature exceeds 400°C, the critical bending moment difference between high strength steel beam and mild steel beam becomes larger, and this can be attributed to the high-temperature mechanical properties of Q460 steel, especially the elastic modulus, degrading slowly than that of mild Q235 steel. It is evident from the figure that the critical bending moments of both steel beams decrease as the increase of span length. This can be attributed to the fact that lateral—torsional buckling can more easily occur for a longer beam. Also shown in the figure is that the lateral torsional buckling resistance of the steel beams in elevated temperature subjected to concentrated load is greater than that of steel beams under uniformed distributed load, which is consistent with that of in ambient temperature.

Based on the ESM, the obtained critical temperatures of Q460 steel beam are compared to those of Q235 steel beam with different overall stability coefficients and the comparisons are shown in Fig. 6.64. It is shown that the critical temperature decreases as the increase of the load response ratio. Also shown in the figure is that the influence of overall stability coefficient on the critical temperature is not significant. In addition, with identical load response ratio and overall stability coefficient, the critical temperature of Q460 steel beam is considerably higher than that of Q235 steel. Overall, there is a significant difference in fire resistance between high strength steel beam and mild steel beam.

6.5.5.2 Temperature distribution pattern

The effects of temperature distributions, namely, uniform temperature distribution and nonuniform temperature distribution, on critical moment

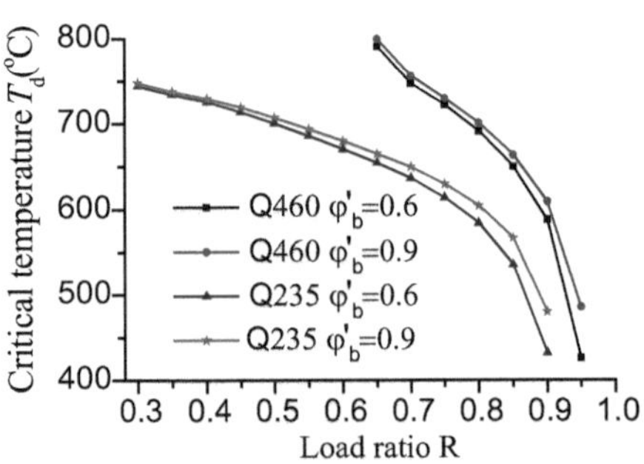

Figure 6.64 Critical temperature of Q235 and Q460 steel beam with nonuniform temperature distribution.

and critical temperature are studied herein by analyzing steel beams subjected to concentrated load and distributed load with different span lengths. Lateral torsional buckling resistances of high strength Q460 steel beams with different temperature distributions are plotted in Fig. 6.65. It can be seen from Fig. 6.65 that lateral torsional buckling resistances of high strength Q460 steel beam under uniform load and concentrated load decrease as the rise of steel temperature. This can be attributed to the degradation of mechanical properties of steel in the elevated temperature. It is also shown in Fig. 6.65 that lateral torsional buckling resistance of high strength Q460 steel beam with nonuniform temperature distribution is greater than that of high strength Q460 steel beam with uniform temperature distribution for the same web temperature. In addition, the deviation is more apparent with the rise of web temperature, and the maximum difference is obtained at the web temperature reaches 650°C, then the deviation starts to reduce with the further increase of web temperature. This can be attributed to the fact that the variation of thermal gradient in the steel beam under fire condition will result in slight difference in the lateral torsional buckling resistance between beams with uniform and nonuniform temperature distribution when the effect of thermal gradient is not significant at a lower temperature. However, the thermal gradient greatly influences lateral torsional buckling resistance of high strength Q460 steel beam with nonuniform temperature distribution with the gradual rise of temperature. Thus, the difference of lateral torsional buckling resistance decreases with the degradation of mechanical properties of steel under high temperature, even though that the effect of thermal gradient is still significant.

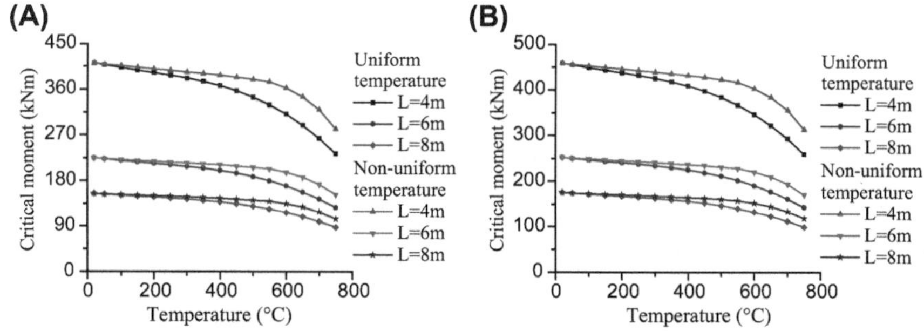

Figure 6.65 Critical bending moment of high strength Q460 steel beam under different temperature distribution patterns. (A) Under distributed load and (B) under concentrated load.

To study the effect of temperature distribution on critical temperature of high strength Q460 steel beam, the critical temperatures of Q460 steel beam with different temperature distributions associated with overall stability coefficients being 0.6 and 0.9 are evaluated by the ESM and shown in Fig. 6.66. It is evident from the figure that the critical temperature of high strength Q460 steel beam subjected to both uniform and nonuniform temperature distribution decreases as the increase of load response ratio with the fact that the decrease of the critical temperature in the latter case is more significant. It should be noted that, for the same overall stability coefficient, the critical temperature of high strength Q460 steel beam with the nonuniform temperature distribution is higher than that with the uniform temperature distribution. Taking the thermal gradient into account, the difference can be resulted from the elastic modulus of top flange and bottom flange of high strength Q460 steel beam with the nonuniform temperature distribution is larger than that of the beams with the uniform temperature distribution at the same web temperature. At the same load ratio, there is a decrease in critical temperature of high strength Q460 steel beam with uniform temperature distribution with the increase of overall stability coefficient. However, the increase of stability coefficient causes a slight increase in the critical temperature of high strength Q460 steel beam subjected to the nonuniform temperature distribution.

6.5.6 Simplified design approach

Compared with the proposed ESM, the finite element analysis is laborious and requires large amount of calculations. For evaluating fire resistance of high strength Q460 steel beams subjected to the nonuniform temperature

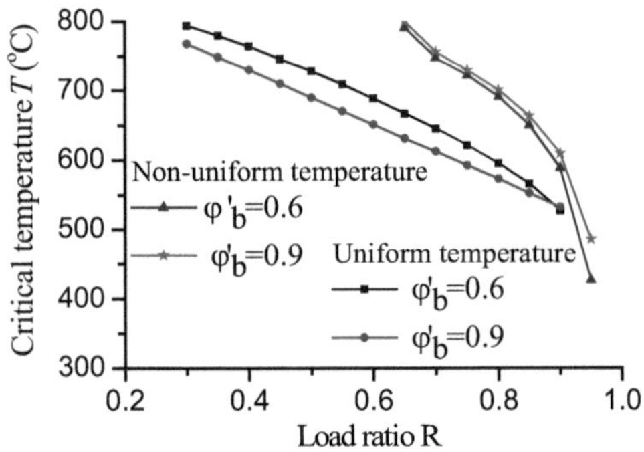

Figure 6.66 Critical temperature of high strength Q460 steel beam with different temperature distribution pattern.

distribution in practice, a simplified design approach (SDA) for evaluating the critical temperature and the overall stability coefficient for high strength Q460 steel beams exposed to fire are presented. The simplified approach is derived by utilizing numerical surface fitting to the results obtained from ESM for high strength Q460 steel beams with a large range of sectional dimensions (flange widths within the range of 175–200 mm and the heights within the range of 350–500 mm).

6.5.6.1 Overall stability coefficient

As the aforementioned analysis, the overall stability coefficient of high strength Q460 steel beam is influenced by the beam span length and web temperature. Then overall stability coefficients of Q460 steel beams with different cross sections and different load types (concentrate load and uniform load) are evaluated by the ESM and subsequently simplified based on numerical surface fitting. Eq. (6.44) presents the proposed simplified equation for the stability coefficient of high strength Q460 steel beams subjected to nonuniform temperature distribution. The validity of the approach is demonstrated by comparing predicted results with the results obtained by the ESM as shown in Fig. 6.67.

$$\varphi_{bT}^{eq'} = 1.896 - 0.333L - 0.00152T_g + 0.0189L^2 + 2.7 \times 10^{-6}T_g^2 - 3.434 \times 10^{-5}T_gL \leq 1.0 \tag{6.44}$$

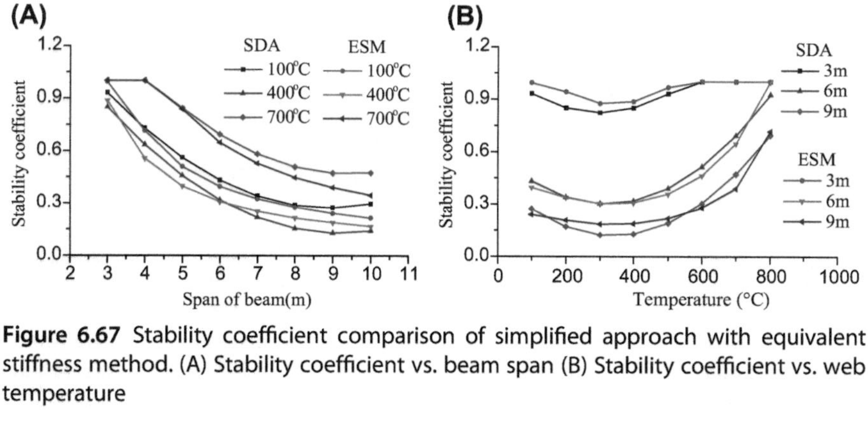

Figure 6.67 Stability coefficient comparison of simplified approach with equivalent stiffness method. (A) Stability coefficient vs. beam span (B) Stability coefficient vs. web temperature

where L and T_g are the span length and web temperature of the steel beam, respectively.

The result comparison shown in Fig. 6.67 demonstrates that the overall stability coefficients predicted by Eq. (6.44) are in good agreement with that of the ESM. Therefore, the proposed simplified equation is appropriate for evaluating the overall stability coefficients of high strength Q460 steel beam with flange widths of 175–200 mm and heights of 350–500 mm.

6.5.6.2 Critical temperature

In consideration of the load type and cross section properties, the critical temperature of high strength Q460 steel beam with different span lengths under different load ratios are evaluated with the ESM, and the obtained results are used for surface fitting. Consequently, the following equation for critical temperature of high strength Q460 steel beams subjected to the nonuniform temperature distribution is proposed.

$$T_d = 790.615 - 0.30406 \times 10^{13}\left[1 + erf\left\{\frac{R - 5.330}{1.018}\right\}\right] \times \left[1 + erf\left\{\frac{L - 102.217}{81.889}\right\}\right] \quad (6.45)$$

where T_d is critical temperature; R is the load ratio; and L is the span of Q460 steel beam.

Shown in Fig. 6.68 is the comparison the critical temperatures calculated from the ESM and that of the simplified design equation (SDA). Both the results are in good agreement with an average difference of 3.59%. It is

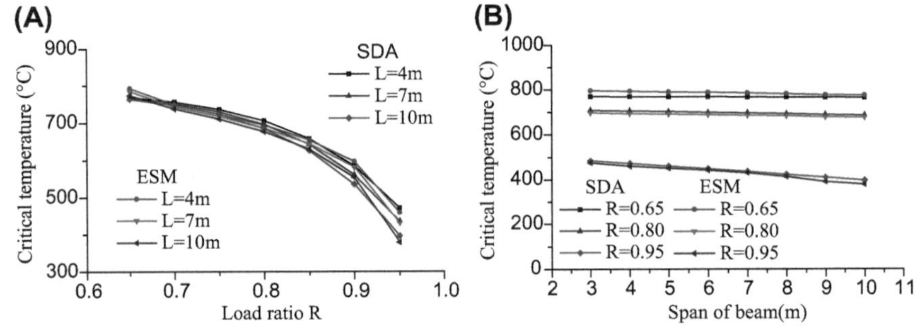

Figure 6.68 Critical temperature comparison of simplified approach with equivalent stiffness method. (A) Critical temperature versus load ratio and (B) critical temperature versus beam span.

concluded that the simplified design equation for calculating the critical temperature is feasible and accurate for high strength Q460 steel beam with flange widths of 175–350 mm and heights of 250–500 mm.

6.6 Effect of creep on fire resistance of high strength Q460 steel beams

To investigate the effect of creep on fire response of restrained Q460 steel beams, the finite element software ANSYS is adopted, considering thermal gradients and high nonlinear structural problems.

6.6.1 Finite element model

A typical Q460 steel beam with axial and rotational restraint (shown in Fig. 6.69) is established in ANSYS. A uniform distributed load is applied on the beam. SHELL181 element, used to model the steel beam, has four nodes with six degrees of freedom (three translations and three rotations) per node. This element is applicable for large strain and can account for creep effect at elevated temperatures. COMBIN14 element was chose to simulate the rotational and axial restraint at two ends of the beam. The NLGEOM is set on to deal with the geometric nonlinear analysis because large displacement occurs in restrained steel beam at high temperature. To simplify the analysis process, the temperatures in steel are calculated by increment method and the temperatures are applied as a special load on column.

Initial imperfections were incorporated into the beam for the buckling analysis. Modal analysis at ambient temperature was made to gain the first buckling mode, which was scaled to that maximum displacement equals to 1‰ beam length, and the scaled deformation was used to update the geometry of the beam. Initial residual stresses were also taken into account, and the maximum value of the initial residual stresses was adopted to be 30% of steel yield stress at room temperature (Kodur et al., 2010). The values of initial residual stresses were assigned at relevant element integration points.

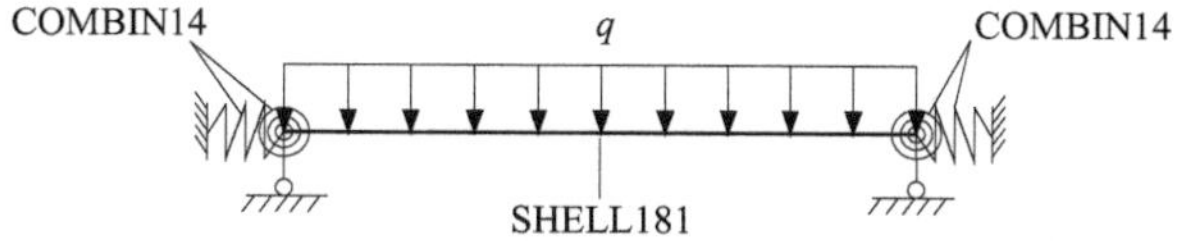

Figure 6.69 Finite element model of restrained steel beam

6.6.2 Model validation

Data from fire tests on restrained steel beam conducted by Li and Guo (2008) are choose for verifying the validation of the established model. The fire tests were performed on two restrained steel beams of H250 × 250 × 8 × 12 section with length of 4.5 m, and two-point loading of 130 kN was applied on both beams. The axial and rotational stiffness of the beams are 39.54 kN/mm and 1.09×10^8Nm/rad, respectively. The beams are made of Q235 steel (measured average yield strength is 291 MPa). However, the coefficients of Norton creep model derived from test data (Wang et al., 2017) are based on Q460 steel (measured average yield strength is 492 MPa). Study by Luecke et al. (2005) indicated that stress can be modified when calculating creep strain to accommodate the creep properties of restrained beams and the coefficients for stress modification are ratios between yield strength in creep model and yield strength in calculation. After this modification, creep strain decreases with higher yield strength under similar stress level. The expression of creep model for Q235 steel can be written as

$$\dot{\varepsilon}_{cr} = d_1 \left(\sigma \frac{492}{291} \right)^{d_2} e^{-d_3/T} = d_1 \left(\frac{492}{291} \right)^{d_2} \sigma^{d_2} e^{-d_3/T} \tag{6.46}$$

In the analysis, elastic modulus is 2.06×10^5 MPa and Poisson's ratio is adopted as 0.3 at room temperature. The value of thermal expansion coefficient is 1.4×10^{-5}. Different with web and bottom flange, top flange can be seen as 3-sides fire exposure due to the protection of concrete slab connected top flange of beam. Because of large dispersion in measured temperatures, the average temperatures of two components were employed for simplicity: one is average value of measured temperatures of web and bottom flange and the other is average measured temperature of top flange.

In Fig. 6.70, predictions obtained from ANSYS considering creep effects are compared with test data. Results shown in Fig. 6.70 indicate that the creep has significant influence on the fire response of beams, especially after 15 min fire exposure. Overall, predictions from the model reach a good agreement with test data when high-temperature creep is incorporated in the analysis. When ignoring the creep effect, the calculated deflections and thermal induced axial forces are smaller in the later stage of heating phase. The predicted deformation of test specimen by the finite element model is similar to photo taken after test, as shown in Fig. 6.71. It can be seen that large deformation occurs in restrained steel beam (maximum approximately is $l/16$), and local buckling happens at the ends of bottom flange.

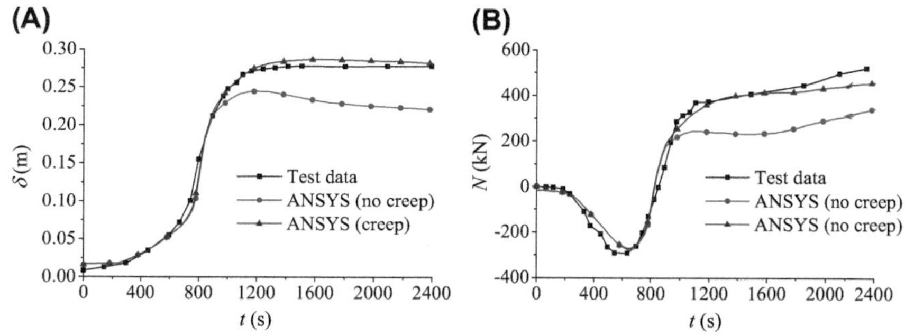

Figure 6.70 Comparison of response of beam between prediction and test. (A) Deflection and (B) axial force.

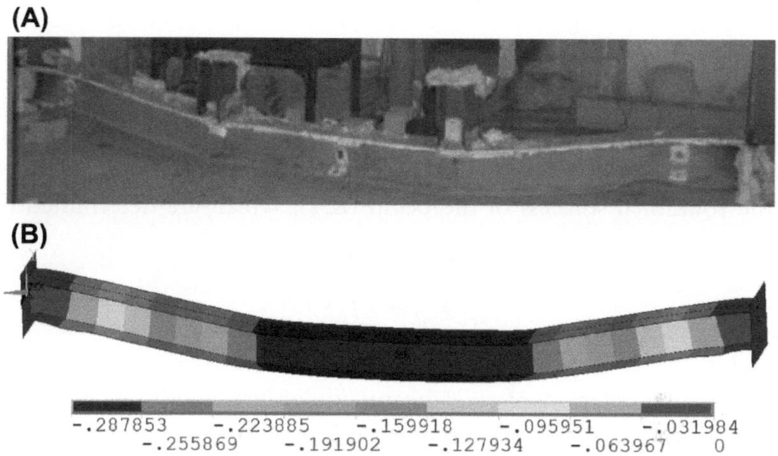

Figure 6.71 Deformation of restrained steel beam. (A) Test and (B) finite element model.

Therefore, taking high-temperature creep into consideration can lead to better predictions of fire response.

6.6.3 Parametric studies

To understanding the response of restrained high strength Q460 steel beams in fire conditions including creep effect, the established finite element model was utilized to carry out parametric studies on the influencing factors, including load level, axial restraint stiffness, rotational restraint stiffness, load type, heating rate, and temperature distribution pattern.

The load level in this chapter is represented as a nondimensional factor *LR*. It is a ratio of calculated moment and plastic limit moment capacity of cross section. The calculated moment is determined by $ql^2/8$ based on simply supported beam theory, where q is the distributed load on the beam. The plastic limit moment capacity of cross section decreases when the beam is under the effect of rotational stiffness, resulting in that *LR* can be greater than 1.0.

As to the axial restraint imposed by the supporting system, the level of axial restraint (α) is often represented as the ration between axial stiffness of support ($K_{a,s}$) and axial stiffness of beam ($K_{a,b}$), which are determined by

$$\alpha = K_{a,s}/K_{a,b} \tag{6.47}$$

$$K_{a,b} = AE/l \tag{6.48}$$

where A is cross-sectional area of beam; E is elastic modulus of steel; and l is the span of beam.

Similar to axial restraint stiffness level, the level of rotational restraint (γ) may be expressed as the ratio between rotational stiffness of support ($K_{r,s}$) and the rotational stiffness of the beam ($K_{r,b}$), which are determined by

$$\gamma = K_{r,s}/K_{r,b} \tag{6.49}$$

$$K_{r,b} = EI/l \tag{6.50}$$

where I is inertia moment of cross section of beam.

The failure criterion of steel beams under fire condition proposed in literature (Dwaikat, 2010) indicated that $l/20$ (l is the span of the beam) could be taken as limit state of restrained steel beam as the restrained steel beam will experience large deflection with low deformation rate due to restraint of slab. Therefore, the temperature at which deflection reaches $l/20$ is used to determine critical temperature of restrained steel beam (T_{cr}).

6.6.3.1 Load level

The fire resistance analysis was performed on unprotected restrained Q460 steel beams with section of H400 × 200 × 8 × 13. Uniform distributed load was applied on these beams. The length of beam is 8 m. These beams were exposed to ISO-834 standard fire, and nonuniform temperature distribution in cross section was neglected. In the parametric studies, the values of axial and rotational stiffness ratio are both taken 1.0 as examples.

The effect of load level (*LR*) on fire resistance of restrained Q460 steel beam is illustrated in Fig. 6.72, where the deflection (δ) and axial force (N)

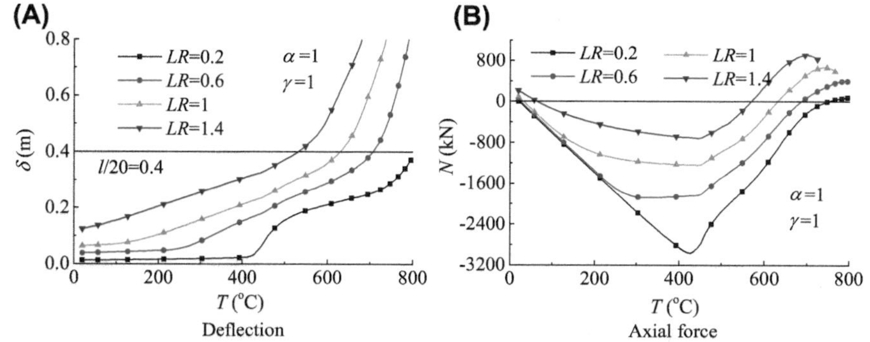

Figure 6.72 Effect of load ratio on the fire response of the beam.

are plotted as function of temperature for various load levels. It is shown in Fig. 6.72 that the fire resistance decreases with increased load level, and the axial forces for different load levels are close to zero when deflection reaches $l/20$. Thus, it could be concluded that load level has significant effects on fire resistance of restrained Q460 steel beams.

6.6.3.2 Axial restraint stiffness

The variation of deflection and axial force of restrained Q460 steel beam with temperature (T) for different axial restraint stiffness is plotted in Fig. 6.73. It is indicated in Fig. 6.73 that at a given temperature below 720°C, the higher the axial restraint stiffness is, the larger the deflection is. When temperature exceeds 720°C, deflections for various axial restraint stiffness are similar. From Fig. 6.73B, it is shown that axial force curves for α equals 1 and 5 follow similar trend after temperature reaches more than

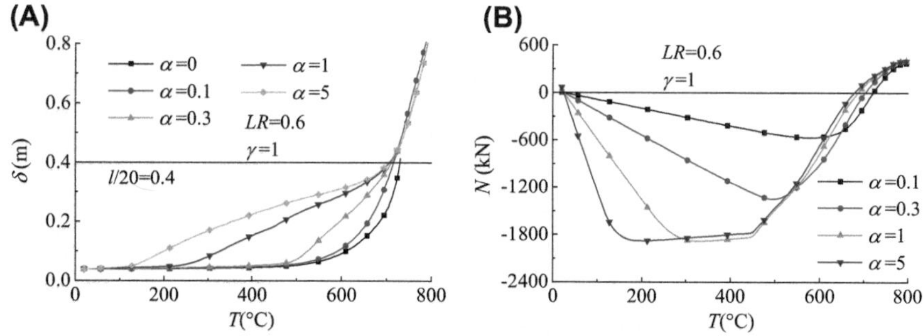

Figure 6.73 Effect of axial stiffness on the fire response of the beam. (A) Deflection and (B) axial force.

300°C. The reason is that the cross section in those two cases both reach yielding limit after 300°C, and the axial force decreases when temperature beyond 450°C due to lower yield strength of Q460 steel. In addition, the axial forces of restrained Q460 steel beam for various axial restraint stiffnesses are close to zero when the deflection reaches $l/20$.

The relationships of critical temperature (T_{cr}) of restrained Q460 steel beam and axial restraint stiffness ratio for different load levels are shown in Fig. 6.74A, indicating that the higher the axial restraint stiffness is, the lower the critical temperature is. It can be found that axial restraint stiffness has slight influence on critical temperature at a given load level due to the maximum variation of temperature 25°C only. The variation of critical temperature with load level for different axial restraint stiffness is plotted in Fig. 6.74B, which shows slight difference in critical temperatures for different axial restraint stiffness at same load level. Hence, the effect of axial restraint stiffness on fire resistance of restrained Q460 steel beam can be neglected.

6.6.3.3 Rotational restraint stiffness

The effect of rotational restraint stiffness on the fire resistance of restrained Q460 steel beam is illustrated in Fig. 6.75, where the deflection and axial force are plotted as a function of temperature for different rotational restraint stiffness. It is shown that in Fig. 6.75, the higher the rotational restraint stiffness is, the smaller the deflection is, and axial force tends to be compressive due to higher compressive capacity of restrained steel beam provided by rotation restraint. When deflection reaches $l/20$, the

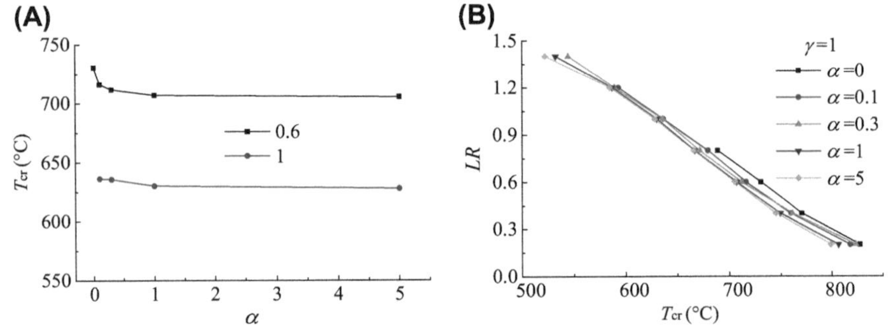

Figure 6.74 Critical temperature of the beams with various load level and axial restraint stiffness. (A) Critical temperature–axial stiffness curve and (B) load ratio–critical temperature curve.

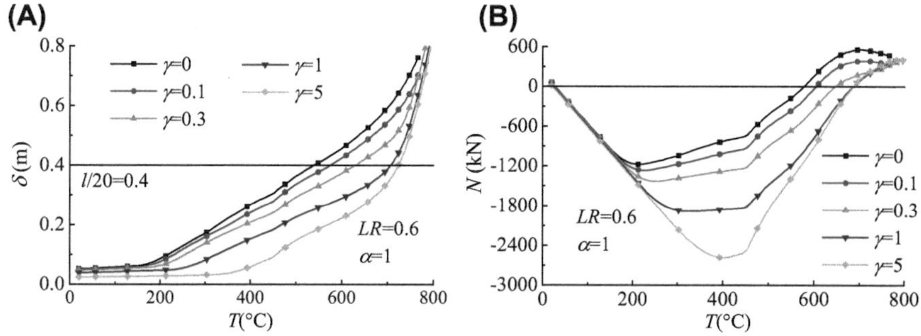

Figure 6.75 Effect of rotational stiffness on the fire response of the beam. (A) Deflection and (B) axial force.

temperature is in a range of 540–720°C, and the axial force is close to zero, for different rotational restraint stiffness.

The variation of critical temperature of restrained Q460 steel beam with rotational restraint stiffness ratio for various load levels is shown in Fig. 6.76A. The effect of rotational restraint stiffness on critical temperature decreases with increased rotational restraint stiffness ratio. The LR-T_{cr} relationship for various rotational restraint stiffnesses is shown in Fig. 6.76B. Results from Figs. 6.75–6.76 indicate that rotational restrained stiffness has considerable influence, namely, higher fire resistance capacity with higher rotational restraint stiffness, on fire resistance of restrained Q460 steel beam.

6.6.3.4 Span–depth ratio

The span–depth ratio means the ratio of beam span and cross-sectional height. The effect of span–depth ratio on the deflection and axial force

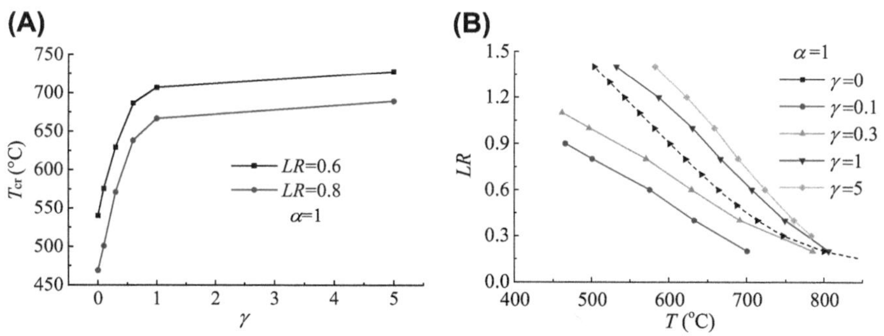

Figure 6.76 Critical temperatures of the beams with various load level and rotational restraint stiffness. (A) Critical temperature–rotational stiffness curve and (B) load ratio–critical temperature curve.

of restrained Q460 steel beam is plotted in Fig. 6.77. It indicates that span–depth ratio has notable influence on restrained Q460 steel beam. The deflection increases with increased span–depth ratio, but temperatures at which deflection reaches $l/20$ for different span–depth ratio are similar.

The relationships of critical temperature and span–depth ratio for various load levels are plotted in Fig. 6.78A, indicating that critical temperature is relatively low when span–depth ratio is large or small and the peak critical temperatures at various load levels are different. The LR-T_{cr} relationships for various span–depth ratios are given in Fig. 6.78B, which indicates that span–depth ratio has considerable influence, which varies with load levels, on fire resistance of restrained Q460 steel beam. Too large or small span–depth ratio can lead to degradation of load bearing capacity;

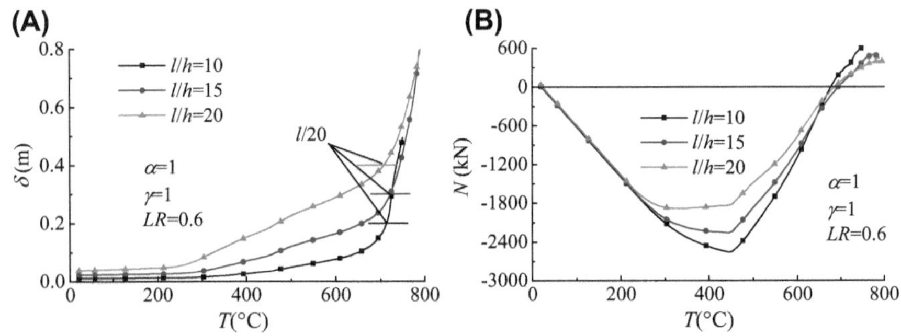

Figure 6.77 Effect of span–depth ratios on the fire response of the beam. (A) Deflection and (B) axial force.

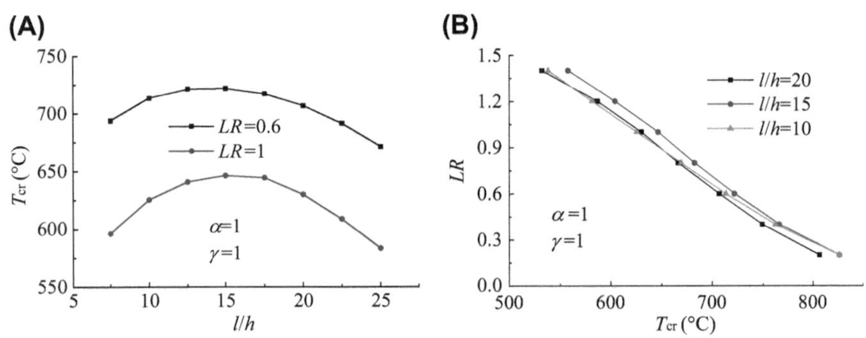

Figure 6.78 Critical temperature of the beams with various load levels and span–depth ratios. (A) Critical temperature–span–depth ratios curves and (B) load ratio–critical temperature curves.

therefore, an appropriate span—depth ratio has beneficial effect on capacity of restrained Q460 steel beam under fire condition.

6.6.3.5 Load type

Three load types, namely, uniformly distributed load, concentrated loads at $l/3$, and $2l/3$ of beam and concentrated load at midspan, were selected. The effect of load type on deflection and axial force and LR-T_{cr} relationships are plotted in Fig. 6.79. It can be seen that load type has slight influence on fire resistance of restrained Q460 steel beam. The deflections of beams under uniform load and two-point load are almost same because they have same moment diagram area. Likewise, the beam under concentrated load has relatively small deflection resulting from smaller moment diagram area. Hence, the effect of load type could be neglected.

6.6.3.6 Heating rate

Temperature curves of restrained steel beam are derived based on the ratios between thermal conductivity and thickness of fire insulation (d_i/λ_i). At a given temperature, the creep increases with decreased heating rate, and this is because creep is dependent on time, resulting in larger deformation. Four restrained Q460 steel beams with different insulation thickness were designed to determine temperature curve.

The effect of heating rate on the deflection and axial force is plotted in Fig. 6.80. Heating rate decreases with increased d_i/λ_i, especially when temperature exceeds 500°C, resulting in larger creep at same temperature. In addition, the axial force of restrained Q460 steel beam is close to zero when deflection reaches $l/20$.

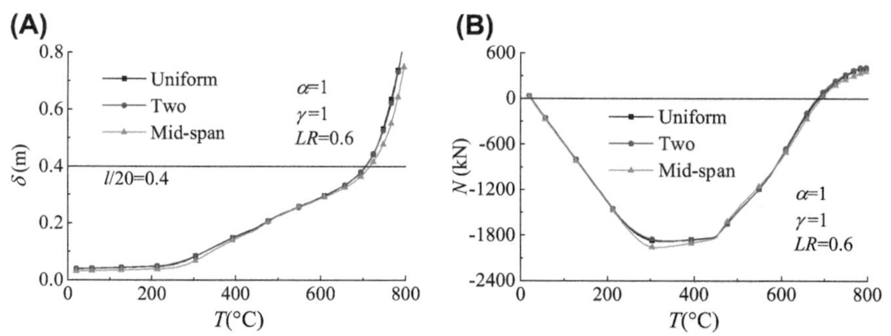

Figure 6.79 Effect of load type on the fire response of the beam. (A) Deflection and (B) axial force.

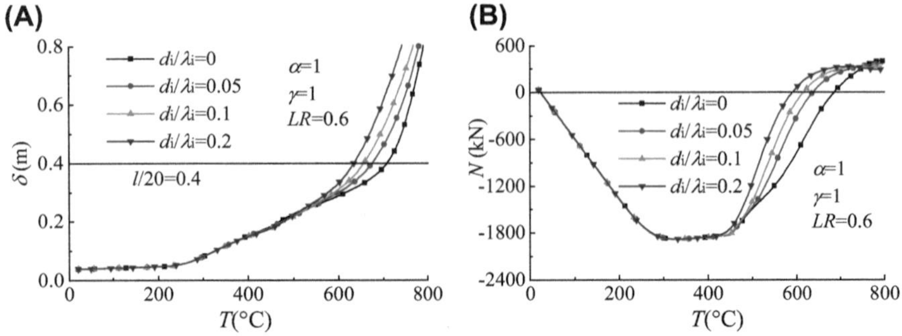

Figure 6.80 Effect of heating rates on the fire response of the beam. (A) Deflection and (B) axial force.

The variation of critical temperature with d_i/λ_i is shown in Fig. 6.81A with different heating rates using ANSYS and calculation method specified in EC3 (2005). It can be seen that the larger the d_i/λ_i is, the lower the critical temperature is, at same load level. This can be attributed to the fact that larger creep deformation occurs at lower heating rate. It should be noted that the critical temperature obtained from EC3 (2005) is relatively high when d_i/λ_i exceeds 0.1 approximately, which can lead to unconservative design in restrained steel beam. The LR-T_{cr} relationships for different heating rates are described in Fig. 6.81B. Thus, a conclusion can be drawn that heating rate has significant influence on fire resistance of restrained Q460 steel beam, that is to say the lower the heating rate is, the lower the load-bearing capacity under fire condition is.

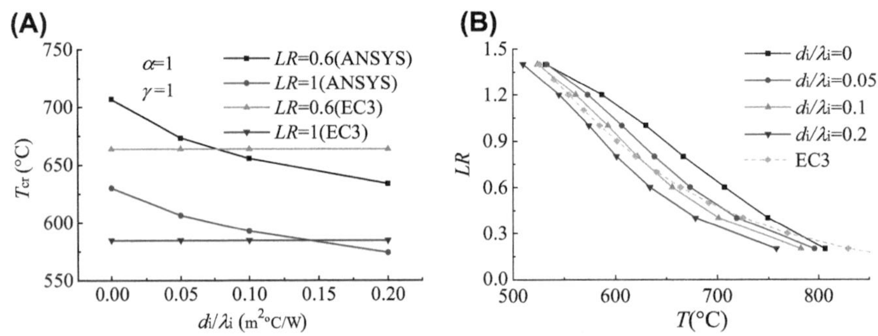

Figure 6.81 Critical temperature of the beams with various load ratios and heating rates. (A) Critical temperature—heating rates curve and (B) load ratio-critical temperature curve.

6.6.3.7 Temperature distribution pattern

Generally, temperature in cross section of restrained steel beam is not uniformly distributed due to fire protection of concrete slab on top flange. According to the experimental results of Liu et al. (2002), the sectional temperature of steel beam can be divided into two parts, one of which is temperature of top flange and the other is web and bottom flange. Adopting above nonuniform and uniform distribution, deflection and axial force of restrained Q460 steel beam are described in Fig. 6.82 as a function of temperature. It can be seen that the effect of temperature distribution pattern on fire response of restrained Q460 steel beam is slight. Nonuniform temperature distribution leads to thermal bending of steel beam, resulting in larger deflection. The LR-T_{cr} relationships for different temperature distributions are plotted in Fig. 6.83, indicating the effect of temperature distribution pattern on fire resistance of restrained Q460 beam in a range ($0.3 < LR < 0.9$).

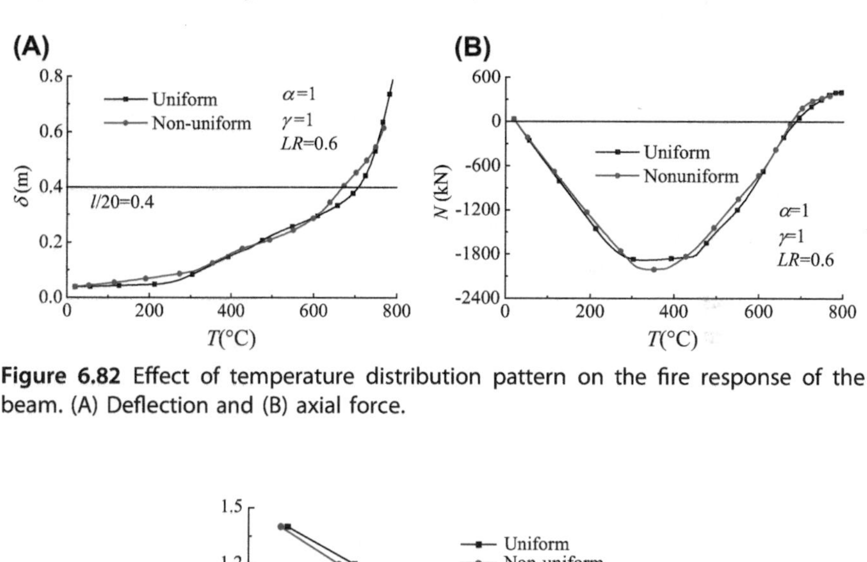

Figure 6.82 Effect of temperature distribution pattern on the fire response of the beam. (A) Deflection and (B) axial force.

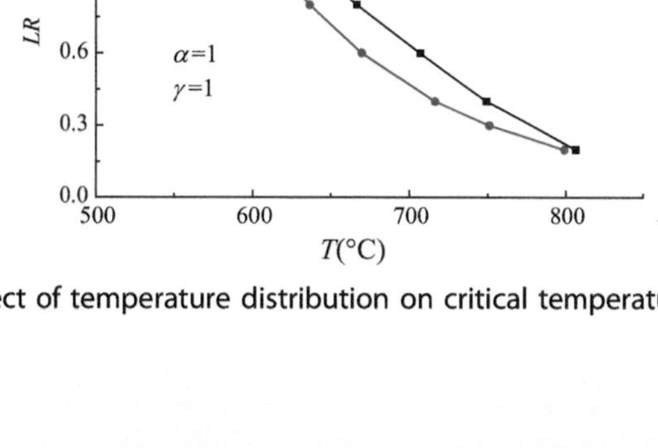

Figure 6.83 Effect of temperature distribution on critical temperature of the beam.

6.6.4 Design approach of fire resistance of restrained Q460 steel beams

The thermal axial force is generated and moment changes when the steel beam is exposed to fire. Thus, it is difficult to predict the evolutions of axial force, moment, and midspan deflection at elevated temperature. In current design codes (EC3, 2005, CECS200, 2006), simple approaches are provided but neglecting creep effect (Fig. 6.83).

For simplicity, the critical temperatures of restrained Q460 steel beams under fire condition could be taken as critical temperature of restrained Q460 steel beam for fire design. The above parametric study definitely specifies that rotational restraint stiffness, span–depth ratio, heating rate, and temperature distribution play critical roles in determining fire resistance of restrained Q460 steel beam. Using regression analysis, a simplified calculation method was developed accounting for these factors as well as creep. The governing equation for fire resistance of restrained Q460 steel beam is

$$M_q \leq M_{pT} \tag{6.51}$$

where M_{pT} is resistant moment of restrained Q460 steel beam, which can be calculated as

$$M_{pT} = LRf_y W_p \tag{6.52}$$

$$LR = \frac{\beta_r \beta_{T1} a}{1 + bk\beta_{T2}(T_{cr} + T_0)^c} \tag{6.53}$$

$$k = 0.00765(l/h)^2 - 0.224(l\,/\,h) + 2.42 \tag{6.54}$$

where M_q equals to $ql^2/8$; W_p is plastic section modulus; β_r and T_0 are coefficients related to rotational restraint stiffness ratio (γ), presented in Table 6.19; β_{T1} and β_{T2} are coefficients related to temperature distribution pattern, which are both 1.0 for uniformly temperature distribution pattern and 1.05 and 1.65 for nonuniform temperature distribution pattern, respectively; l/h is the span–depth ratio of beam; a, b, and c are coefficients related to d_i/λ_i, presented in Table 6.20. The explicit limitations for the proposed calculation method are as follows: (1) the fire curve follows as ISO 834

Table 6.19 The value of β_r and T_0

γ	0	0.1	0.3	1	>5
β_r	0.636	0.64	0.726	1	1.12
T_0	103	68.2	29.6	0	-5.6

Table 6.20 The value of a b and c

d_i/λ_i(m$^{2\circ}$C/W)	0	0.05	0.1	0.2
a	1.53	1.76	1.82	1.88
b	1.7×10^{-28}	4.18×10^{-25}	2.26×10^{-24}	6×10^{-24}
c	9.82	8.72	8.5	8.4

standard fire; (2) the value of d_i/λ_i is lower than 0.2 m$^{2\circ}$C/W; (3) the critical temperature T_{cr} is within the range of 500–800°C; and (4) no lateral buckling occurs in beams.

Under a predefined applied load, the critical temperature could be calculated by

$$T_{cr} = \left[\frac{1}{bk\beta_{T2}}\left(\frac{\beta_r\beta_{T1}a}{LR} - 1\right)\right]^{\frac{1}{c}} - T_0 \tag{6.55}$$

Critical temperatures for a large number of restrained Q460 steel beams with various load levels, different rotational restraint, span–depth ratio, and heating rate are predicted by ANSYS, and the proposed approach and the comparison are plotted in Fig. 6.84. Good agreement is found and the difference between them is within 5%.

6.7 Knowledge gaps and research needs

6.7.1 Knowledge gaps

Following the foregoing test results on mechanical properties of new-type steels at elevated temperature, this section focuses on the remaining knowledge gaps. Overall knowledge gaps were subsequently identified, discussed,

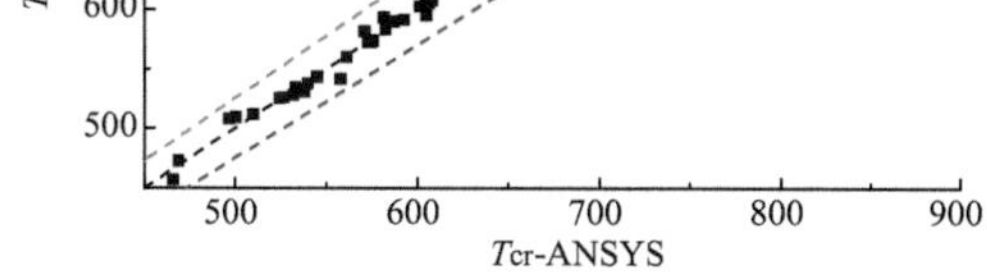

Figure 6.84 Comparison of critical temperature between simplified approach and ANSYS.

and prioritized by researchers and practitioners at the synthesis conference. The most important knowledge gaps are presented in the following subsections. These include (1) the uncertainties of mechanical properties of steels to the chemical composition and heating process. The reasons for these uncertainties are addressed in separate subsections, namely, (2) the lack of data sources and (3) shortcomings of existing methods. Further important knowledge gaps relate to (4) the accepted creep models of steel at elevated temperatures, (5) the variation of Poisson's ratio of steel at elevated temperatures, and (6) using material properties for fire resistance design.

6.7.1.1 Uncertainties of mechanical properties of steels

The mechanical properties of steels at elevated temperatures are probabilistic due to the following influencing factors, such as chemical composition, melting technique, temperature distribution, testing method, testing setup, and testing technician. As we know, at room temperature, the design value of mechanical properties of steels is determined by considering most of influencing factors and structural reliability. Therefore, the design values of mechanical properties of steels at elevated temperatures should be determined by probabilistic method. In addition, the requirement of structural reliability and the effect of these influencing factors are different with that at ambient conditions. In the current fire safety design guidance, the design values of mechanical properties were adopted as the reduction factor multiplies the values at room temperature. In the future, the design values of mechanical properties of steels at elevated temperatures are worthy of investigating.

6.7.1.2 Lack of test data

Even there are numerous test data available in the published paper and books, test data are still needed as new types of steel are produced every year. Because of the difference in chemical composition and rolling process, the reduction factor of strength and elastic modulus for different type steel at elevated temperature are not similar with each other. The reliable method is, for each type of steel, carrying out tension testing and deriving the design values of mechanical properties based on abundant of test data and probabilistic methods.

6.7.1.3 Shortcomings of tensile methods

As discussed in Chapter 2, some specifications provide detailed testing procedures for tension test of steels at elevated temperatures, but there was a freedom to select the tension strain rate and other tension parameters.

These selections will affect the results of test data. In addition, the precision accuracy of different tension setup is not similar, and this accuracy also affects the test results. Finding a stable and reliable test method to obtain the actual mechanical properties is also necessary.

6.7.1.4 Accepted creep models

Although some widely used creep models are available in literatures, all these models are quite dependent of creep test data and are only applicable to one special steel. This is because the creep models are derived based on the characteristic of creep curve, and the parameter values are fitted data without any physical meanings. Reliable creep models based on the microstructure of steel and deformation mechanism are worthy of investigating.

6.7.1.5 Variation of Poisson's ratio

Few literatures investigated the Poisson's ratio of steel at elevated temperatures. In current fire safety analysis, the value of 0.3 is often adopted as Poisson's ratio of steel at elevated temperatures. There is some fluctuation of Poisson's ratio of steel at elevated temperatures, and the value adopted also affects the analysis results. The accurate value of Poisson's ratio of steel as a function of temperatures is needed to conduct comprehensive modeling of steel structures in fire conditions.

6.7.1.6 Fire resistance design of structures made in new steels

Because of the variation of mechanical properties of different type steel, the corresponding design methods on fire safety of steel structures are also different, especially the design values of mechanical properties, stability coefficient of steel member under compression, or bending. The current design codes on fire safety of steel structures are all proposed based on research finding on mild steel and mild steel structures. In the recent years, new type steels, such as high strength steel, stainless steel, aluminum alloy, corrosion resistance steel, and fire resistance steel, are used more and more in new high-rise or long-span buildings. Unfortunately, the fire resistance research on new type steel and structures is very limited.

6.7.2 Research needs

According to the previously mentioned knowledge gaps, in the next few years, the following topics need to be investigated:

(1) Determination of design values of mechanical properties of steels at elevated temperatures,

(2) Tension tests on new type steels at elevated temperatures,
(3) Reliability of and requirement of tension test method,
(4) Creep models based on material microstructure and deformation mechanism,
(5) Poisson's ratio of various steels at elevated temperatures,
(6) Performance-based design approach on fire safety of steel structures made in new type steels.

By completing the previous research topics, the safety and reliability of steel structures made of new-type steels can be guaranteed.

References

Ali, F., Shepherd, P., Randall, M., Simms, I.W., et al., 1998. The effect of axial restraint on the fire resistance of steel columns. J. Constr. Steel Res. 46 (1–3), 305–306.

ASTM, 2009. Designation: A370-09 Standard Test Methods and Definitions for Mechanical Testing of Steel Products. West Conshohocken, PA.

Ban, H.Y., 2012. Research on the Overall Buckling Behavior and Design Method of High Strength Steel Columns under Axial compression. PhD thesis. Tsinghua University, China.

CECS200:2006, 2006. Chinese Technical Code for Fire Safety of Steel Structure in Buildings, China Engineering Construction Standardization Association Standard. China Plan Press, Beijing, China.

Chen, W.F., Atsuta, T., 1976. Theory of beam-columns. In: In-Plane Behaviour and Design, vol. 1. J. Ross Publishing Classics. MCGRAW-HILL.

Chen, J., 2003. Stability of Steel Structures—Theory and Design. Science press of China, Beijing (in Chinese).

Chen, J., 2003. Stability of Steel Structures-Theory and Design. Science Press of China, Beijing.

Chen, J., 2010. Measurement and Analysis on Residual Stress in GJ Steel Welded I-Shaped Cross-Sectional Member. Master's Degree Thesis Chongqing University, Chongqing.

Dwaikat, M., 2010. Response of Restrained Steel Beams Subjected to Fire Induced Thermal gradients. PhD thesis. Michigan State University, East Lansing.

Franssen, J.M., 2000. Failure temperature of a system comprising a restrained column submitted to fire. Fire Saf. J. 34, 191–207.

GB 50017-2003, 2003. Code for Design of Steel Structures. State Standard of the P.R. China. China Planning Press, Beijing, China.

Huber, A.W., Beedle, L.S., 1954. Residual stress and the compressive strength of steel. Weld. J. 33 (12), 589—s. Research suppl.

Kodur, V.K.R., Dwaikat, M.M.S., 2009. Response of steel beam-columns exposed to fire. Eng. Struct. 31 (2), 369–379.

Kodur, V.K.R., Dwaikat, M.M.S., Fike, R., 2010. High-temperature properties of steel for fire resistance modeling of structures. J. Mater. Civ. Eng. 22 (5), 423–434.

Li, G.Q., Guo, S.X., 2008. Analysis of restrained steel beams subjected to heating and cooling Part I: Theory. Steel Compos. Struct. 8 (1), 1–18.

Li, G.Q., Han, L.H., Lou, G.B., Jiang, S.C., 2006. Fire Resistance Design of Steel Structure and Composite Steel and Concrete Structure. Chinese Architecture and Building Press.

Li, K.X., Xu, W.M., 1989. An improvement on "the inverse calculation segment length method". J. Chongqing Inst. Architect. Eng. 11 (3), 37–43.
Li, W., Guo, Z.H., 1993. Experimental investigation of strength and deformation of concrete at elevated temperatures. J. Build. Struct. 14 (1), 8–16.
Liu, T.C.H., Fahad, M.K., Davies, J.M., 2002. Experimental investigation of behavior of axially restrained steel beams in fire. J. Constr. Steel Res. 58 (9), 1211–1230.
Luecke, W.E., McColskey, J.D., McCowan, C.N., et al., 2005. Federal Building and Fire Safety Investigation of the World Trade Center Disaster: Mechanical Properties of Structural Steels. National Institute of Standards and Technology.
Qiang, X.H., Wu, N.D., Jiang, X., Luo, Y.F., 2018. Experimental and theoretical study on high strength steel extended endplate connections after fire. Int. J. Steel Struct. 18 (2), 609–634.
Qin, S.Q., 2015. The Effect of Elevated Temperature on Welding Residual stress. Master thesis. Chongqing University, China.
Rodrigues, J.P., Neves, I.C., Valente, J.C., 2000. Experimental research on the critical temperature of compressed steel elements with restrained thermal elongation. Fire Saf. J. 35 (2), 77–98.
Tan, K.H., Toh, W.S., Huang, Z.F., Phng, G.H., 2007. Structural responses of restrained steel columns at elevated temperatures, part 1: experiments. Eng. Struct. 29 (8), 1641–1652.
Tao, Z., Wang, X.Q., Uy, B., 2012. Stress-strain curves of structural and reinforcing steels after exposure to elevated temperatures. J. Mater. Civ. Eng. 25 (9), 1306–1316.
Valente, J.C., Neves, I.C., 1999. Fire resistance of steel columns with elastically restrained axial elongation and bending. J. Constr. Steel Res. 52 (3), 319–331.
Wang, Y.C., 1997. Effects of structural continuity on fire resistant design of steel columns in non-sway multi-storey frames. Fire Saf. J. 28 (2), 101–116.
Wang, Y.C., 2004. Post buckling behaviour of axially restrained and axially loaded steel columns under fire conditions. J. Struct. Eng. 130 (3), 371–379.
Wang, P.J., Li, G.Q., Wang, Y.C., 2010. Behaviour and design of restrained steel column in fire: part 2: parameter study. J. Constr. Steel Res. 66 (8–9), 1148–1154.
Wang, Y.B., Li, G.Q., Chen, S.W., et al., 2012a. Numerical analysis of mechanical property of Q460 high strength steel welded box columns under axial compression. Ind. Constr. 42 (1), 26–31.
Wang, Y.B., Li, G.Q., Chen, S.W., et al., 2012b. Experimental and numerical study on the behavior of axially compressed high strength steel columns with H-section. Eng. Struct. 43, 149–159.
Wang, Y.B., Li, G.Q., Chen, S.W., Sun, F.F., 2012c. Experimental study on the ultimate bearing capacity of axially compressed high strength steel of H-section columns. J. Civ. Eng. 45 (6), 58–64.
Wang, W.Y., Liu, B., Kodur, V.K.R., 2013a. Effect of temperature on strength and elastic modulus of high strength steel. J. Mater. Civ. Eng. 25 (2), 174–182.
Wang, Y.B., Li, G.Q., Chen, S.W., et al., 2013b. Parametric study on the ultimate load-bearing capacity of Q460 high strength steel H-shaped columns under axial compression. Prog.Steel Build. Struct. 15 (5), 8–13.
Wang, W.Y., Kodur, V.K.R., Yang, X.C., et al., 2014a. Experimental study on local buckling of axially compressed steel stub columns at elevated temperatures. Thin-Walled Struct. 82, 33–45.
Wang, Y.B., Li, G.Q., Chen, S.W., et al., 2014b. Experimental and numerical study on the behavior of axially compressed high strength steel box-columns. Eng. Struct. 58, 79–91.

Wang, W.Y., Li, G.Q., Ge, Y., 2015. Residual stress study on welded section of high strength Q460 steel after fire exposure. Adv. Steel Constr. 11 (2), 150–164.
Wang, W.Y., Yan, S.H., Liu, J.P., 2017. Studies on temperature induced creep in high strength Q460 steel. Mater. Struct. 50 (1), 68.
Zhang, L.L., Wang, P.J., 2018. Simplified analysis method for catenary action of restrained cellular steel beams at elevated temperature considering strain reversal. Fire Saf. J. 95 (1), 145–159.

Index

Note: 'Page numbers followed by "f" indicate figures and "t" indicates tables'.

D

E

F

G

H

I

R

FSC
www.fsc.org
MIX
Papier | Fördert
gute Waldnutzung
FSC® C083411

Kolibri 360 GmbH
Ferdinand-Jühlke-Straße 7
99095 Erfurt, Deutschland
produktsicherheit@kolibri360.de